1980 SAUNDERS COLLEGE
Philadelphia

PHYSICS AN

To Beverly

PHYSICS AND MUSIC

The Science of Musical Sound

HARVEY E. WHITE
University of California, Emeritus

DONALD H. WHITE
Oregon College of Education

Library of Congress Cataloging in Publication Data

White, Harvey Elliott, 1902–
Physics and music.

Bibliography: p.
Includes index.
1. Music—Acoustics and physics. 2. Musical instruments—Construction. 3. Sound—Recording and reproducing. 4. Architectural acoustics. I. White, Donald H., joint author. II. Title.
ML3805.W44 781'.1 79-22472
ISBN 0-03-045246-5

Printed in the United States of America
0123 032 987654321

PREFACE

Almost everyone enjoys music, whether as a performer or as a listener. This book is written for the student who wants to go beyond the perceptual stage of music to learn how musical sound is created and why it is perceived as it is. The student does not need a background in science to understand the material presented here. However, he or she should have some knowledge of the structure of music, for example, a familiarity with the piano keyboard.

This text, intended to be self-contained, surveys a wide range of topics related to the acoustics of music. We begin with a brief history of the art and science of music, followed by the general principles of sound, musical scales, and the primary ways in which sound can be generated as well as those characteristics common to all instruments. In later chapters we deal with various mechanical and electronic recording devices, the playback of hi-fi stereophonic and quadrophonic sound, and the design of electronic musical instruments. Finally, we explore the acoustics of auditoriums, concert halls, and outdoor theaters in both perceptual and measurable terms so that the student will be able to analyze or modify the acoustical response of the listening environment.

We have tried to make this text as readable and comprehensive as possible. For example, we have included over 300 diagrams, photographs, and tables as illustrative matter. Algebraic equations are often used to express experimentally based relationships. New terms are set in boldface type since the subject matter introduces a sizeable new vocabulary. These terms are also boldface in the index, which serves as a glossary as well. At the end of each chapter we have included (1) questions to aid the student in reviewing and consolidating the material, (2) problems for students who wish to employ the relationships discussed in the chapter quantitatively, (3) student projects, which can be done at home or in the laboratory, and (4) references for students who want to explore specific topics in depth or concentrate on areas of personal interest.

In recent years the field of musical acoustics has been the scene of considerable interest and activity, both in the classroom and in the research laboratory. New findings are continually being published in technical journals as well as in popular magazines. In treating the central concepts of this book, we have often glossed over interesting details which the student will miss unless he or she supplements this text with additional reading. If students are stimulated to explore some of the byways of acoustics by following up some of the references noted, we will consider this book a success.

In drawing together the considerable material required for the range of topics presented in this text, we have relied upon input and review from many musicians, scientists, and specialists in other areas. We wish to thank Isidor Elias, Santa Barbara City College; Sterling Gorril; Edwin Hahn, University of California, Berkeley; Joseph Hirschberg, University of Miami; David Jasnow, University of Pittsburgh; Martin E. Rickey, Indiana University; Thomas Rossing, Northern Illinois University; Larry Rowan, University of North Carolina, Chapel Hill; William R. Savage, University of Iowa; Wave Shaffer, Ohio State University (Emeritus); Daniel Strohl; William E. Vehse, West Virginia University; Gabriel Weinreich, University of Michigan, Ann Arbor; and Robert W. Williams, University of Washington. We are also grateful to Klemi Hambourg, Max Larson, Richard Sorensen, David E. Wallace, and Ronald Wynn of Oregon College of Education for their assistance.

Berkeley, California
Monmonth, Oregon
October, 1979

Harvey E. White
Donald H. White

CONTENTS

Part Two

Hearing and Harmony 77

PART ONE

The Nature of Sound

Chapter One

THE ART AND SCIENCE OF MUSIC

The word *science* stems from the Latin root *scire,* meaning "to know." It is a branch of human endeavor in which nature is studied and analyzed in an attempt to understand its structure and behavior. Music is thought of as an art form, a refinement of the production and reception of sound.

1.1 Music of the spheres

Music and science may seem to many to be quite far apart. Yet these subjects are related in many interesting ways. The study of the relation between nature and musical harmony extends back into antiquity. During the sixth century B.C., Pythagoras and his followers[1] believed both mathematics and music to be expressions of the harmony of nature (see Figure 1 – 1). The findings attributed to Pythagoras can be expressed as follows: Pluck a stretched string and perceive its pitch. If now only one-half the length of the string is allowed to vibrate, it will sound an octave higher. If two-thirds of the string is allowed to vibrate, it will sound a perfect fifth higher. A length of three-fourths will sound a perfect fourth higher; and so on. These relationships were perceived as harmonious, in the broadest sense.

The harmonious relationship was recognized in terms of the vibrating portions of the string, expressed as simple mathematical relationships: 2:1 = octave, 3:2 = fifth, 4:3 = fourth, and so forth. These perfect musical consonances were directly related to simple relationships of whole numbers.

FIGURE 1–1 Demonstration of harmonic proportions from Franchino Gafurio's "Theorica Musice" of 1492. In these diagrams Pythagoras is seen with his follower Philolaus demonstrating the tones of bells of different sizes, glasses filled to different levels, strings that are stretched by different weights, and pipes of different lengths.

The ancient Greeks were strongly oriented toward harmonies in nature, and music was simply one of these natural harmonies. The Pythagoreans carried their concept of natural musical harmony far beyond mathematics. They believed that all regularities in nature are musical. In particular, they believed that the planets produced harmonious sounds as they moved in space. This is the "music of the spheres," a harmonious relationship of the heavenly bodies.[1]

Various aspects of the Pythagorean school affected the understanding of nature and music well into the Middle Ages.[2] Johannes Kepler (1571–1630), a tavern keeper who became an astronomer, was the first to reduce the studies of planetary motion to three laws of motion. Nevertheless, he too was entranced by the concept of a celestial chorus, and he speculated on which planet sang soprano, which tenor, and so on.[3]

1.2 The science of acoustics

As science and music developed from the seventeenth century on, scientists and musicians continued to take an interest in each other's activities. Scientists are often amateur musicians, and musicians are often amateur scientists. As time passed, breakthroughs were made in understanding sound through the studies of acoustical pioneers such as Sauveur, Mersenne, Chladni, and many others.[4] As acoustics developed, its principles were applied to the development of music and musical instruments by designers such as Boehm, Sax, and others.

In recent years, the science of acoustics has developed to the point where its impact on the understanding of music is truly significant. Unfortunately, many people are suspicious of science, tending to equate it with negative aspects of technology. Others fear that an analytical approach may detract from the "mystique" of the aesthetic experience. Some simply avoid science as a result of an unfortunate experience in a science or mathematics class. It is the hope of many musicians and scientists, however, that through cooperation and mutual understanding, music can be a more effective means of enhancing the intellectual and aesthetic experience of everyone. It is in this spirit that this book is written.

1.3 Subjective and objective descriptions

Musicians and scientists have many characteristics in common. They are both creative, they relate emotionally to their activities, and they enjoy communicating their activities with others. A basic difference, however, is that the scientist attempts to describe a situation in **objective** terms. That is, he or she will use instruments capable of taking quantitative measurements. In so doing, the results can be expressed unambiguously, consistently, and often in numerical terms. For example, sound intensity can be expressed in picowatts per square meter or in decibels.

Among musicians it is common practice to use **subjective** or **perceptual** descriptions. For example, a tone can be described as *bright, dark, hollow, nasal, harsh, pure, golden, vibrant, rich, round, tinny, raspy, woody, reedy, mellow, fuzzy,* and so on. Since sensory experience is what music is all about, subjective descriptions seem appropriate. However, it is difficult to try to explain the meaning of subjective terms verbally. To convey what is meant by a "bright sound," for example, it is customary to create several such sounds as examples. To the trained ear, this is usually adequate, although it is never totally unambiguous.

It may seem that an objective description could be found for every sensory experience. For example, one could conceivably express loudness in units of intensity, pitch as vibration frequency, and so forth. Careful study, however, almost always shows the relationships to be much more complex than the examples above, as illustrated in the following diagram:

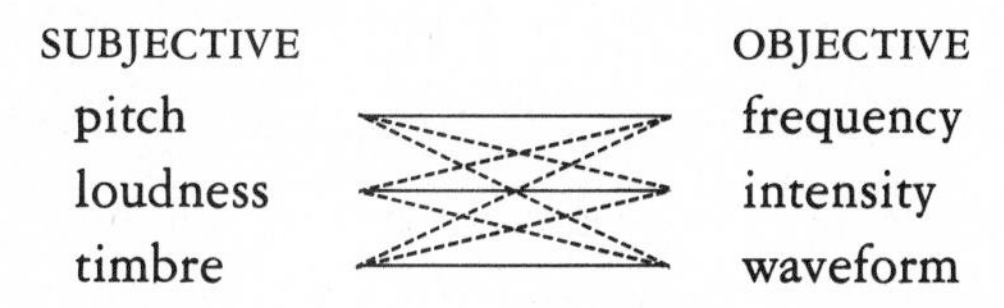

For example, the loudness of a sound is strongly correlated with the intensity of the sound wave. But if its frequency is high enough, it can't be heard at all. Loudness, therefore, depends also upon frequency.

Expressing subjective quantities in objective terms is usually not simple. But it is important to try to make this our objective.

1.4 Sound

It has been said, "If a tree falls in the forest and there is no one to hear it, there is no sound." Whether this is true or not depends upon one's definition of sound. If sound is defined as "the sensation produced by stimulation of the organs of hearing," the statement is true. If sound is defined as "the mechanical vibrations transmitted through an elastic medium," like the air, the saying is false. This is an example of *measurable* versus *perceptual* interpretation. In this text, we will take the word *sound* to be perceptual (subjective). Where the distinction is important, we will use *sound wave* for the measurable (objective) descriptor. In many cases, the distinction is unimportant, and the word *sound* will cover both aspects.

Sound is one of the pollutants of our big cities today, as well as being a source of communication and expression. The purring sound of the automobile, the passing of a truck or a bus, the sounding of a train or a factory whistle, the rat-a-tat-tat of an air hammer or a riveter, the whine of a siren or a fire engine, the passing overhead of a jet plane or a helicopter, and the hum of a group of people talking are but a few examples of sound. Some of these sources are quite objectionable to our hearing, and great effort is made to reduce their sound levels. Some of them, like whistles and sirens, are purposely loud and are made for their attention-getting capabilities. Musical instruments, on the other hand, are designed for entertainment, personal expression, and enjoyment, and they are usually pleasant to hear.

The science of musical sounds involves detailed descriptions and precise measurements of vibrations, the waves they produce, and their reception by the ear. If sounds are heard in an auditorium, we are also interested in the acoustics of the enclosure. If they are recorded on tape or records, we are interested in the fidelity of the recording and in the monophonic, stereo, or quadraphonic playback to a large or small audience. In this book, all these subjects will be taken up in some detail.

1.5 The modern metric system of units

A number of the physical laws of nature are basic to a better understanding of musical sounds. The concepts of *speed, velocity, acceleration, mass, force, work, energy,* and *power* are presented in Appendix I, and each will be referred to at the appropriate place in this book. All the known laws of nature are based upon numerical values, measured in the laboratory as the result of planned experiments and formulated and expressed by the simplest mathematical relationships.

Since precise measurements are essential to the establishment of reliable laws, common units of measurement are essential if they are to be used by others. Over the years, it has been found that all measurable mechanical quantities can be expressed in terms of three fundamental units:

length mass time

All other quantities related to mechanics and sound, such as speed, velocity, acceleration, and force, can be expressed in terms of these three.

In recent years, an international committee has developed the International System of Units (SI) in order to develop a worldwide standard. The modern metric system, adopted in the United States,[5] is essentially a system using the SI units with very minor modifications. The SI units for length, mass, and time are meters (abbreviated as m), kilograms (kg), and seconds (s), respectively.

1.6 New vocabularies

To combine the *art of music* with the *science of music,* there are essentially two different vocabularies that will be used. New words as they appear in this book will be given in boldface type and listed in the index with the page number in boldface type. Each chapter will, as far as possible, include basic principles, classroom demonstrations, laboratory experiments, and one or more solved numerical problems. At the end of each chapter, the student will find several questions pertaining to that chapter's subject matter and several numerical problems, where appropriate. The mathematics involved is confined to simple algebraic relationships, and the problems are given accordingly.[6]

1.7 Projects

A great deal can be learned from doing projects. By exhibiting a musical instrument in class, various principles being studied can be tried out. The size, shape, and weight of an instrument and its component parts can be observed close at hand by all students, and there are always a few present who know how to play one or more instruments and can "put them through their paces."

Nearly every educational institution has a band and an orchestra, and all kinds of instruments can be borrowed for classroom or laboratory study. Some school laboratories have considerable equipment for experimentation. However, if a laboratory is poorly equipped, inexpensive devices can be purchased in a hobby shop, hardware store, or electronics supply house, and do-it-yourself kits can become class projects. It is by taking part in such activities that one learns a great deal about the science of musical sounds.

NOTES

[1]J. Bronowski, *The Ascent of Man* (Boston: Little, Brown, 1974), p. 155.

[2]E. Helm, "The Vibrating String of Pythagoras," *Scientific American* (December 1967):92.

[3]J. Kepler, "The Harmonies of the World," in *Great Books of the Western World,* vol. 16, ed. R. M. Hutchins (Chicago: Encyclopedia Britannica), p. 1005. Kepler is discussed in various biographies such as, M. Casper, *Kepler,* IV 8 (New York: Abelard-Schuman, 1959).

[4]D. C. Miller, *Anecdotal History of the Science of Sound* (New York: Macmillan, 1939) (presently out of print but still available in many libraries).

[5]A wide variety of information on the modern metric system can be obtained by writing to the Office of Technical Publications, National Bureau of Standards, U.S. Department of Commerce, Washington, D.C. 20234.

[6]A variety of additional problems can be found in W. R. Savage, *Problems for Musical Acoustics* (New York: Oxford University Press, 1977).

Chapter Two

VIBRATING SYSTEMS

2.1 Periodic motion

Any motion that repeats itself in equal intervals of time is called **periodic motion.**[1] The swinging of a clock pendulum, the dancing of a mass on the end of a coiled spring, the vibration of a weighted metal strip clamped at one end, and the vibration of the prongs of a tuning fork are four examples of periodic motion. See Figure 2–1.

Consider a mass m hanging freely from the end of a coiled spring, as shown in Figure 2–2. If we raise the mass to point A, a distance $+a$, and then release it, the mass will vibrate up and down with periodic motion. If the mass

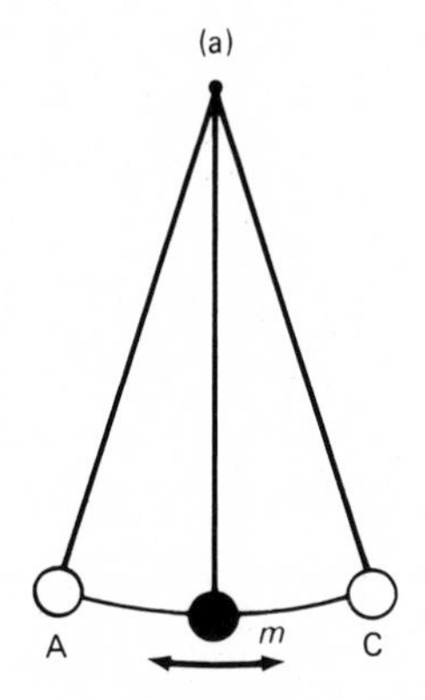

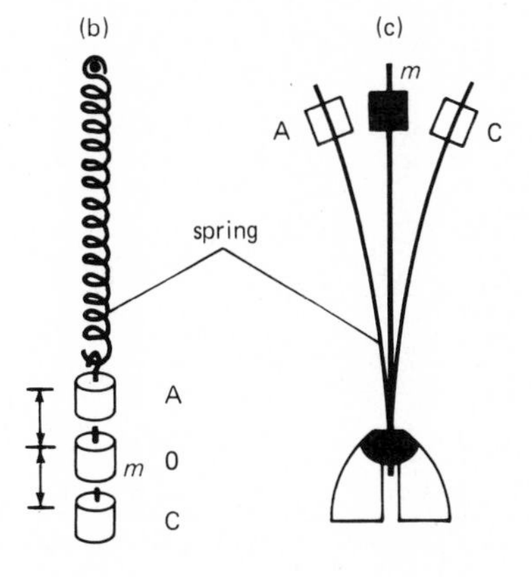

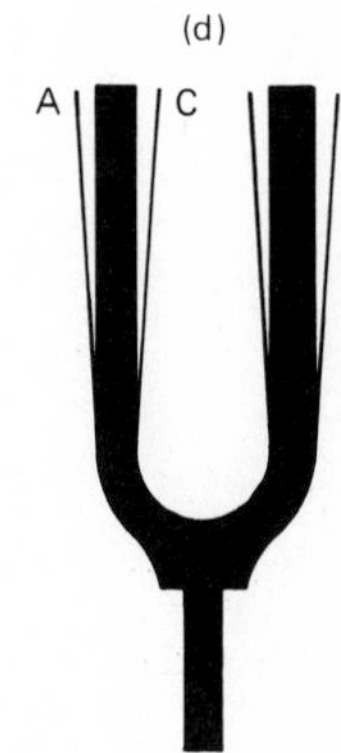

FIGURE 2–1
Four common vibrating objects found in the average laboratory: (a) a simple pendulum, (b) a spring pendulum, (c) a vibrating strip, and (d) a tuning fork.

is pulled down to the point C, and then released, it will move up to A, then back to C, back to A, back to C, and so on. In either case, the motion repeats itself in equal intervals of time. Such a single vibration is called a **cycle.**

The time T it takes to vibrate through one complete cycle is called the **period**, measured in seconds:

$$T = \text{period in seconds} \quad [2a]$$

If the mass m requires exactly *one second* to make one complete cycle, we say the period is exactly one second, and we write

$$T = 1.0 \text{ s} \quad [2b]$$

If the mass m requires exactly two seconds to complete a single cycle, we say the period $T = 2.0$ s. If it requires 0.65 second to make one complete cycle, the period $T = 0.65$ s. (Note that we abbreviate the unit of second by the symbol s.)

Note that if the mass starts at A, the cycle is not completed until the mass m moves down to C and back to A. If the mass starts at C, the cycle is not completed until the mass moves up to A and down to C. If we start counting time as the mass passes the midpoint, or equilibrium position 0, and it moves to C and back to 0, only half a cycle has been completed.

The **displacement** y of a mass point at any given time is given by the distance from its equilibrium. The **amplitude** a is a position value expressing the maximum displacement from the equilibrium point 0 to the highest

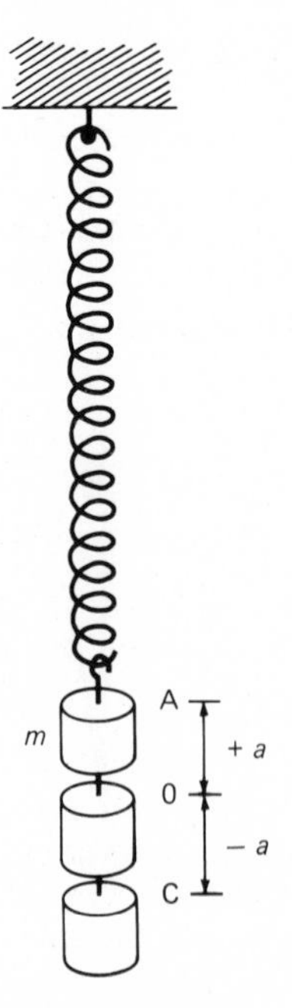

FIGURE 2-2
A mass m suspended from the end of a coiled spring vibrates up and down with periodic motion.

point A or the lowest point C. During vibration, the displacement varies between $-a$ and $+a$:

$$-a \leqslant y \leqslant +a \qquad [2c]$$

2.2 Frequency of vibration

The **frequency** of a vibrating body is defined as the number of complete cycles per second. For example, if the mass on the end of a spring completes a cycle in one-half second, the period $T = \frac{1}{2}$ s, and it will make two cycles in one second. We can therefore write

frequency = 2.0 cycles per second

In abbreviated notation, this is

$$f = 2.0 \text{ c/s}$$

If a vibrating object completes a vibration in one-quarter second, $T = \frac{1}{4}$ s, it will make four vibrations in one second, and we can write

$$f = 4.0 \text{ c/s}$$

Observe that the period and the frequency are reciprocals of one another:

$$\text{period} = \frac{1}{\text{frequency}} \qquad \text{frequency} = \frac{1}{\text{period}} \qquad [2d]$$

or in symbols,

$$T = \frac{1}{f} \qquad f = \frac{1}{T} \qquad [2e]$$

If the period is $\frac{1}{2}$ s/c, the frequency is 2 c/s. If the period is $\frac{1}{50}$ s/c, the frequency is 50 c/s.

The international (SI) unit of frequency is the *hertz,* abbreviated Hz.

$$\frac{1 \text{ cycle}}{\text{second}} = 1 \text{ hertz} \qquad [2f]$$

$$1 \text{ c/s} = 1 \text{ Hz} \qquad [2g]$$

This unit is named in honor of Heinrich R. Hertz, a German physicist who became known during the latter part of the nineteenth century for his discovery of electromagnetic waves.*

*Heinrich Rudolph Hertz (1857–1894), German physicist, was born at Hamburg on 22 February 1857. He studied physics under Helmholtz in Berlin, at whose suggestion he first became interested in Maxwell's electromagnetic theory. His research with electromagnetic waves was carried out at Karlsruhe Polytechnic between 1885 and 1889. As professor of physics at the University of Bonn, after 1889, he experimented with electrical discharges through gases.

Example

The string of a violin vibrates with a frequency of 250 Hz. Find its period.

Solution

Direct substitution of $f = 250$ Hz in the first equation of Equation (2e) gives

$$T = \frac{1}{f} = \frac{1}{250 \text{ c/s}} = 0.0040 \text{ s/c}$$

The answer is read, "T equals four thousandths of a second per cycle."

2.3 Time graphs

Figure 2–3 shows a series of instantaneous positions of the vibrating mass suspended at the end of a spring as it moves up and down with periodic motion. These diagrams are like the individual photographs of a motion picture film. In other words, the vibrating system is shown at *equal intervals of time.*

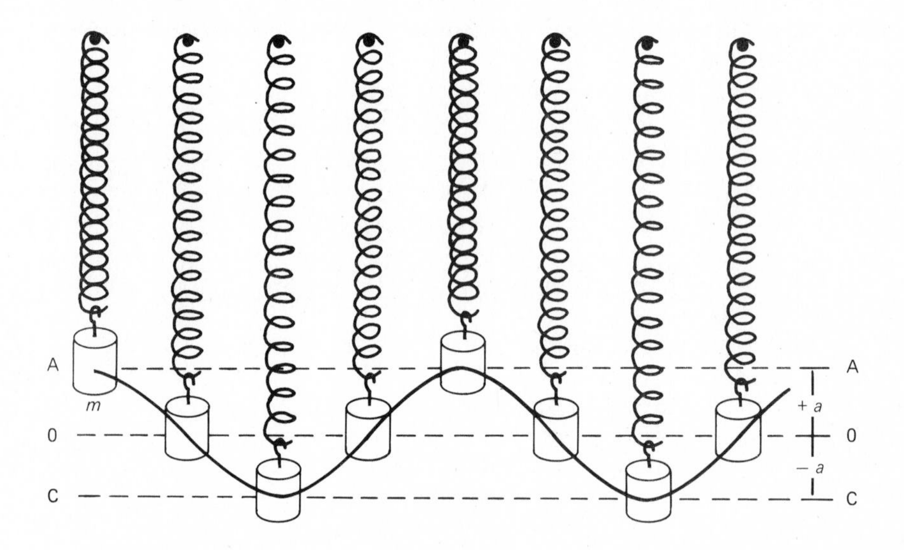

FIGURE 2–3
Positions of a vibrating mass at equal time intervals.

If we plot a time graph of the motion, we obtain a diagram similar to that shown in Figure 2–4. Vertical distances are measured in centimeters (cm) or meters (m) and horizontal distances in seconds (s). The curved line shows the displacement y at all instants of time. Observe that the curve is smooth and shows a type of symmetry about the time axis. The displacement is continually changing, is periodic, and has a maximum of $+a$ and a minimum of $-a$.

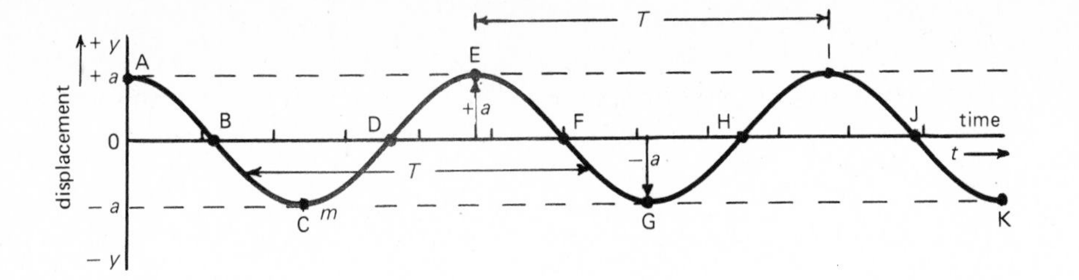

FIGURE 2–4
Time graph of the periodic motion of a mass *m* suspended from the lower end of a coiled spring.

2.4 Simple harmonic motion

If the mass in Figure 2–2 is displaced from its equilibrium position, the spring acts to return the mass to its equilibrium position. The further the mass is pulled, the stronger is the returning force of the spring. This force, then, has two features: (a) its direction is toward the equilibrium point 0, and (b) its magnitude is proportional to its displacement from equilibrium. A system with such a force is said to obey **Hooke's law.**

Different vibrating bodies may produce a variety of time graphs. If the vibrating system obeys Hooke's law, it will result in a simple time graph, such as is shown in Figure 2–4. Such bodies are said to execute **simple harmonic motion.** The four objects shown in Figure 2–1 are good examples of such motion, provided that the amplitudes are not too large.

Simple harmonic motion (abbreviated SHM) can be illustrated by the projection, on any diameter, of a graph point moving in a circle with **uniform circular motion.** See Figure 2–5. The *graph point* p moves around the circle of radius a with uniform speed v. If, at every instant of time, a perpendicular is

FIGURE 2–5
As p moves around the circle with uniform speed, the projection p moves up and down the *y*-axis with simple harmonic motion.

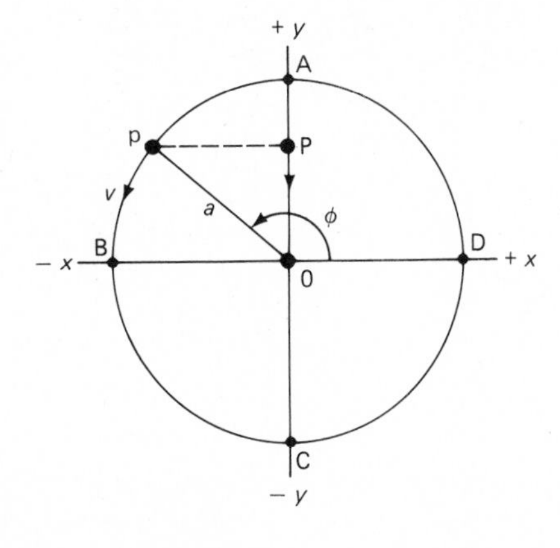

drawn from p to the diameter AC, the intercept P, called the **mass point,** moves with simple harmonic motion.

Moving up and down along the y-axis, the speed of the mass point P is continually changing. At the center point 0 it has its greatest speed, while at A and C it is momentarily at rest. Starting from either end of its path, the speed increases until it reaches 0. From there it slows down, coming to rest at the opposite end of its path. The return trip is a repetition of this motion in the reverse direction.

2.5 Phase angles

The position of the mass point P on a time graph is shown in Figure 2-6. The period of a vibrating body is seen as the time interval between any point P_1 and the next corresponding point P_2, as shown. In this example, $T = 5.0$ s, and the frequency $f = 0.20$ Hz.

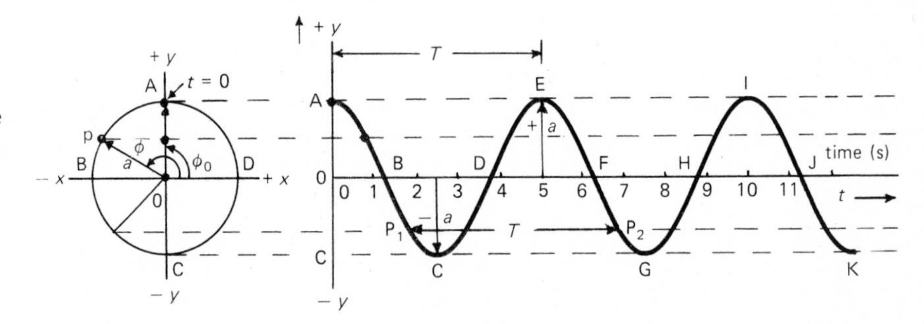

FIGURE 2-6
Time graph for simple harmonic motion showing the circle of reference ABCD, the phase angle ϕ, the initial phase angle ϕ_0, the period T, and the amplitude a.

The angle ϕ between the $+x$-axis, line 0D, and the line 0p is called the phase angle. At point A on the circle of reference, and on the time graph, the displacement y is equal to $+a$, and the phase angle $\phi = 90°$. At point B, the displacement $y = 0$, and the phase angle $\phi = 180°$. At point C, the displacement $y = -a$, and the phase angle $\phi = 270°$, while at point D, the displacement $y = 0$, and the phase angle $\phi = 360°$. In radians, these same phase angles are $\phi = \frac{1}{2}\pi$ at A, $\phi = \pi$ at B, $\phi = \frac{3}{2}\pi$ at C, and $\phi = 2\pi$ at D. (For *radian measure of angles,* see the Appendix, A1.19.)

If we start a cycle from any point on the circle of reference, we call the angle it makes with the $+x$-axis, at time $t = 0$, the initial phase angle $\phi = \phi_0$. In Figure 2-6, for example, the time graph is drawn with the initial phase angle $\phi_0 = 90°$. Starting at A at time $t = 0$, the point p moves counterclockwise around the circle once in the time of the period $T = 5$ s, and the time graph traces out one complete cycle. Now, at position A once again, the total phase angle $\phi = \phi_0 + 360°$, or a total of 450° from the starting point. Once more around the circle of reference and the total time is 10 s, the graph point and the mass point are again at A, the time graph is at I, and the total phase angle that has been turned through is $\phi = 450° + 360° = 810°$.

2.6 Damped vibrations

When any object is set into vibration, and no means of sustaining its motion are applied, the amplitude diminishes continuously until it finally comes to rest. Such vibrations are called **damped oscillations,** or **damped vibrations.** See Figure 2–7. With most musical instruments, the decrease in vibration amplitude is attributed largely to friction in the moving parts of the instrument and to radiation of energy in the form of sound waves. Later we will see that some musical instruments provide the means for continuously increasing, decreasing, or maintaining constant amplitude and **sound intensity.** We will also find that sound intensity is proportional to the *square of the amplitude.*

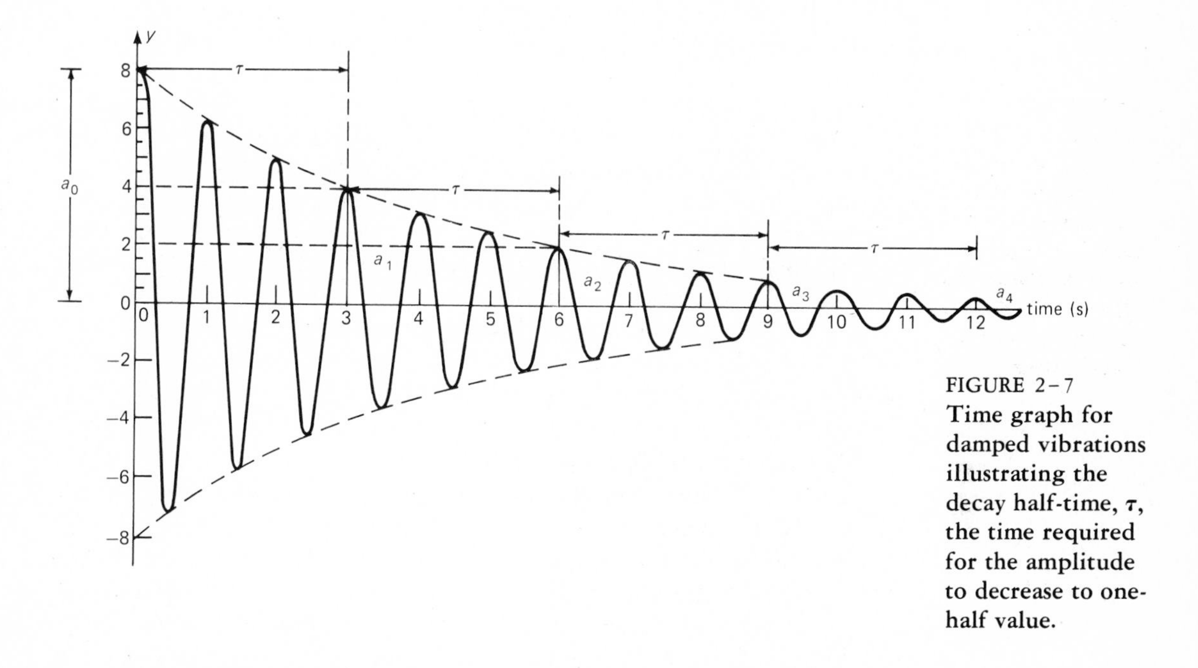

FIGURE 2–7
Time graph for damped vibrations illustrating the decay half-time, τ, the time required for the amplitude to decrease to one-half value.

A characteristic of damped vibrations is the so-called **decay half-time.** This can be defined as the time required for the amplitude to decrease to half its initial value. Interestingly enough, this half-time is usually independent of the initial amplitude and has the same value for all portions of its time graph.

As shown in Figure 2–7, the decay half-time τ is 3 s. In the first 3 s, the amplitude drops from 8 to 4. During the next 3 s, the amplitude drops from 4 to 2, and so on. Later in other chapters we will see how these principles apply to the strings of musical instruments like the piano, harp, banjo, guitar, and violin.

2.7 Laboratory experiments

Any one of the objects in Figure 2 – 1 lend themselves to excellent laboratory experiments.

1. A simple pendulum is easily made by using a piece of solid metal as a bob and a piece of string about a meter long as a support. Pull the bob 5 cm to one side, and with the sweep hand of an ordinary watch or stopwatch, find the time it takes to make 20 complete oscillations. Repeat the experiment a number of times by starting the bob with larger and smaller amplitudes, and determine the period for each value. Make a table, using as headings (a) *initial amplitude* and (b) *period.* Does the period change with amplitude?

Similar experiments can be performed by using the same bob and varying the pendulum length and tabulating *period* and *length.*

Still another experiment can be performed by varying the weight of the bob and keeping the length constant. Tabulate *period* and *weight.* Does the period change with the length of the pendulum? How? Does the period change with the weight of the bob?

2. As a second experiment use a lightweight coil spring, and suspend a known mass to the lower end. Pull the mass down 5 cm, release it, and determine the time it takes to make 20 oscillations. Change the initial amplitude to larger and smaller values and record the time. Make a table and use as headings (a) *period,* and (b) *Initial amplitude.* How does the period vary with amplitude? Vary the mass and determine how the period varies with mass.

3. Clamp a strip of bronze or steel in a vise, as shown in Figure 2 – 1(c). With a piece of tape, fasten a known mass to the top end. Pull the mass 5 cm to one side and release it. Measure the time required to make 20 vibrations. How does the period vary with (a) the length of the spring, (b) the initial amplitude, and (c) the attached mass?

4. Select a tuning fork with a pitch of $G_2 - C_3$(99 Hz to 132 Hz), and fasten a felt-tip pen to one tine. Clamp the fork above a tabletop, so that the pen presses on a strip of paper that lies on the table. Set the tuning fork in motion, and draw the paper along at a measurable rate. The pen will trace out a damped vibration pattern. The speed of the paper can be determined by timing the duration of the pull with a stopwatch or by having a second vibrating source draw a much slower vibration curve (a vibragraph). Find the frequency of the tuning fork.

QUESTIONS

1. Define or briefly explain, in your own words, each of the following: amplitude, period, frequency, displacement, decay half-time, and phase angle. Use examples you have observed.

2. What are damped vibrations? Make a diagram and briefly explain.

3. What is a time graph? Make a diagram to illustrate simple harmonic motion.

4. Make a list of three systems which (a) undergo simple harmonic motion, (b) undergo harmonic (periodic) motion but not simple harmonic motion, (c) move but are not harmonic.

5. Does Hooke's law apply to each of the systems in Figure 2 – 1? Explain.

PROBLEMS

1. The A string of a violin, properly tuned, has a frequency of 440 Hz. Find its period.

2. The E string of a violin has a period of 0.001515 s. Find the frequency with which it vibrates.

3. The mass suspended from a coiled spring is set vibrating up and down with an initial amplitude of 5.0 cm. After 30 s, the amplitude has decreased to 2.5 cm. What would be its amplitude 2.0 minutes (min) after starting?

NOTES

[1]H. E. White, *Modern College Physics,* 6th ed. (New York: Van Nostrand, 1972), pp. 277 – 317.

TRANSVERSE WAVES

In nature there are many kinds of waves. In our *submicroscopic* world, atoms and molecules are composed of electrons, protons, neutrons, and mesons that move about as waves within their atomic and molecular boundaries. With appropriate stimulation, these same atoms and molecules emit waves we call *gamma rays, X rays, visible light waves, infrared waves, microwaves, and radio waves.*

In our *macroscopic* world of the earth, waves are produced as the result of moving masses of considerable size. Earthquake waves are produced by sudden shifts in land masses, water waves by the wind or motions of bodies in water, and sound waves by the quick and sudden movement of objects in and through the air. In this chapter we will discuss some of the characteristics of waves.

3.1 Classification of waves

Nearly all types of waves can be classified under one or both of the following two headings: **transverse waves** and **longitudinal waves.** Light waves, and the waves on the vibrating strings of musical instruments, are transverse in character. Sound waves through the air, on the other hand, are longitudinal in character. Earthquakes produce both transverse and longitudinal waves which travel over and through the earth, respectively, while water waves are a combination of transverse and longitudinal waves. It is important, therefore, that we consider transverse and longitudinal waves in general and then as they apply to sound and music, and that we become familiar with the vocabulary involved in their description. We will start with the simplest of all forms of waves, namely, transverse waves. (Longitudinal waves will be discussed in the next chapter.)

3.2 Traveling transverse waves

Transverse waves are defined as those in which all parts of the medium through which they are traveling are vibrating perpendicular to the waves' direction of propagation. All musical instruments belonging to the classification of "strings" involve transverse waves.

As a classroom demonstration, consider a long rope stretching across the room, with the far end fastened to a hook in the wall and the other end held firmly in the hand. If one gives a quick up-and-down flip to the end of the rope, a disturbance moves out along the rope toward the fixed end and is reflected back again toward the hand. See Figure 3–1. This kind of disturbance is called a **transverse wave pulse**. Observe that the initial pulse is "up," as shown in diagram (a), and the reflected pulse is "down," as shown in diagram (b).

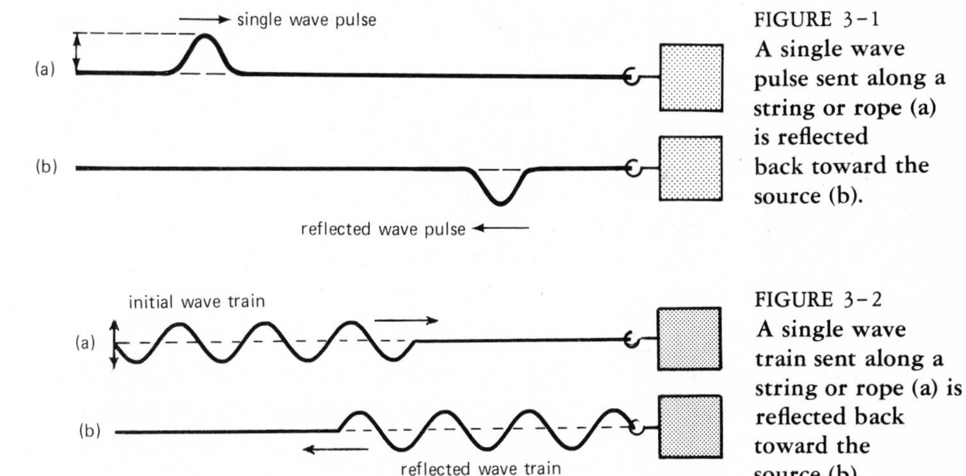

FIGURE 3–1
A single wave pulse sent along a string or rope (a) is reflected back toward the source (b).

FIGURE 3–2
A single wave train sent along a string or rope (a) is reflected back toward the source (b).

If, instead of a sudden flip, the hand is moved down and up several times with simple harmonic motion, a **wave train** like those shown in Figure 3–2(a) will travel along the rope. Reflecting back from the fixed end, the wave train travels in the opposite direction, as shown in diagram (b). Both the initial wave and the reflected wave vibrate in the same plane and constitute transverse waves. Such a wave generated by simple harmonic motion is called a **sinusoidal** or **sine wave**. The wave shape is called a **sinusoid**.

Transverse sine waves are readily demonstrated by a machine illustrated in Figure 3–3. As the handle H is turned clockwise, the small wooden balls at the front face move up and down with simple harmonic motion. As each ball moves vertically along its own line, the waveform ABCDEFG moves to the right, keeping its same shape. Each ball, like each section of the string or rope, moves in a direction transverse (at right angles) to the direction of wave propagation. If the handle is turned counterclockwise, the wave will appear to move to the left. In each case, each ball performs exactly the same motion along its line of vibration, the difference being that each ball is slightly ahead or behind the motion of its nearest neighbor. (The construction of this machine is described in Section 3.7.)

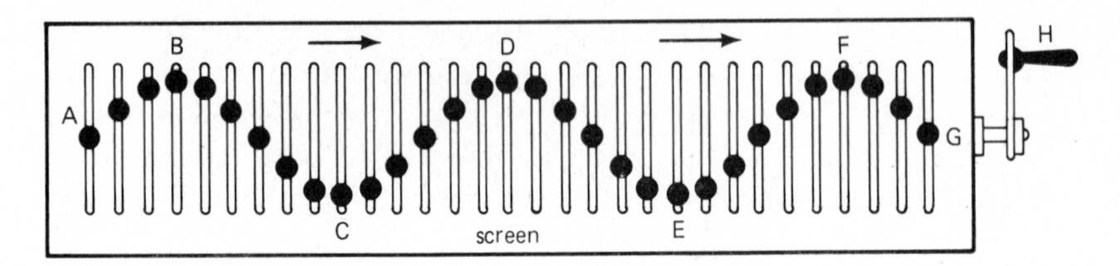

FIGURE 3–3 A mechanical device for demonstrating transverse waves traveling to the right or to the left.

Transverse waves, as well as many other wave effects, can be nicely demonstrated on a **Shive Wave Machine.**[1] This device consists of a series of horizontal bars, each tied onto a common central flexible ribbon. See Figure 3–4. A twist at one end will propagate slowly down the array. Strictly speaking, this wave is neither longitudinal nor transverse but a **torsional** (twisting) wave. The ends of the rods, however, will have essentially a transverse motion.

3.3 Wavelength

When a vibrating source sends out a **transverse wave train** on a string or rope, the waves travel with constant speed. If the source vibrates with a frequency f, the waves are all the same length, as shown in Figure 3–5. The **wavelength** is defined as the distance between two similar points of any two consecutive waves* and is represented by the Greek letter λ. The distance between two consecutive **wave crests**, or two consecutive **wave troughs,**† is

*We refer here to each wavelength of displacement as a *wave* in the same sense as one refers to individual waves of water. A series of such waves is strictly called a *wave train,* although this is shortened to just *wave* if such a meaning is clear by context.
†The word *trough* is strictly appropriate only to waves with vertical displacement, such as water waves. In general, such a trough is simply a crest in the negative direction.

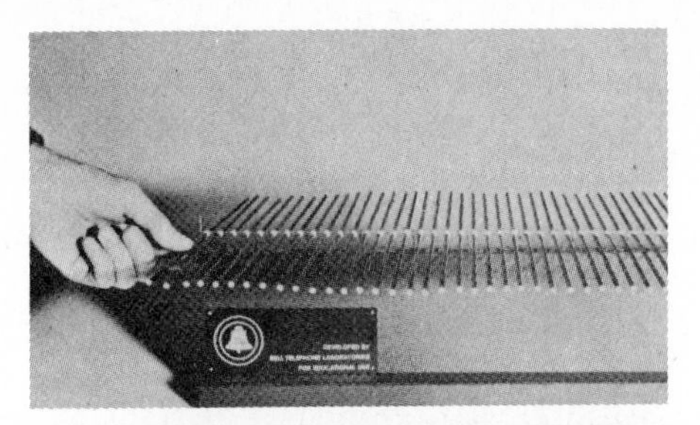

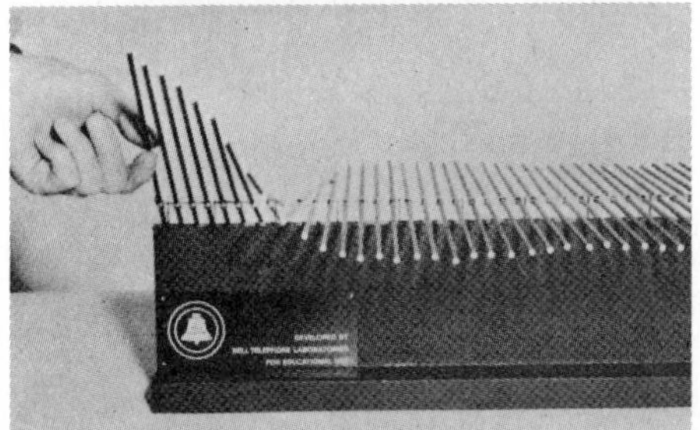

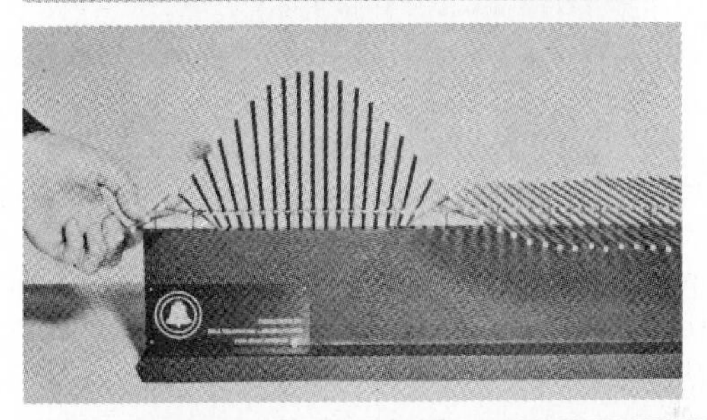

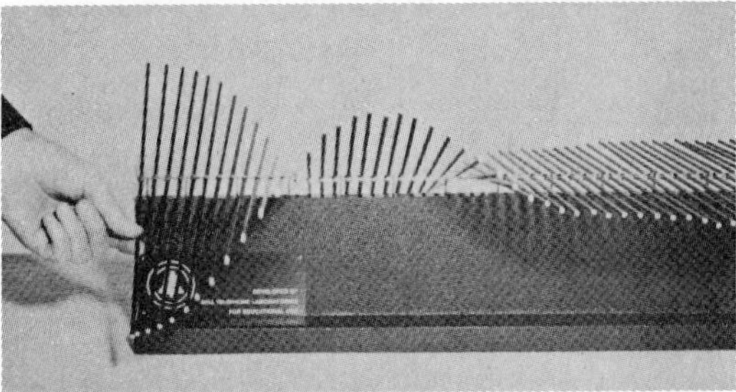

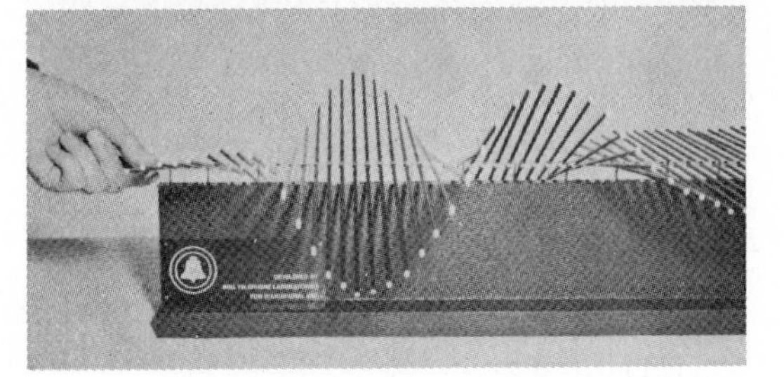

FIGURE 3–4
Five photos of the generation of a single wave on a Shive Wave Machine. The motion of the rods represents a torsional wave, while the white ends of the rods represent a transverse wave. (Copyright 1968 Bell Telephone Laboratories. Reprinted by permission.)

FIGURE 3-5
A transverse wave showing crests, troughs, wavelength λ, amplitude *a*, displacement *y*, and speed *V*.

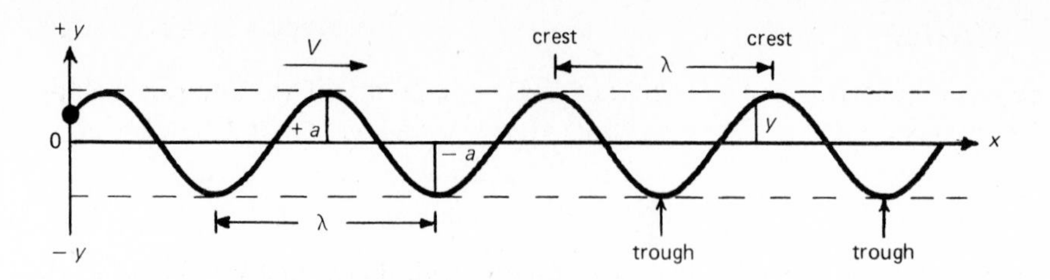

equal to one wavelength. One wavelength is the distance the wave travels in the time of one complete cycle of the source:

$$\lambda = \frac{\text{distance}}{\text{cycle}}$$

In the metric system of units, λ is measured in meters per cycle and is abbreviated m/c, or just m.

The **displacement** y of any given point along a transverse wave, at any given instant of time, is given by the lateral distance of that point from its equilibrium position. The displacement is continually changing from positive to negative with each half cycle. The **amplitude** of any wave, like that of any vibrating mass point, is given by the letter a and does not vary, unless otherwise specified, as shown in the diagram.

The **frequency** of a train of waves is defined as the number of waves passing any given point per second. This is the same as the frequency of the source, since waves do not pile up nor disappear as they travel from the source. Like the frequency of a source, the frequency of waves is universally expressed in hertz, Hz.

3.4 The wave speed

From the definition of speed V, the frequency f, and the wavelength λ, the following simple relation exists between them:

$$V = f\lambda \qquad [3a]$$

This equation follows from the fact that the length of one wave times the number of waves per second equals the distance traveled in one second. The **period** of a wave is the time for one complete wave to pass by any given point, and it is exactly the same as the period of the source. The frequency of a wave and the frequency of its source are equal, and both are measured in hertz.

If any two of the three quantities speed, frequency, and wavelength are known, the third quantity can be calculated by means of Equation (3a).

Example

One end of a long rope is fixed, and the other end is moved up and down with a frequency of 15.0 Hz. Waves travel outward toward the fixed end with a speed of 24.0 m/s. Find the wavelength.

Solution

The given quantities are $f = 15.0$ Hz and $V = 24.0$ m/s. The unknown quantity is the wavelength λ. Dividing both sides of Equation (3a) by f, we obtain

$$\frac{V}{f} = \lambda \quad \text{or} \quad \lambda = \frac{V}{f} \qquad [3b]$$

where we put the *known* quantities on the right-hand side of the equation and the *unknown* quantity on the left.* Direct substitution of the known quantities gives

$$\lambda = \frac{24.0 \text{ m/s}}{15.0 \text{ c/s}} = 1.60 \text{ m/c}$$

The units of time cancel, and the wavelength is 1.60 m/c, or simply 1.60 m.

3.5 Standing transverse waves

Nearly all sounds emanating from stringed musical instruments originate from **standing transverse waves**. Such waves can be illustrated by combining two transverse wave trains traveling in opposite directions along the same rope or string. One of the ways of illustrating this is shown in Figure 3–6. The two ends of the string or rope are anchored. Consider a wave train moving to the right and its reflection moving to the left, as shown at the top of Figure 3–6. If there are an integral number of half wavelengths between the ends, it is seen that the two waves reflect into each other at each end. The rope will sustain the sum of such waves, since the amplitudes at the endpoints add

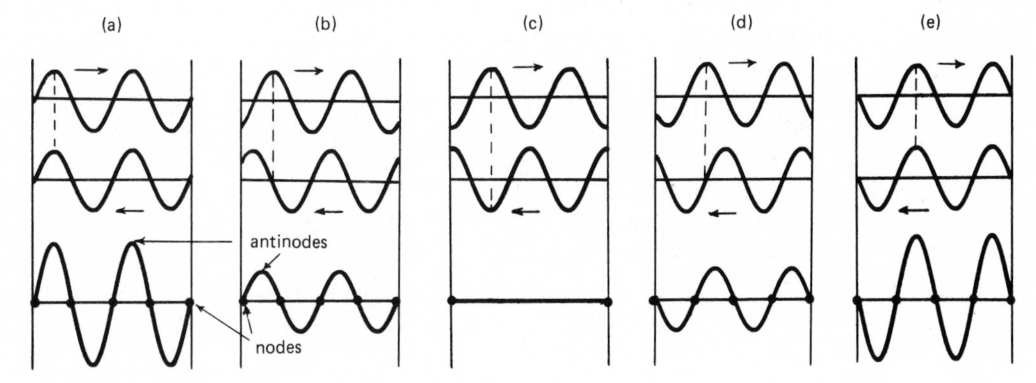

FIGURE 3–6
Two waves of equal amplitude and wavelength traveling along the same string or rope will add to produce a standing wave. In (a) and (e), where crest meets crest, we have a large resultant amplitude. In (b) and (d), where crest meets zero displacement, we have a smaller resultant. In (c), where crest meets trough, we have complete cancellation and no displacement.

*In solving numerical problems, and before any substitutions of known quantities are made, symbols representing known quantities are customarily transferred to the right-hand side of the equation and the unknown quantity to the left, as is done here.

to zero, as they must. The amplitude at the other points along the rope are also arrived at by simply adding the amplitudes of the two traveling waves, as shown at the bottom of Figure 3-6. The general appearance of the resulting standing wave will be that of dividing the rope into stationary sections of equal length, as shown. The dark points, where the rope has no up-and-down motion, are called **nodes**, and the points halfway between, where the motion has the greatest amplitude, are called **antinodes**. An entire wave section between two consecutive nodes is called a **loop**. Note carefully that each loop has a length of $\frac{1}{2}\lambda$:

$$L = \frac{1}{2}\lambda \qquad [3c]$$

A good classroom demonstration may be performed with the long rope shown in Figure 3-7. One end of the rope is fastened, and the other end is held firmly in the hand. The hand is moved up and down continuously with simple harmonic motion. As the waves reach the fixed end of the rope, they are reflected back to meet succeeding waves coming up. If the waves have just the right frequency, the rope will sustain both waves by dividing into sections, as shown.

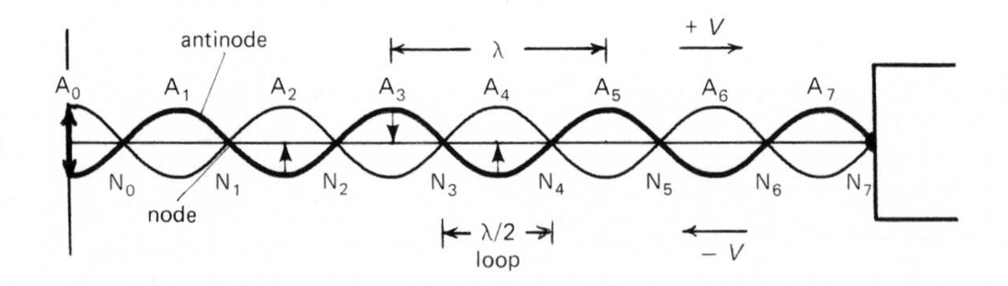

FIGURE 3-7 **A classroom demonstration of standing waves on a string or rope showing nodes $N_1, N_2, N_3, \ldots$, antinodes $A_1, A_2, A_3, \ldots$, and wavelength λ.**

When the string of a musical instrument vibrates in such a simple pattern of nodes and loops, each small section of the string moves up and down with SHM. The amplitude of the motion is zero at the nodes and increases continuously in both directions to a maximum at the antinodes. A wave machine revealing these characteristics in slow motion is illustrated in Section 3.10.

3.6 Laboratory experiment

A good laboratory experiment on standing transverse waves is shown in Figure 3-8. One end of a thread or string is fastened to one prong of an electrically driven tuning fork, and the other end passes over a pulley to a weight holder. By adding and removing weights in small increments, the tension in the string can be varied, and, at certain specific values, nodes and loops can be obtained. Greater tension will produce fewer loops, and decreased tension will produce more loops. If the frequency of the fork is known, one can meas-

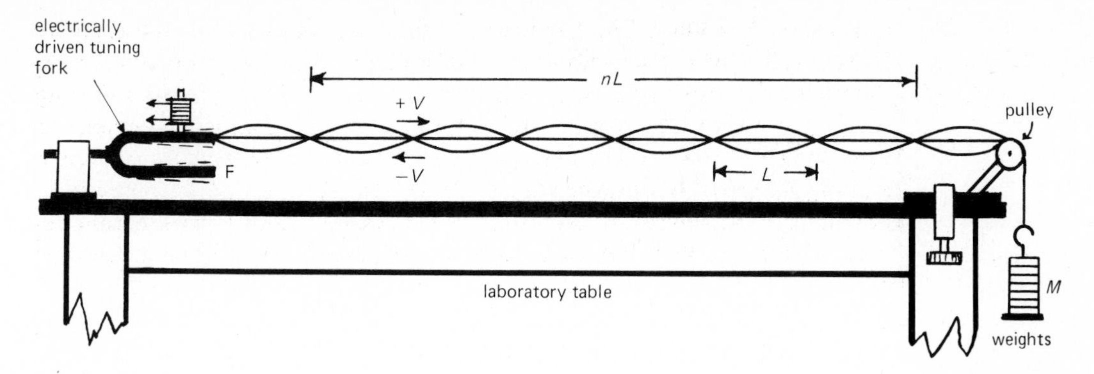

FIGURE 3–8 A laboratory experiment for determining the speed of waves on a stretched string, set vibrating in what is called a ***standing wave pattern.***

ure the length of several of the equal-length loops, and, using Equation (3a), calculate the speed of the transverse waves along the string. The values shown in Table 3–1 are typical for this kind of experiment, where the particular string used, from fork to pulley, was varied between 90 cm and 160 cm, and the fork had a stated frequency of 95.0 Hz. If apparatus of this kind is available, students should make their own set of measurements. Record the data under column headings 1, 2, 3, and 4, and then calculate values for column headings 5, 6, and 7.

TABLE 3–1: Typical Measured Values for the Vibrating String Experiment

1	2	3	4	5	6	7
TRIAL	M (kg)	nL (m)	NO. OF LOOPS n	L (m)	λ (m)	V (m/s)
1	0.033	1.113	7	0.159	0.318	30.2
2	0.130	1.148	4	0.287	0.574	54.5
3	0.190	1.038	3	0.346	0.692	65.7
4	0.250	1.302	3	0.434	0.868	82.5
5	0.425	1.158	2	0.579	1.158	110.0

3.7 Shop projects

Four different machines for generating different kinds of waves and wave motion have been designed and constructed by one of the authors in the school machine shop. Details of the first two, described in Sections 3.2 and 3.5, are given below, and the other two are given in the next chapter.

Inside the box in Figure 3–3 are a set of 31 identical levers, with offset disks D like the one shown complete in Figure 3–9. The grooved wheels A mounted on each rod PQ move up and down with SHM by identical disks of radius R. All 31 disks are mounted off-axis by the same amount d, and they are

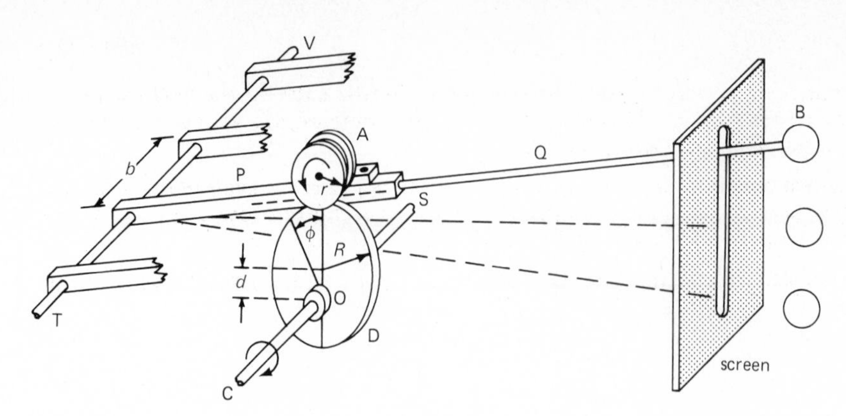

FIGURE 3-9
Details for the construction of the wave machine shown in Figure 3-3.

rigidly mounted perpendicular to their common crankshaft. Each disk lags by an angle of 30° behind the one preceding it. For producing reasonably accurate SHM for each ball B, to within 5 percent of the correct displacement, the offset distance should not be greater than 30 percent of the radius R.

The following dimensions are recommended for a traveling wave display 1.50 m long, containing 2.5 wavelengths: lever spacing $b = 5.0$ cm, offset distance $d = 2.0$ cm, P = 8.0 cm (aluminum), $r = 2.0$ cm (aluminum), Q = 28 cm (aluminum), $R = 7.5$ cm (aluminum), B = 3.0 cm in diameter (wood), CS = 1.0 cm in diameter (steel), and TV = 1.0 cm in diameter (steel).

The second machine is for demonstrating transverse standing waves and is identical to the first machine, but with one difference. The off-axis distances are variable, and the following are recommended: $d_1 = 0$, $d_2 = 1.0$ cm, $d_3 = 1.73$ cm, $d_4 = 2.0$ cm, $d_5 = 1.73$ cm, $d_6 = 1.0$ cm, $d_7 = 0$, $d_8 = -1.0$ cm, $d_9 = -1.73$ cm, and so on. See Figure 3-10.

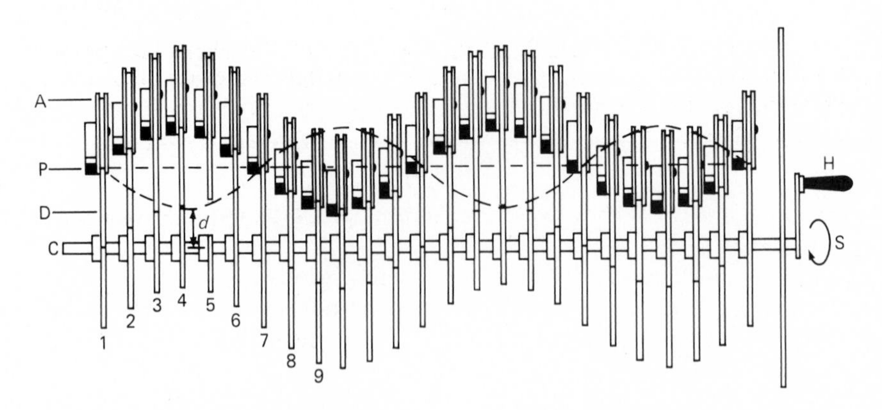

FIGURE 3-10
Mechanical details for a machine to demonstrate standing transverse waves.

All machines may be reduced in size by proportionately reducing all dimensions.

QUESTIONS

1. Define, or briefly explain in your own words, each of the following: (a) transverse waves, (b) reflected wave train, (c) wavelength, (d) wave equation, and (e) standing transverse waves.

2. Make a diagram of standing transverse waves, three wavelengths long. Label the nodes, antinodes, loops, and one wavelength. How does the wavelength compare with the length of one loop?

3. What kind of waves are (a) earthquake waves and (b) waves on a violin string?

PROBLEMS

1. A harp string has a length of 30.5 cm and vibrates with a node at each end and an antinode in the center. If its frequency is 440 Hz, find (a) the wavelength and (b) the speed of the waves on the string.

2. A string is set vibrating with nodes and loops by means of an electrically driven tuning fork. If five loops are measured and found to have a length of 0.66 m, and the fork has a frequency of 85 Hz, find (a) the wavelength, in centimeters, and (b) the speed of the waves on the string, in meters/second.

PROJECT

A simple transverse wave,[2] similar to the Shive Wave Machine* (Section 3.2), can be made by suspending about 2 m of tape recorder tape from the ceiling and attaching plastic drinking straws horizontally at even intervals.[2] Any twist applied at the bottom will propagate up the ribbon and reflect from the top. The speed of propagation will be slow enough to be easily observed and controlled.

NOTES

[1]J. N. Shive, *Similarities in Wave Behavior* (New York: Bell Telephone Laboratories Inc., 1961). A film by the same title is also available from Bell Telephone Laboratories or through local telephone business offices.

[2]M. D. Levenson, "Wave Motion Demonstration," *Physics Teacher* 12 (1974): 47.

*Various devices, such as a Shive Wave Machine and film loops, are readily available from scientific apparatus supply houses.

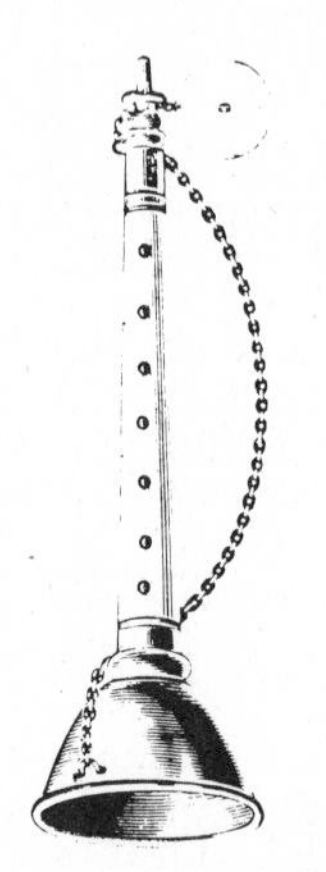

Chapter Four

LONGITUDINAL WAVES

When a string of a cello, guitar, or piano is set vibrating, the periodic motion sets the whole instrument vibrating. The vibrating string, as well as the instrument surfaces, strikes the neighboring air molecules, and these in turn strike others, transmitting the periodic disturbances outward. Traveling in all directions from the source, such periodic vibrations constitute compressional sound waves, as well as energy flow.

4.1 Traveling longitudinal waves

Sound waves transmitted through gases are *longitudinal* in character. Longitudinal waves are defined as those disturbances in which all parts of the me-

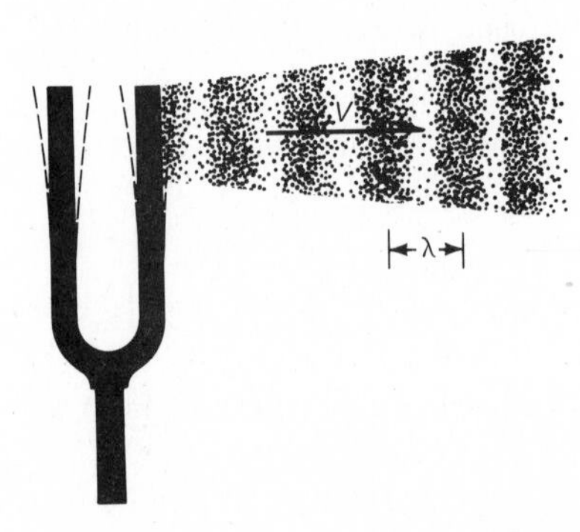

FIGURE 4-1
Sound waves emitted by the vibrating prong of a tuning fork are longitudinal in character.

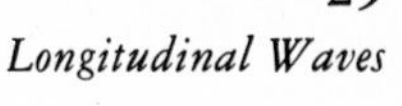

dium through which the waves are traveling are vibrating, about their equilibrium positions, parallel to the waves' direction of propagation.

In Figure 4–1 the prongs of a tuning fork are shown vibrating back and forth with simple harmonic motion. By periodically bumping into air molecules, each prong sends out pressure waves through the air at approximately 350 m/s, or 1150 ft/s (ft is the abbreviation for foot or feet). Although the waves are shown radiating from one section of the fork, waves travel outward in all directions and with greater intensity in some directions than in others.

Longitudinal waves are readily demonstrated in slow motion by a coiled spring, or "slinky," every few turns of which are suspended by strings or threads, as shown in Figure 4–2. When one end of the coil is given a single

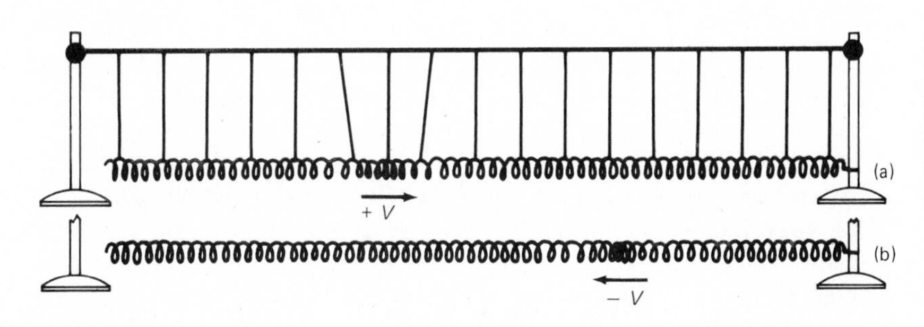

FIGURE 4–2
(a) A longitudinal wave pulse travels along a coiled spring with a speed $+V$. After reflection from the other end, it travels in the opposite direction with the same speed $-V$.

push or pull, a **longitudinal wave pulse** is sent along the spring. Traveling with constant speed to the right, this pulse arrives at the far end and is reflected back to the left with the same speed. A return pulse will occur whether the far end is rigidly fastened, as shown, or is free to move.

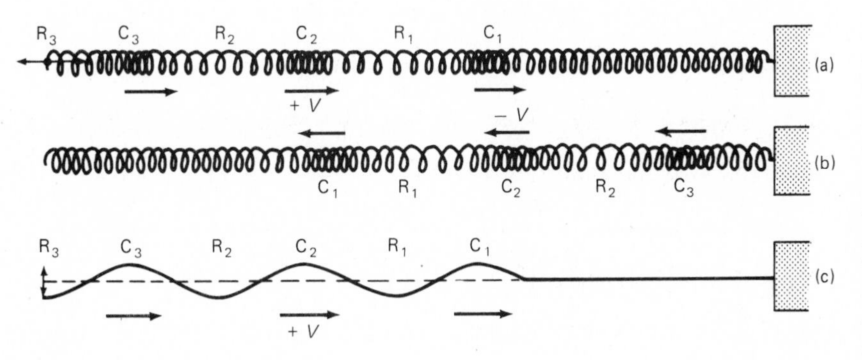

FIGURE 4–3
A longitudinal wave train is shown traveling to the right in diagram (a), and a reflected wave train is shown traveling to the left in diagram (b). A transverse wave train traveling to the right along a rope is shown for comparison in diagram (c).

If the left end of the spring is moved back and forth with simple harmonic motion, a train of longitudinal waves is sent along the spring, as shown in Figure 4–3(a). Reflected from the far end, the wave train reverses its direction and travels toward the source.

Because such waves are difficult and tedious to draw, it is customary to represent them by transverse graphs, as shown in Figure 4–3(c). The crests correspond to points of compression (high pressure) C_1, C_2, C_3, . . ., and the troughs correspond to points of rarefaction (low pressure) R_1, R_2, R_3, ... Similar graphs were made for transverse waves (Figure 3–5) and, in those cases, corresponded visually to the transverse wave itself.

An excellent classroom demonstration of the vibrations of each small part of the medium (air) can be illustrated by a longitudinal wave machine, as shown in Figure 4–4. As the crank H is turned at constant speed, each ball

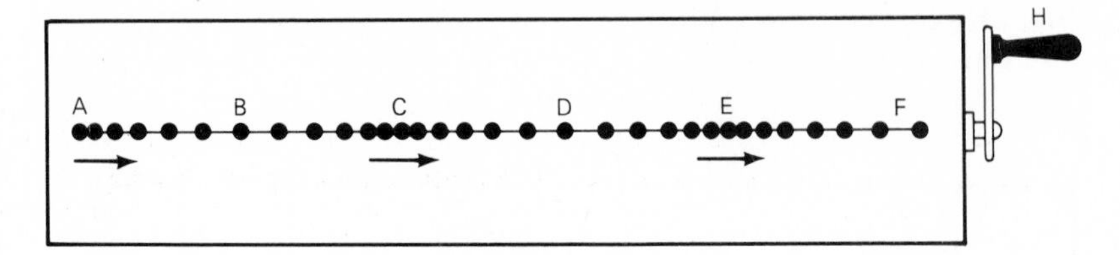

FIGURE 4–4 Side view of a longitudinal wave machine used for demonstrating the vibrations of air molecules as sound waves pass through them (see Figures 4–10 and 4–11 for construction details).

at the end of a rod moves back and forth about its equilibrium position with SHM. Each ball is slightly ahead of or behind the motion of its immediate neighbor, and this produces regions of compression and regions of rarefaction moving along from left to right. If the crank is reversed in direction, the waves appear to move from right to left. A time-lapse series of drawings of the balls only is shown in Figure 4–5. Construction details are given in Section 4.5.

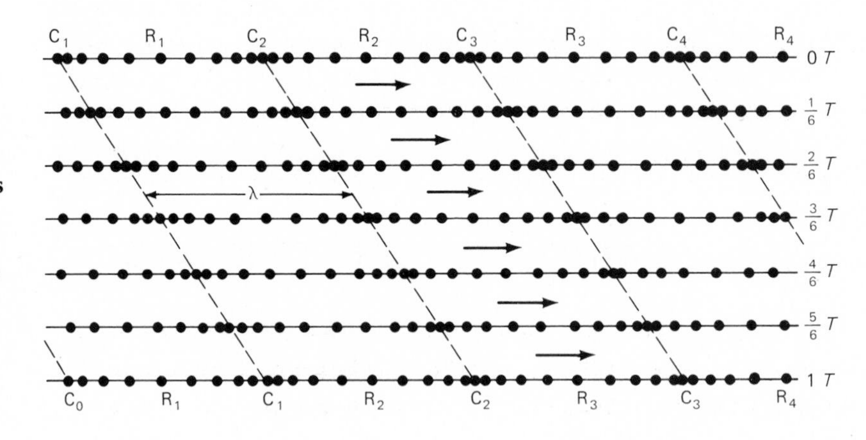

FIGURE 4–5 Time-lapse drawings of the positions of air molecules as longitudinal waves travel through them to the right. Time intervals are one-sixth of a period apart.

4.2 Standing longitudinal waves

All wind instruments, woodwinds as well as brasses, sustain longitudinal waves in their enclosed air columns. The vibrating air mass within each instrument is the source of its sound. To demonstrate standing waves, we start with the coiled spring or slinky described in Figure 4–2 and move the left end back and forth continuously with SHM. The continuous wave train moving to the right is reflected back on itself from the other end. The two wave trains, traveling with the same speed, one to the right and the other to the left, interfere and set up displacement nodes and antinodes. See Figure 4–6. At the nodes N_1, N_2, N_3, . . ., the spring remains practically at rest, whereas at the antinodes A_1, A_2, A_3, . . ., the spring is in a state of vibration with maximum amplitude. For example, when the particles at A_1 are moving to the right, those at A_2 are moving to the left, those at A_3 are moving to the right, those at A_4 are moving to the left, and so forth.

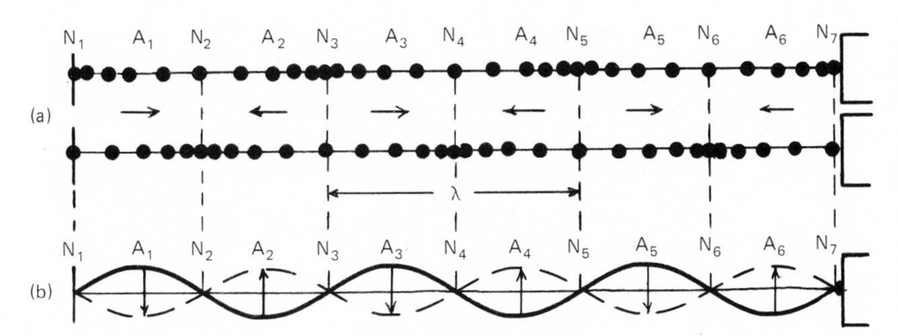

FIGURE 4–6 Comparison of the vibrations in (a) the air molecules in standing longitudinal waves of an air column with (b) the standing transverse waves in a rope.

The time-lapse drawings of standing longitudinal waves shown in Figure 4–7 are reproduced for comparison purposes. They show the relative motions of particles in the medium. The drawings represent conditions at regular intervals of one-eighth of a period *T*. With standing transverse waves, each small part of the string moves at right angles to the direction of the waves, while in longitudinal standing waves, the motions of the air molecules are along the line of propagation.

The wave machine shown in Figure 4–8 makes a good classroom demon-

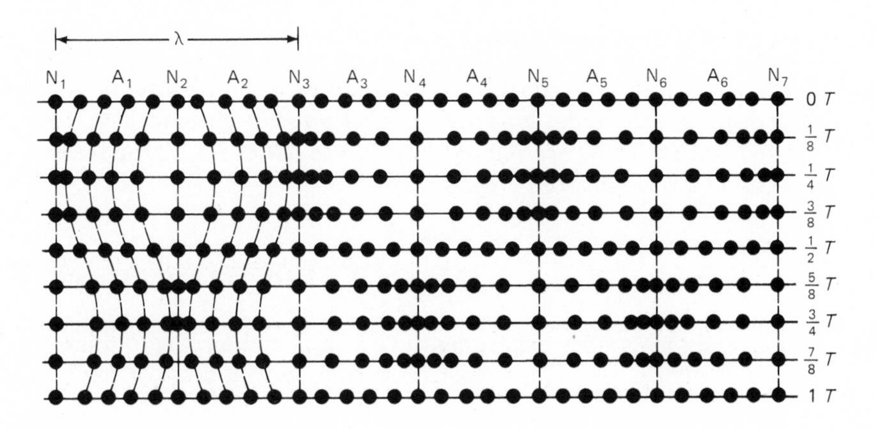

FIGURE 4–7 Time-lapse diagrams of the positions of air molecules in longitudinal standing waves in an air column showing their behavior at nodes, antinodes, and in between.

stration. White balls on the ends of vertical rods move back and forth horizontally with amplitudes that vary between zero at the nodes and a maximum at the antinodes.

While sound waves in air, and standing waves in wind instruments, are longitudinal in character, it is customary to draw them as transverse plots. Such graphs have already been seen to be visually similar to the shape of transverse waves. This is part of the reason for making the one-to-one comparisons shown in Figures 4–3 and 4–6. In the following chapters, sound waves will therefore be represented by transverse graphs.

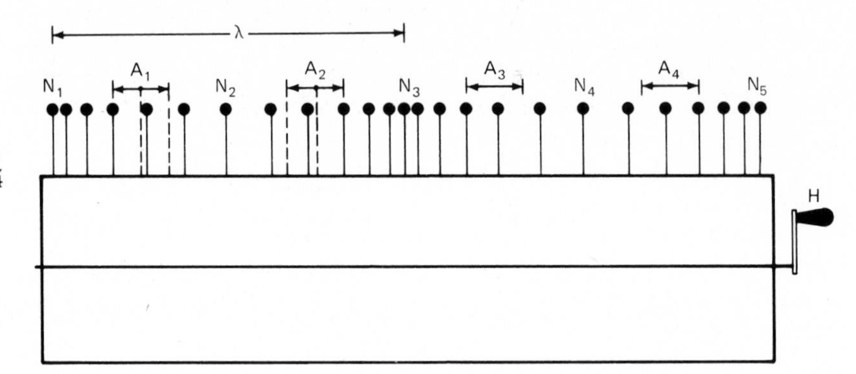

FIGURE 4–8 Machine for demonstrating standing longitudinal waves. (See also Section 4.4 and Figures 4–10 and 4–11.)

4.3 Pressure nodes and antinodes

An examination of Figure 4–6 will show that the nodes N_1, N_2, N_3, . . . are regions where the displacement of air molecules is zero, and they are continuously at rest. At one instant, alternate nodes N_1, N_3, N_5, . . . find the neighboring molecules closest together, and one-half of a period later they are farthest apart. When they are close together, the gas pressure is high, and when they are far apart, the gas pressure is low. Therefore, these are pressure antinodes.

At the displacement antinodes A_1, A_2, A_3, . . ., the motion is a maximum. Although the molecules have their greatest amplitude here, their average distance apart remains constant, which means the pressure is constant. Hence displacement nodes N_d are regions of pressure antinodes A_p, and displacement antinodes A_d are regions of pressure nodes N_p. For more details, see Section 12.3 and Figure 12–3.

4.4 Laboratory experiment

An interesting laboratory experiment involving standing longitudinal waves is shown in Figure 4–9. This experiment consists of a brass or steel rod R, about 60 to 70 cm long and 6 mm (abbreviation for millimeter) in diameter, clamped rigidly at its center N. At one end is a small lightweight disk, serving as a plunger P, in a glass tube G about 4 cm in diameter. A movable plug P′ at the other end permits one to vary the length of the entrapped air column.

The left end of the rod is stroked longitudinally as indicated, using a piece of cloth previously coated with rosin or soaked in alcohol. Properly held and stroked from right to left, the rod will sing out, with a node at N. The disk plunger P sends longitudinal waves through the column of air, and the waves reflected back from P′ may set up standing waves in the air column. With successive trials, moving the plunger P′ to different positions along the tube, positions can be found where fine cork dust particles, lying in the bottom of the tube, will divide up into equally spaced piles, as indicated. The cork piles will be more pronounced if the glass tube is turned slightly before stroking, as the particles will then settle to the bottom at the antinodes where the air vibrations are most active.

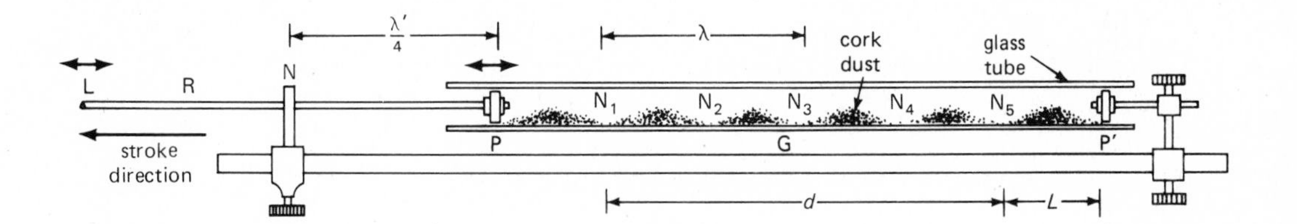

FIGURE 4–9
Laboratory experiment for measuring the wavelength of standing longitudinal waves in an air column and in a metal rod.

If the total length of a number of equally spaced cork piles is measured, the frequency of the vibrating rod, which is also the frequency of the waves in the air column, can be calculated. Assume the speed of sound in air to be 350 m/s, and use Equations (3a) and (3c).

Example

Air is enclosed in a glass tube as shown in Figure 4–9. If the vibrating brass rod sets up standing waves in the air column that produces cork dust piles 6.0 cm long, find (a) the wavelength of the sound waves in the tube and (b) the frequency of the sound waves. Assume the speed of sound in air to be 350 m/s.

Solution

The given quantities are $L = 6.0$ cm and $V = 350$ m/s.

(a) Using Equation (3c), we solve for λ and obtain

$$\lambda = 2L$$

which, upon substituting of the known quantity L, gives

$$\lambda = 2 \times 6.0 = 12.0 \text{ cm}$$

(b) Using Equation (3a), we solve for f and obtain

$$f=\frac{V}{\lambda}$$

which, on substitution of the known speed V, in meters/second, and the wavelength λ, in meters, yields

$$f=\frac{350 \text{ m/s}}{0.12 \text{ m/c}}=2917 \text{ Hz}$$

4.5 Shop projects

Two machines for generating traveling and standing longitudinal waves can be constructed in the average college student's workshop. The principal machine tools needed are a machine lathe, a drill press, a milling machine, and a metal band saw.

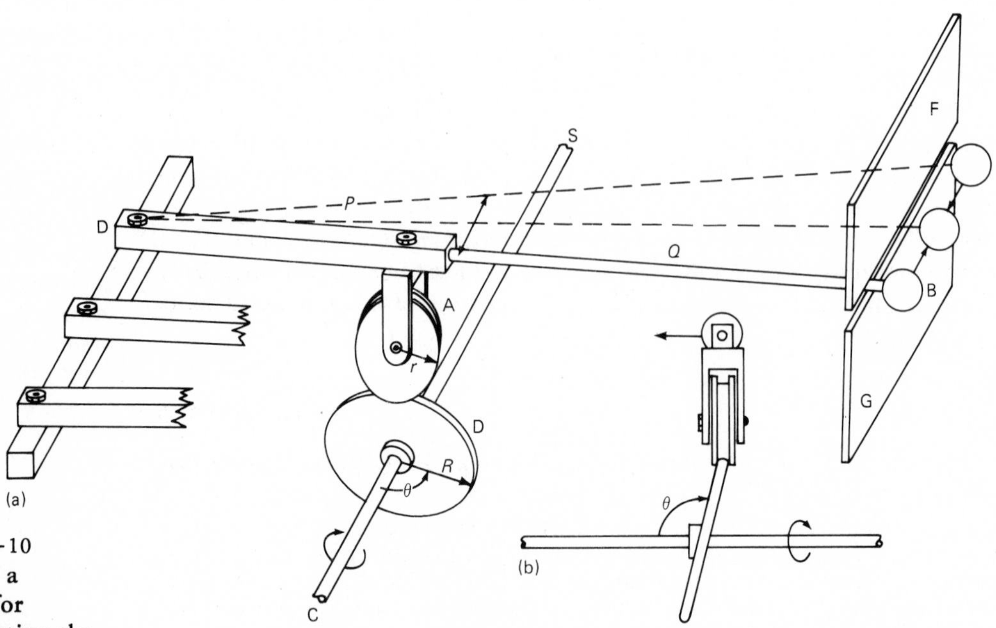

FIGURE 4–10 Details of a machine for demonstrating the vibrating motion of air molecules in traveling longitudinal waves: (a) pivoting arm, (b) canted guide wheel.

Inside a wooden box, similar to the one shown in Figure 4–4, is a set of 25 levers and disks, like the one shown in Figure 4–10(a). Each grooved wheel A, mounted so it swivels at the top, moves back and forth as it rolls on the top of the tilted disk D below. In the machine shown here, all 25 disks are mounted at their centers, tilted at the same angle θ with the axle rod on which they rotate. See detail in diagram (b). Each disk lags on the axle by an angle of 30° behind the one on its left.

With each lever pivoted at D on a rigid support, rotation of the axle CS will cause each ball at B to move back and forth horizontally with SHM. The screen FG, with the horizontal slot, represents the front face of the box. Turning the axle handle H in Figure 4–11 will cause compressed regions of balls to move to the right, while reversing the direction of H reverses their direction of motion.

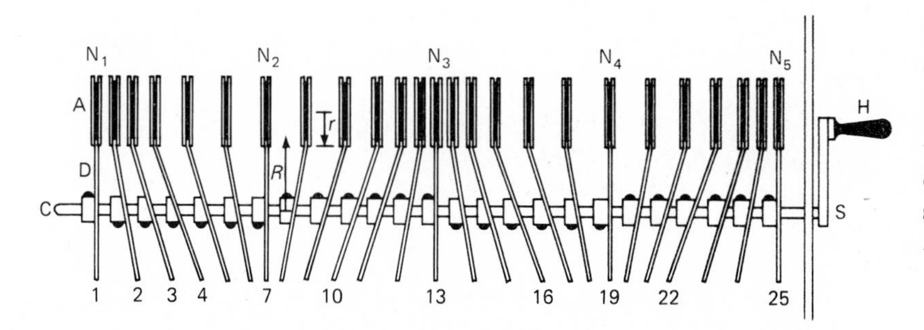

FIGURE 4–11 Working diagram of a machine for demonstrating the vibrating motion of air molecules in standing longitudinal waves.

In the second machine demonstrating standing longitudinal waves, the tilt angles of the disks vary from one lever to another, as shown in Figure 4–11. Disks 1, 7, 13, 19, and 25 are at nodes N_1, N_2, N_3, . . ., where there is no motion, and can be replaced by fixed balls on the front face of the machine. Disks 4, 10, 16, and 22 are at antinodes and have the greatest tilt angles, and the corresponding balls on the front face have the greatest amplitudes. Rotation of the handle H in either direction will give rise to the same standing wave patterns.

QUESTIONS

1. Define or briefly explain each of the following: (a) traveling longitudinal waves and (b) standing longitudinal waves.

2. Make a diagram showing longitudinal waves traveling along a coiled spring or slinky suspended in a horizontal position by threads or strings. Show the positions of the individual turns of the spring at one instant only. Also make a diagram of the corresponding transverse wave immediately below, and label the crests C and the troughs T.

3. Make a diagram of standing longitudinal waves on a coiled spring or slinky suspended as shown in Figure 4–2. Show the positions of the individual turns of the coil at one instant only. Also make a corresponding standing transverse wave pattern immediately below, and label the nodes N and the antinodes A.

4. If the nodes of the transverse wave (previous question) correspond to the slinky displacement nodes, how does this differ from comparing nodes in the

transverse wave to the slinky compression nodes? Make a drawing to illustrate the difference.

PROBLEMS

1. Air is enclosed in a glass tube as shown in Figure 4–9. The metal rod is 62.0 cm long, is clamped at its center, and is set vibrating with standing longitudinal waves. The length of five cork dust piles are measured and found to be 26.0 cm long. Assuming the speed of sound in air is 348.0 m/s, find (a) the frequency of vibration, (b) the wavelength of sound waves in the metal rod, and (c) the speed of sound in the rod.

2. A glass tube containing air and some cork dust is shown in Figure 4–9. A metal rod 55.4 cm long is used to produce standing longitudinal waves. (a) If the speed of sound in the metal is 3500 m/s, find the frequency of vibration. (b) If six cork dust piles are measured and found to be 33.6 cm long, what is the speed of sound in air?

PROJECT

If a longitudinal wave machine is not available in your laboratory, secure a coil slinky and suspend it by a number of threads, as shown in Figure 4–2. A slinky can often be purchased at a relatively low price at a local toy or hobby shop. Try sending waves along the coiled spring to study the motion. Try clamping the far end, and compare the returned waves in both cases. What differences can you detect?

Chapter Five

SOUND TRANSMISSION

Our present knowledge and treatment of sound is usually divided into three parts. These are

sources transmission reception

This chapter is devoted principally to the *transmission* of sound and its *speed* of travel through a medium. For sound to be transmitted from one place to another, a material medium—solid, liquid, or gas—is required. The medium itself is not transported from source to receiver. Only the waves themselves are transmitted, and these constitute a flow of sound energy.

5.1 Sound waves

That sound is transmitted by air, or any other gas, may be demonstrated by suspending a small bell in a jar, as shown in Figure 5-1. As the air is slowly removed from the jar, the ringing of the bell becomes fainter and fainter. When a good vacuum is obtained, no sound can be heard. As soon as air is admitted to the jar, however, the ringing becomes clearly audible again. The vibrating bell strikes air molecules, knocking them away from the metal surface. These molecules strike the adjacent air molecules, and they in turn strike others. Upon reaching the side of the jar, the glass walls are periodically bombarded by the molecules and set vibrating. The walls in turn set the outside air vibrating. Arriving at the observer's ear, the disturbance strikes the eardrum, setting it into motion. Without air to transmit the vibrations from the bell to the inside surface of the glass jar, no sound waves can leave the jar.

The transmission of sound by liquids may be illustrated by the experiment shown in Figure 5 – 2. A tuning fork with a thin aluminum disk attached to its base is set vibrating and then touched against the surface of a dish of

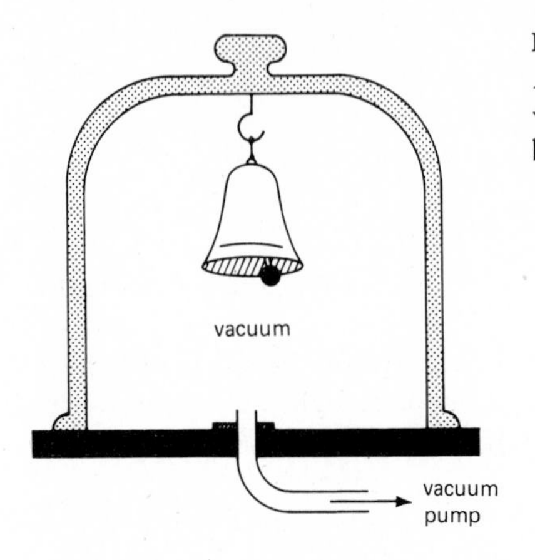

FIGURE 5 – 1
A bell ringing in a vacuum jar cannot be heard.

water. The vibrations of the fork and disk travel through the water to the bottom of the dish and to the tabletop. The tabletop itself is set into vibration with the same frequency as the fork, thus acting as a soundboard to make the sound louder.

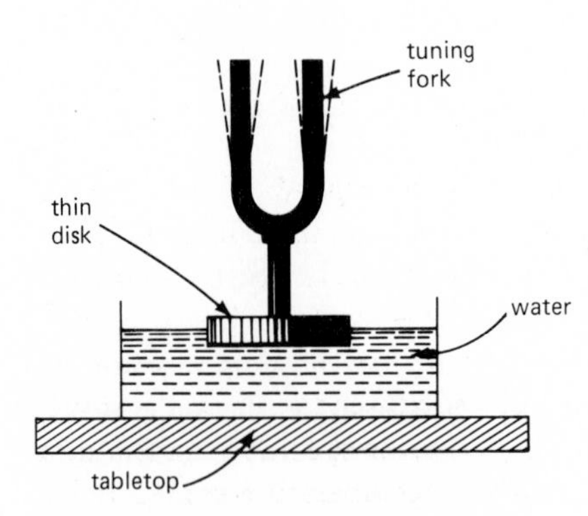

FIGURE 5 – 2
An experiment that demonstrates that sound waves travel through water, a liquid.

The transmission of sound by solids is illustrated in Figure 5 – 3. A vibrating tuning fork is brought into contact with the end of a long wooden rod. The longitudinal vibrations travel down the length of the rod, which causes the hollow wooden box at the other end to vibrate. Sound is clearly heard coming from the box.

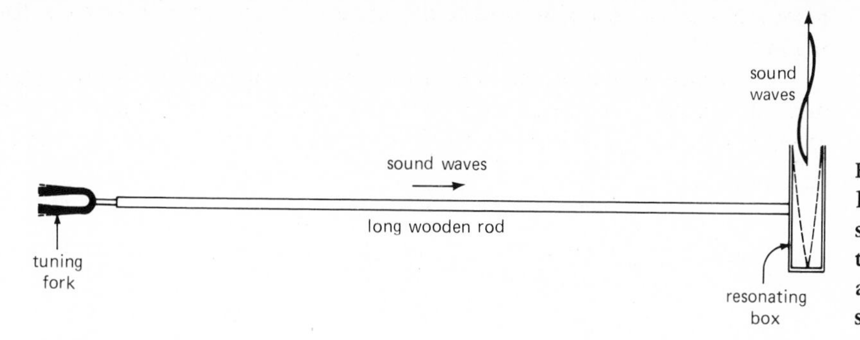

FIGURE 5–3
Demonstration of sound waves traveling through a wooden rod, a solid.

5.2 The speed of sound

Although light and sound both travel with a finite speed, the speed of light is so great in comparison that an instantaneous flash of light may often be regarded as taking no time to travel several miles. When we see the light from a distant lightning flash and hear its sound (thunder) several seconds later, we know that the difference in time is due to the relatively slow speed of sound. Knowing that sound requires 3 full seconds to travel 1 kilometer (km) [5 seconds per mile (mi)], the distance of a passing thunderstorm can be determined by timing the arrival of the thunder. Similarly, when a distant carpenter strikes a nail with his hammer, the arrival of the accompanying sound is not heard until an appreciable time later. For comparison, at normal room temperature and pressure, the speed of sound and the speed of light are approximately as follows:

$$\text{speed of sound} = 350 \text{ m/s}$$
$$\text{speed of light} = 300{,}000{,}000 \text{ m/s}$$

The earliest successful attempts to measure the speed of sound were made in 1640 by Marin Mersenne, a French physicist, and in 1656 by Giovanni Borelli and Vincenzo Viviani, Italian physicists. Since that time many experimenters have improved upon these earliest measurements by using various methods. The most accurate measurements up to 1934 were made by Miller.* With coast defense guns as a source of sound and a set of electronic receivers located at certain distances apart, very accurate speed determinations were made. The results gave a speed of 331 m/s at a temperature of 0°C and at **standard atmospheric pressure,** which is 1.013×10^5 newtons/meter^2 (N/m^2). More refined measurements were made and reported eight years lat-

*Dayton C. Miller was an American physicist noted for his experiments on the quality of musical sounds and on the ether drift. He collected and had in his possession the largest collection of flutes in the world. He turned over these instruments to the Smithsonian Institute in Washington, D.C., where they are now on exhibit.

er,[1] and they gave the value 331.5 m/s. This is equivalent to 1193.4 km/hour (h), or 741.6 mi/h.

As a general rule, sound travels faster in solids and liquids than it does in gases. This is illustrated by the data shown in Table 5-1, which gives the speed of sound as measured when traveling through a few common materials.

TABLE 5-1: The Speed of Sound in Common Materials at 0°C

SUBSTANCE	SPEED (m/s)
GAS	
Air	331
Carbon dioxide	258
Carbon monoxide	337
Hydrogen	1,269
LIQUID	
Alcohol	1,213
Benzine	1,166
Turpentine	1,376
Water	1,435
SOLID	
Aluminum	5,104
Brass	3,500
Diamond	14,000
Glass	5,500
Ivory	3,013
Nickel	4,973
Oak (with grain)	3,850
Steel	5,130

To find the speed of sound, we make use of the simple relation that speed equals the total distance traveled divided by the time of travel:

$$\text{speed} = \frac{\text{distance traveled}}{\text{time of travel}} \qquad V = \frac{x}{t} \qquad [5a]$$

If, for example, it takes sound 1.0 min to travel a distance of 21 km, we can use Equation (5a) to find the speed of sound. If we substitute the known quantities directly into Equation (5a), we obtain

$$V = \frac{21{,}000 \text{ m}}{60 \text{ s}} = 350 \text{ m/s}$$

Or, as another example, suppose sound waves in a steel rod have a speed of 5130 m/s. How far will a train of waves travel in 2.0 min? Solving Equation (5a) for the unknown quantity x, we obtain

$$x = Vt \qquad [5b]$$

and substituting the known quantities, we obtain

$$x = 5130 \text{ m/s} \times 120 \text{ s} = 615{,}600 \text{ m}$$

or

$$x = 616 \text{ km}$$

It is well known that temperature has a small but measurable effect upon the speed of sound. For example, for each degree rise in temperature, the speed increases by 0.610 m/s. Written as an equation, this is

$$V = V_0 + 0.610t \qquad [5c]$$

where V_0 is the speed in meters/second at 0°C and t is the temperature in degrees Celsius.

Example

Light reflected from a jet stream emerging from the blast of a boat whistle is noted by an observer on a distant hilltop. If the temperature of the air is 25°C, find (a) the speed of sound and (b) the time it will take the sound to reach the observer if he is 2.50 km away.

Solution

The given quantities are $t = 25°\text{C}$, $V = 331.5$ m/s, and $x = 2500$ m. (a) To find the speed of sound at 25°C, we can use Equation (5c). Upon substitution of the known quantities, we obtain

$$V = 331.5 + 0.610 \times 25 \text{ m/s}$$
$$V = 331.5 + 15.25 \text{ m/s}$$
$$V = 346.8 \text{ m/s}$$

(b) To find the time it takes for the sound to reach the observer on the hill, we use Equation (5b). Solving this equation for the unknown t, we obtain

$$t = \frac{x}{V} = \frac{2500 \text{ m}}{346.8 \text{ m/s}}$$

$$t = 7.21 \text{ s}$$

5.3 Reflection of sound waves

This phenomenon may be demonstrated in many different ways. One arrangement is shown in Figure 5–4, in which a Galton whistle* sounding a high-

*A Galton whistle can be purchased at almost any apparatus supply house. It is quite small in size. By turning a screw in its base, very high frequencies, some of them beyond the audible range, can be produced.

pitched note acts as a source of sound and a sensitive gas flame acts as a receiver. (See Project 5B.) A solid screen located between the two casts a **sound shadow**, thus permitting only the reflected waves from the wall of the room to reach it. If the whistle, blown by compressed air, is sounded continuously, the flame will be noticeably unstable. If, under these conditions, the experimenter walks close to the wall through the sound path, the flame will remain unstable until his body blocks the path at B or D, at which time the flame burns smoothly.

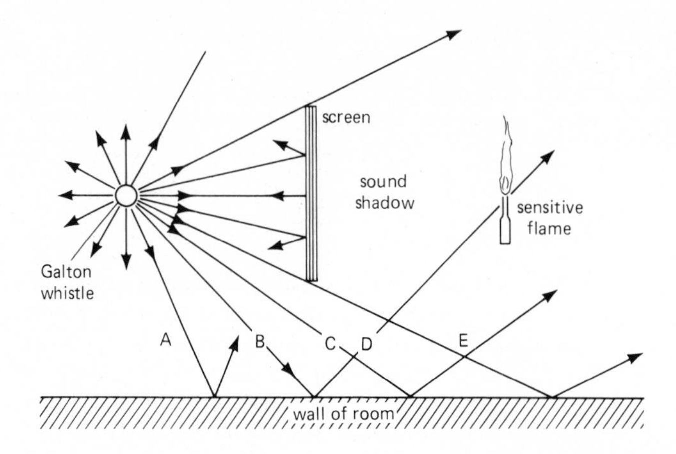

FIGURE 5–4 **Demonstration of the reflection of sound waves from the wall of a room.**

Sound shadows of this kind are characteristic of high-pitched notes and demonstrate that short waves tend to travel in straight lines. The longer waves of lower-pitched notes tend to bend around corners. The latter effect, known as **diffraction,** is quite noticeable where a carillon of large bells is being played. On walking around the corner of a nearby building, one notices that the sharp cutoff in the intensity of the high-pitched bells is quite marked, while the sounds from the low-pitched bells continue with good intensity. High-pitched notes also show diffraction, but to a lesser degree. This phenomenon of diffraction is due to the wave nature of sound and will be taken up in Chapter 7.

Oftentimes, in the hills or in large rooms, one can shout several words and hear a well-pronounced echo. This is another example of the reflection of sound waves. The echo is the return by reflection of the sound waves from a large obstructing surface some distance away.

5.4 Refraction of sound waves

The bending of sound waves by layers of air at different temperatures is called **refraction.** This phenomenon, which can be demonstrated in various ways, is due to the greater speed of sound in warm air than in cold. (See Section 5.2.)

If you have ever been boating on a lake or river, you probably have noticed that you are able to hear music from a quite distant radio or sound system at nightime but not in the daytime. This is an example of the refraction of sound waves. The situation is shown in Figure 5-5. At night, the air near the water is colder than it is higher up, so that the higher speed in the warmer air bends the waves back down. During the day, the air close to the water is warmer, and the sound waves bend up away from the water as shown.

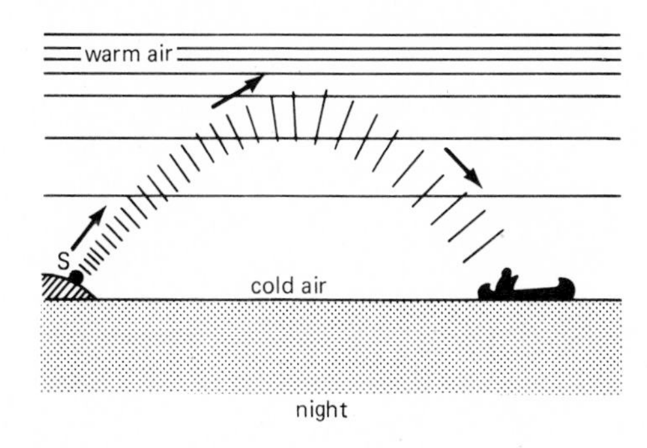

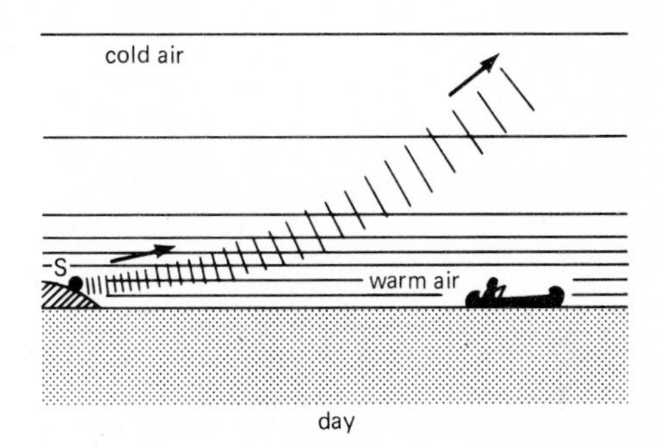

FIGURE 5-5
The lower speed of sound waves in cooler air causes sound waves to be refracted up in daytime and down at nighttime.

Recent experiments of this kind have been performed with the very loud sounds from big guns. Sound waves refracted back from high up in the stratosphere indicate with some degree of certainty the existence of very warm layers of air at altitudes of 40 to 64 km. Refraction in cases such as this is similar to reflection from a mirror surface, since the waves travel in more or less straight lines, going up and back, but bend over when they enter, more or less abruptly, a warmer layer.

5.5 Acoustic impedance

We have seen in Figures 3-1 and 3-2 how a wave pulse or wave train traveling along a rope is reflected from the fixed end back toward the source. Suppose a rope is suspended from one end so that the other end hangs free, and a wave pulse is sent downward by a sudden flip of the top end. See Figure 5-6. The pulse will be reflected, but with one difference from that shown in Figure 3-1. Instead of returning on the opposite side of the rope, it will return on the same side, as shown.

Let us now imagine that we fasten two ropes A and B together, as shown in Figure 5-7, and send a wave train along one rope toward the junction. What would happen at the junction if the ropes are of different size, and how would the result compare with two ropes of the same size?

In the top graphs of all four pairs of curves, a wave train is shown being sent to the right along rope A. The lower diagram in (a) shows that when rope

B is thicker and more massive than rope A, it will hardly move at all under the periodic force of the impressed waves arriving at the junction. Most of the wave energy will be reflected back to the left on rope A, with little loss of amplitude. Rope B is said to have a *larger impedance* than rope A. This situation is similar to the case of a uniform rope with a fixed end as shown in Figure 3–2.

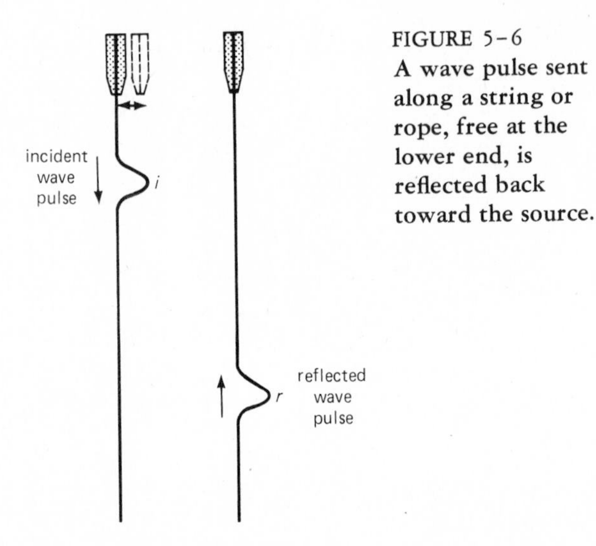

FIGURE 5–6
A wave pulse sent along a string or rope, free at the lower end, is reflected back toward the source.

In diagram (b), ropes A and B are the same and act as a single uniform rope with an imaginary junction J. Here the impedances of the two ropes are equal, and all the energy is transmitted. This is a case of *matched impedance.*

In diagram (c), rope B is less massive than rope A, and some energy is transmitted and some is reflected. Rope B is said to have a *smaller impedance* than rope A.

In diagram (d), rope B is absent, and the open end of rope A reflects all the energy back toward the source. In this case, the second medium is said to have *zero impedance.*

These diagrams serve to illustrate a general definition of **impedance,** the ratio of the force on a system to the response of that system. This general definition will allow us to apply the concept to a wide variety of examples.

Acoustic impedance concerns the response of a mechanical system to an applied vibratory force. Its quantitative form depends upon the specific situation: a sound wave traveling through air, the response of a specific system of moving parts, such as a violin bridge, and so on. We will see that the concepts of impedance and impedance matching come up in a variety of places throughout the remainder of this text.

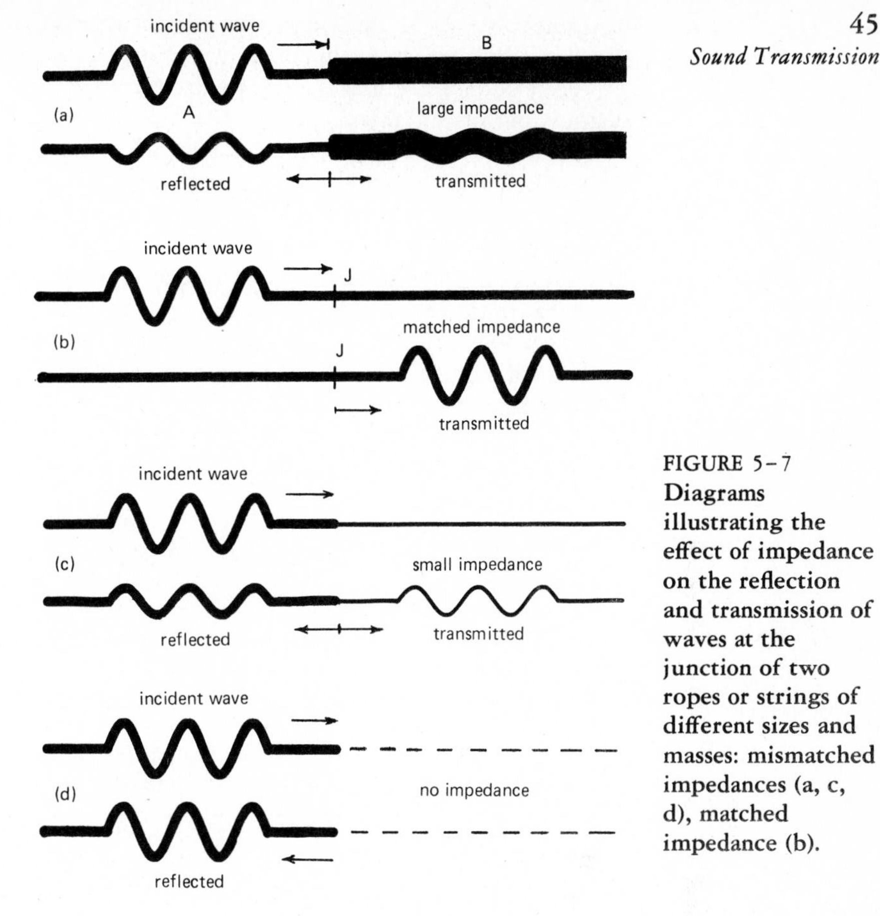

FIGURE 5-7 Diagrams illustrating the effect of impedance on the reflection and transmission of waves at the junction of two ropes or strings of different sizes and masses: mismatched impedances (a, c, d), matched impedance (b).

5.6 Impedance matching

To transmit the maximum amount of energy from one medium to another, the two media should be connected to one another and ideally have the same impedance. See Figure 5-7(b). If the impedance of two media are quite different, some intermediate system should be provided to transfer energy from one to the other with little reflection at the junction. Such a process is referred to as **impedance matching**, and the system for doing this is called an **impedance-matching transformer.**

Impedance matching is involved in our sensation of hearing, in optical and electrical devices, and in everyday sports like baseball, tennis, golf, handball, billiards, and bicycling. See Figure 5-8. Impedance matching is also involved in nearly all musical instruments. Oftentimes the impedance of a source of sound is quite different from the impedance of the surrounding air.

Instrument design, therefore, involves improving the match between the impedance of the vibrating source and the impedance of the surrounding air.

For example, the impedance of a drumhead is fairly well matched to the surrounding air, and this accounts for the loudness of the audible sound from a struck drum. In the case of a string, however, the string is rigidly fastened at both ends, and little sound will be heard when it is set vibrating. This results from the poor impedance match between the air and the string, due mostly to the small area of contact between the two. Stringed instruments, like the banjo, violin, cello, and guitar, are therefore designed to improve the impedance match by transforming the strings vibrations through a bridge to a drumhead, or large wooden surface, where the impedance match to the air is reasonably good. Such a device, which is common to most stringed instruments, can be called an impedance-matching transformer.

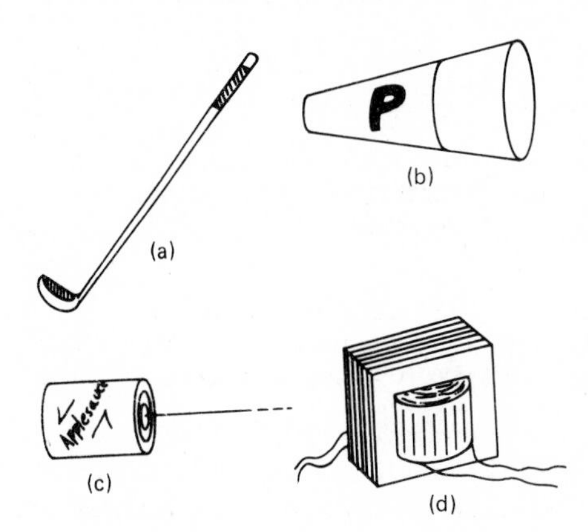

FIGURE 5–8. Examples of impedance-matching transformers include (a) a golf club, (b) a megaphone, (c) a tin can telephone, and (d) an electric transformer.

A second example that involves impedance matching is the *megaphone*, in which the size of the mouth opening is slowly and uniformly widened through the conically shaped walls to a much larger opening, thereby providing a better impedance match with the open air. This device too can be called an impedance-matching transformer.

5.7 Laboratory experiment

The experiment described in Section 4.4 should be performed, and measurements made to determine the speed of sound in metal and in air. See Figure 4–9. Measure the length of several cork dust piles in the glass tube, the length of the metal rod that is set vibrating, and the temperature of the room at the time the experiment is performed. Using Equation (5b) and the observed temperature, calculate the speed of sound in air. Using Equation (3c), calculate (a) the wavelength of the sound in the air column and (b) from the measured length of the metal rod, the wavelength of sound in the metal. Using Equation (3a) and the calculated frequency and wavelength in the metal rod, determine the speed of sound in the metal. Check your answer with the value given in Table 5 – 1.

QUESTIONS

1. Briefly describe an experiment demonstrating the transmission of sound by a long metal rod. Make a diagram and label the essential components.

2. Make a diagram and briefly describe an experiment demonstrating the transmission of sound by a liquid. Label the important components.

3. Describe briefly a demonstration of the reflection of sound waves. Make a diagram and label the important components in the experiment.

4. Explain how sound waves are refracted by the atmosphere. Make a diagram.

5. Define or briefly explain each of the following: (a) impedance, (b) impedance matching, and (c) impedance-matching transformer.

6. Why do you think the term *impedance-matching transformer* can be applied to (a) the swinging of a golf club in hitting a golf ball, (b) a tennis racquet hitting a tennis ball, (c) one billiard ball hitting another, and (d) the chain-and-sprocket wheels of a bicycle?

7. Prior to the invention of hearing aids, people with hearing deficiencies often used an "ear trumpet," a small horn with the narrow end placed into the ear canal. Explain how this device increases the sound transmission into the ear.

PROBLEMS

1. If the room temperature is 32°C, find the speed of sound in air.

2. The speed of sound is measured in the laboratory and found to be 348.0 m/s. What is the temperature of the air?

3. In performing the experiment described in Section 4.4 and shown in Figure 4 – 9, the following measurements were made: the length of five cork dust piles was 33.80 cm; the length of the metal rod was 72.0 cm; the temperature of the room was 30°C. Calculate (a) the speed of sound in air, (b) the wavelength of sound in air, (c) the frequency of sound in the glass tube, (d) the frequency of sound in the metal rod, (e) the speed of sound in the metal rod.

PROJECTS

Project A. These projects, which involve sending musical sounds through gases, liquids, and solids, are very effective in their performance and make excellent classroom demonstrations. Figure 5 – 9 shows two long glass tubes about 4 to 5 cm in diameter and 1.5 to 3.0 m in length. The wooden rod is comparable in length.

The first tube, diagram (a), has an electric doorbell D suspended by rubber bands at end A and a microphone M similarly suspended at the other.

Thin, flexible, stranded and coiled wires, leading through glass seals to a battery and push button at one end and an audio amplifier and loudspeaker at the other end, provide a demonstration of sound waves transmitted by a gas. The end plates are metal.

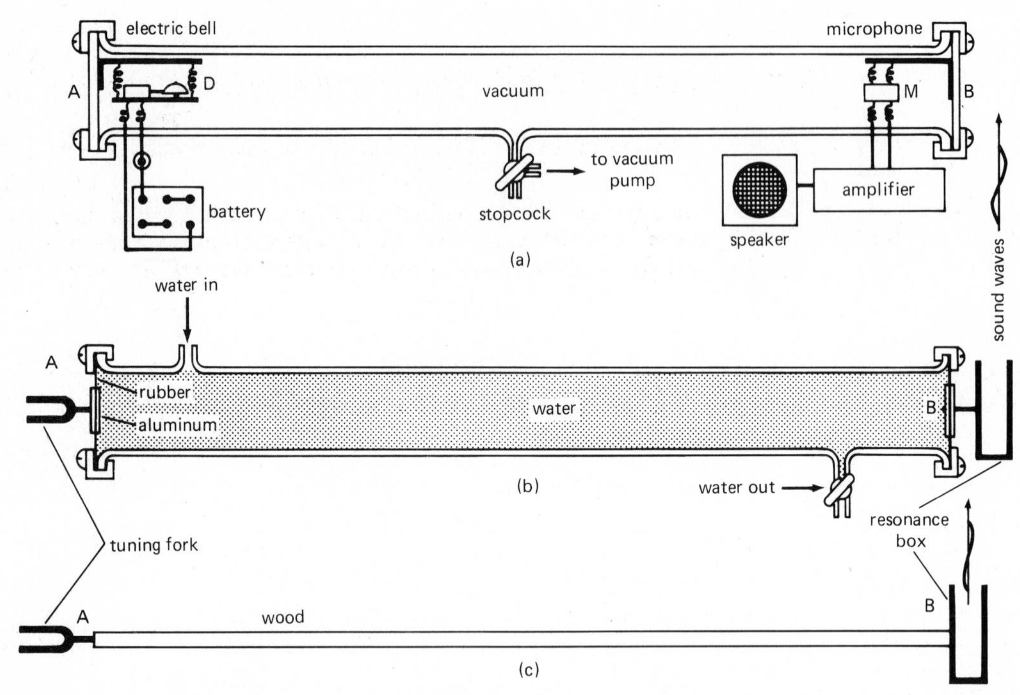

FIGURE 5–9 Student projects for demonstrating the transmission of sound through (a) gases, (b) liquids, and (c) solids.

The second tube, diagram (b), contains a water inlet near one end and an outlet with a stopcock for emptying at the other. The clamped-on ends are rubber membranes with lightweight aluminum disks of smaller diameter to act as plungers. A tuning fork at one end and a resonating box at the other act as source and receiver of sound waves. Each aluminum disk can be made of two thin aluminum sheets riveted together at several points near the edges.

The third demonstration, diagram (c), is like that shown in Figure 5–3. Two or three wooden dowels, purchased at a lumberyard and about 1 cm in diameter, are spliced and glued together to make one long rod.

More sophisticated demonstrations for students who have some knowledge of electronics can be constructed which are even more spectacular. For the water and wood experiments, the tuning forks and resonator boxes can be replaced by the vibrating voice coils of small loudspeakers. A diagram of these auxilliary units is given in Figure 5–10. Three aluminum wires, in the form of a tripod, are glued to the periphery of the voice coil with epoxy glue. A large part of the paper cone can be carefully cut away, leaving three or more ribs to keep the voice coil centered in the magnetic gap. One apex is then glued to the center of each disk. Music from a tape deck or record player is then fed into the modified speaker unit at A, and the results are heard through an audio amplifier and loudspeaker attached to the modified speaker unit at the receiving end B.

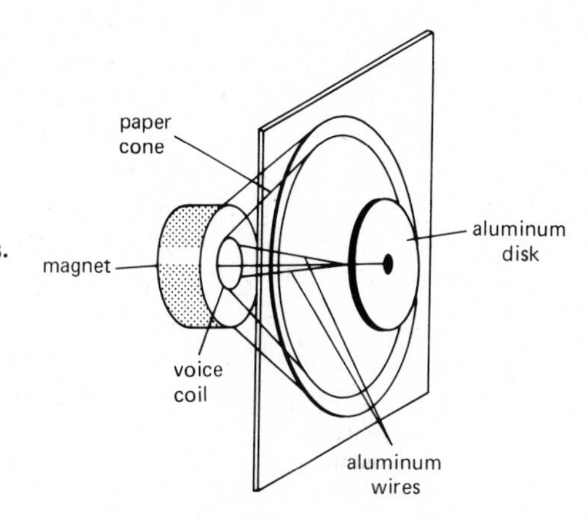

FIGURE 5-10
Details of a small loudspeaker, modified to send sound waves through gases, liquids, and solids.

Project B. The reflection of sound waves from a hard flat surface can be demonstrated by means of equipment similar to that shown in Figure 5-4. A Galton whistle will serve well in this experiment, and it can be blown continuously by compressed air or nitrogen from a pressure tank. For best results, a very high frequency of several thousand hertz should be used. If a Galton whistle is not available, shaking a bunch of keys will suffice.

A nozzle for the sensitive flame can be made of glass tubing 6 to 7 mm in outside diameter. A piece of tubing about 15 cm long is heated in its center section and pulled out to form a restricted section, as shown in Figure 5-11. A file scratch is then made at the center of the restriction and a clean break of the glass will produce two nozzles.

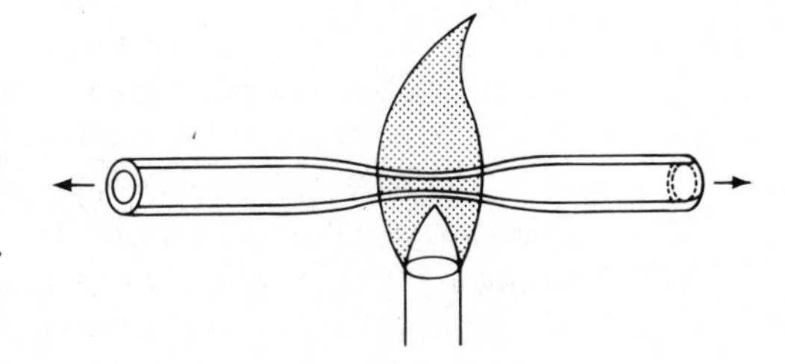

FIGURE 5-11
Construction diagram for making a glass nozzle for a sensitive flame.

With one nozzle connected by a flexible tube to a gas line, a flame is lighted and the pressure adjusted to produce a tall, narrow, smoothly burning jet. When the gas pressure is turned up too high, the jet will become unstable. It will function best as a detector when the pressure is just slightly under this unstable condition and the flame is burning smoothly. When the high-frequency source of waves is sounded, the direct waves passing the tip of the nozzle will cause the flame to become unstable. When the sound source is turned off, the flame should again burn smoothly. Waves reflected from a hard wall, such as a chalkboard, will markedly affect the flame. When the waves are shielded from the flame by any large obstacle, the flame will burn smoothly.

If a sound generator, a microphone, and an amplifier are available, more sophisticated experiments with reflected sound waves from flat and concave surfaces can be performed.

NOTES

[1]H. C. Hardy, D. Telfair, and W. H. Pielemeir, "The Velocity of Sound in Air," *Journal of the Acoustical Society of America* 13 (1942):226.

Chapter Six

RESONANCE, BEATS, AND DOPPLER EFFECT

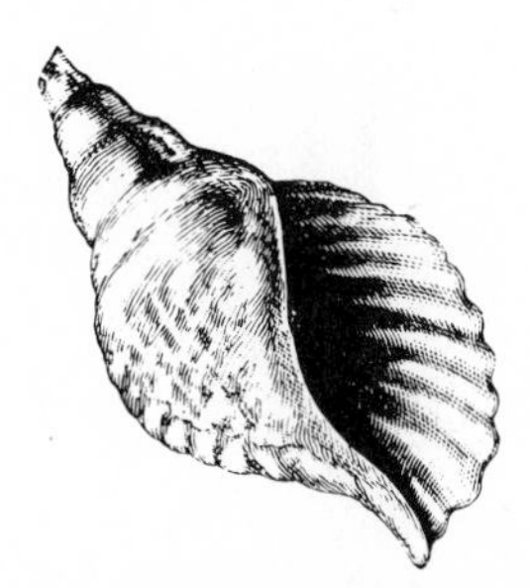

If two sources of sound having the same natural frequency are located some distance apart, and only one of them is sounded, the other will absorb energy from the waves and start vibrating with the same frequency. This is called **resonance**. If the two sources have quite different natural frequencies, this will not happen.

If two sources of different frequencies are sounded simultaneously, they periodically get in and out of step with each other. This produces periodic variations in loudness called **beats**.

If the source of a musical sound and an observer are not moving with respect to each other, the frequency that is heard is identical to that of the source. If, on the other hand, the source or the observer is moving with respect to the other, the observer hears a frequency that is different from that of the source. This phenomenon is called the **Doppler effect**.

These three subjects—resonance, beats, and the Doppler effect—are to be treated in this chapter.

6.1 Resonance

Suppose the corresponding strings of two violins are tuned to exactly the same frequency. If one string is set vibrating, the corresponding string of the other violin, even though it may be located some distance away, will sing out with the same frequency. This phenomenon is an example of **resonance.***

An experiment often used to demonstrate resonance is illustrated in Figure 6–1. Two tuning forks having exactly the same pitch are mounted on separate hollow boxes. Fork A is set vibrating for a moment and then stopped by touching the prongs with the fingertips. Fork B can then be heard vibrating, indicating it has been set into motion with the same frequency. If we examine the hollow boxes, whose function is to serve as resonating cavities, we find the explanation quite simple. The vibrating fork causes the box walls to vibrate, generating acoustical waves. The sound waves emerging from the box A travel across the room and enter box B, pushing out the walls at just the right moments to make the second fork B vibrate with the same natural frequency. This is similar to a person pushing someone in a swing. If she pushes periodically at the right times, the amplitude of the swing increases, whereas if she pushes at the wrong frequency, there is little response.

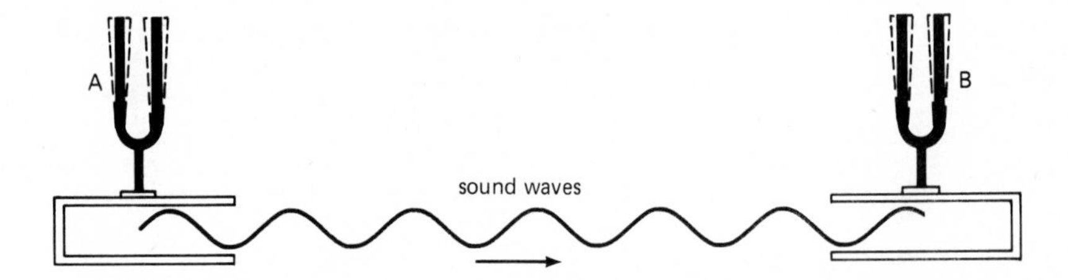

FIGURE 6–1
Two identical tuning forks mounted on resonator boxes and used for demonstrating the phenomenon of resonance.

Another experiment demonstrating resonance, and one that is easily set up in the laboratory, is shown in Figure 6–2. A set of simple pendulums is suspended from a flexible support. This support can be made of thin strips of wood or metal. The center bob E is relatively heavy compared to the others, and it can be made of metal. The others should be light in weight and can be made of wood. It is known that the period of a simple pendulum depends on its length l and the acceleration due to gravity g and is given by

$$T = 2\pi\sqrt{\frac{l}{g}} \qquad [6a]$$

where T is the period in seconds, l is the length of the pendulum in meters, $\pi = 3.14$, and $g = 9.80$ m/s^2. Observe in this relation the fact that the mass does not appear. This is due to the experimental observation that the period and frequency are independent of the mass for the special case of the pendulum.

When the heavy bob E is pulled back and set swinging, the horizontal support responds to the motion and swings back and forth in step with E. The vibrating support in turn acts forcibly on the other pendulums, and they re-

*Musicians often use the word *resonant* when referring to an acoustical enclosure where successive reflections cause an after ring, such as an auditorium, a cathedral, or a cave. Acousticians refer to such a setting as *reverberant.* (See Section 26.5.) To avoid confusion, we reserve the word *resonance* to refer to those strongly frequency-dependent responses that are described in this chapter.

spond with various amplitudes. The short and long pendulums A and I show very little response, while pendulums D and F show a maximum response. The closer the natural frequency of any pendulum to the forced vibration of the support, the greater is the amplitude of the response. See Figure 6–3.

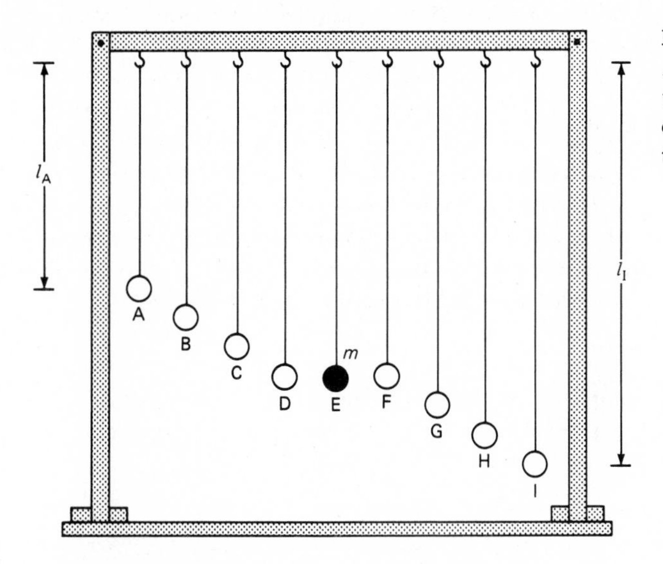

FIGURE 6–2
A demonstration using pendulums of different lengths to show resonance.

Since D, E, and F have the same length, their natural frequencies are equal, and D and F respond to the swaying support and swing with a large amplitude. They respond in *sympathy,* or *resonance,* to the *driving pendulum* E. Resonance may be defined as the state of a system in which a large vibration at the natural frequency is produced in response to an external vibrating stimu-

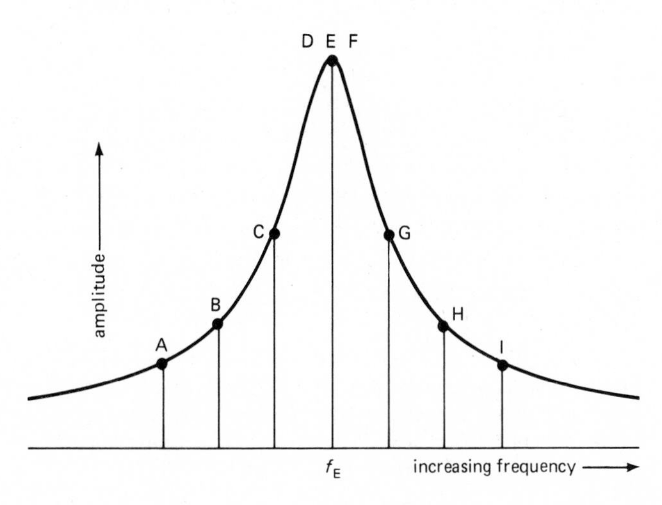

FIGURE 6–3
Resonance curve showing the amplitude response of the pendulums in Figure 6–2.

lus of the same, or nearly the same, frequency. The frequency of maximum response, the **resonant frequency**, is always equal to the natural frequency.

As an experiment, try singing a sustained note directly at the strings of a piano and then listening, in all possible quietness. One of the strings will be heard sounding with sympathetic vibrations of the same frequency.

6.2 Cavity resonators

The first cavity resonators were developed centuries ago and consisted of large ceramic vases standing as decorations in large meeting rooms, cathedrals, and auditoriums. Their function was to absorb certain sounds that were objectionable to an audience. To see how this comes about, we will first consider what are called **Helmholtz resonators.***

The shape of a Helmholtz resonator is shown in Figure 6–4. It is a hollow metal sphere, often made of brass, and it has two openings, which are opposite one another. The larger opening A, called the *neck,* serves as the entrance aperture for sound waves, while the small opening B is held to the ear.

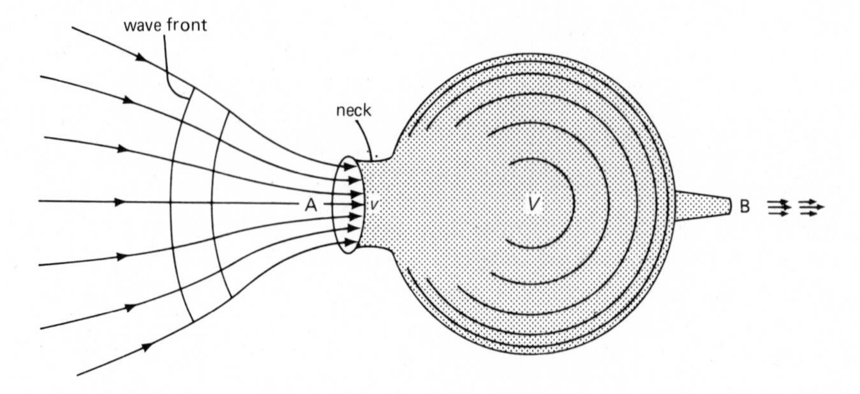

FIGURE 6–4
Typical Helmholtz resonator showing sound energy from a large area being drawn into the cavity neck.

As the sound wave strikes the open neck, the air in the neck moves in and out, causing periodic compressions and rarefactions of the air in the cavity. The air in the cavity behaves like a spring, acting on the mass of the air in the neck. Such a system, consisting of a spring and a mass, will have a natural frequency for vibration. When the incident sound wave matches this natural frequency, resonance occurs. Each resonator responds to waves of a particular frequency, and the ear hears this same note greatly enhanced.

*Hermann Helmholtz (1821–1894) was a noted German physicist who, during his lifetime, made outstanding contributions to the subject of light, sound, and electricity. Probably his greatest contribution to acoustics was his explanation of tone quality in musical notes and the response of the ear. He demonstrated that quality depends upon the number and intensity of the partials present in the musical tone.

In the presence of a resonating cavity, the incident waves are modified, causing the wave energy to be drawn from a considerably larger area, as shown in the figure. Experimentally, a similar attraction of wave energy from a large area occurs with radio and television antennas when the receiver is tuned to the frequency of the passing waves. Certain types of spiral sea shells of the conch family, used by the early Polynesians as horns, show similar response to noise if held to the ear. It is often said erroneously that what one hears is the sound of the sea.

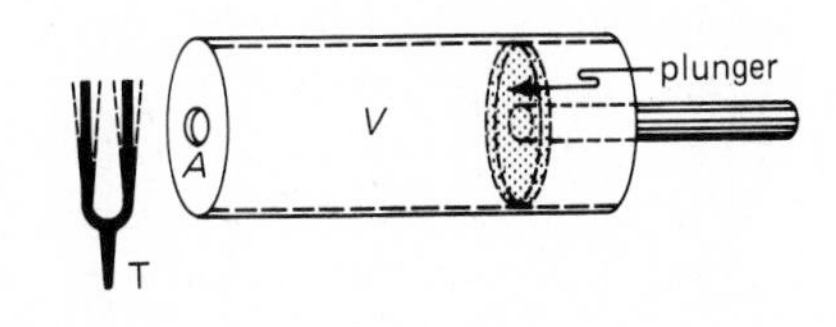

FIGURE 6–5
Diagram of a modern resonance cavity having an adjustable plunger for changing, or tuning, its response frequency.

A modern form of cavity resonator is shown in Figure 6–5. Sound waves from a source enter a small circular hole at one end, while a movable plunger near the other end changes the volume. This permits the resultant frequency to be changed at will, and the plunger is marked with a scale for reading directly the value of the frequency. Cavities of this shape have a very sharp resonance frequency—that is, they have a very narrow frequency resonance band. See Figure 6–3.

6.3 Acoustical filters

Since acoustical resonators are capable of drawing in sound energy in a narrow frequency band, they can be used to absorb those same frequencies from any surrounding sound field. See Figure 6–4. The absorption is attributed to frictional losses at the neck, as large variations in pressure take place within the cavity. Acoustical resonators have been used in buildings to reduce the sound level of certain frequencies. For example, a number of large cavities have been built into the ceiling of one of the studios of the Danish Broadcasting Company. These cavities absorb sounds in the region of 100 Hz.

If sound is to be transmitted some distance through a pipe, a series of cavity resonators can be inserted along the path to absorb certain objectionable frequencies. By tuning each of the resonators to a different frequency, a wide band of frequencies can be suppressed or eliminated. Such an arrangement is called an acoustical wave filter.

6.4 Resonance damping

We have seen in Section 2.6 that real vibrating systems exhibit **damping.** For various reasons, friction enters all such phenomena, bringing about acoustical energy losses in the form of heat production.

As an example, consider a mass m suspended from the lower end of a coiled spring. To the lower side of such a mass we rigidly fasten a small piece of sheet metal D, which we will call a *dasher.* See Figure 6–6(a). If the upper

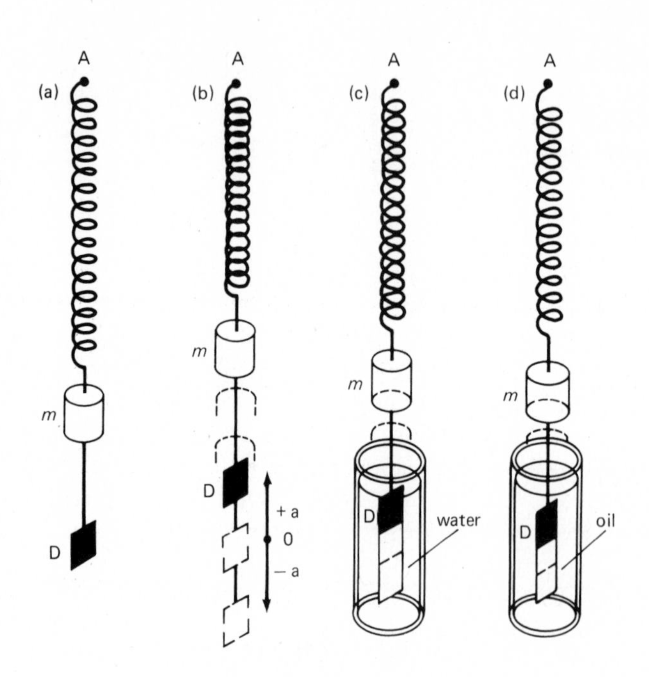

FIGURE 6–6
(a) Stationary spring pendulum and dasher D, (b) vibrating bob with dasher in air, (c) vibrating bob in air with dasher in water, and (d) vibrating bob in air with dasher in oil.

end of the spring is moved up and down with a frequency equal to the natural frequency of the system, the amplitude of the driving force at A can be quite small, yet a large resonant amplitude of the mass m can be gradually built up. See Figure 6–6(b). With higher and lower driving frequencies, the response amplitude will be reduced, and, when plotted as a graph, the amplitude will yield the top curve in Figure 6–7. This curve should be compared with Figure 6–3.

We now place a vessel below the mass so that water surrounds the dasher, as shown in Figure 6–6(c), and repeat the motion of A with the same amplitude as before. Again the system will vibrate, but with smaller amplitude. If the experimenter changes the frequency, the data for curve (b) in Figure 6–7 is obtained. If the dasher is surrounded with oil, the response to all the same impressed frequencies at A will be still smaller, as in curve (d).

The friction between the dasher and fluid provides damping to the motion through energy losses. The maximum amplitude attained under the conditions above will depend on the amplitude of the driving force at A. Small energy losses in the top curve of Figure 6–7, and the pendulum curve in Figure 6–3, due largely to air friction on the moving masses, prevent the maximum amplitude from becoming infinite at the resonant frequency *f*. Otherwise, the continuous supply of small energy increments to the vibrating system could not be dissipated, and the amplitude would continue to rise. Such damping effects are common to all musical instruments, as well as the human voice, and will be referred to frequently in later chapters.

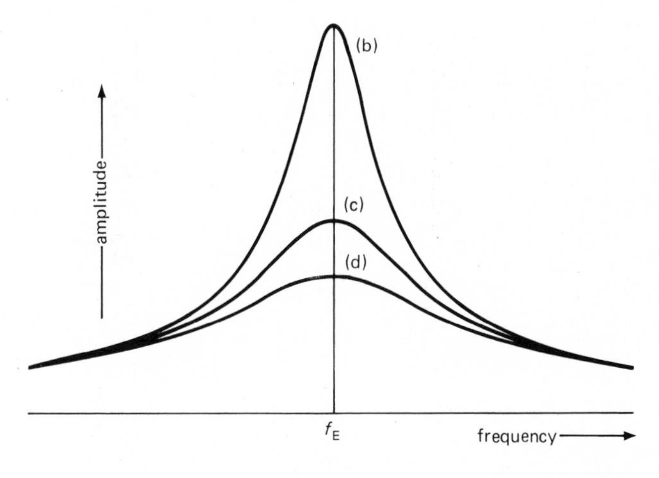

FIGURE 6–7
Amplitude curves of response to damping for the demonstration shown in Figure 6–6. Observe (b) the tall sharp peak in the response with little damping, (c) the medium response with medium damping, and (d) the low response and broad frequency range with large damping.

6.5 Beats

When two notes of slightly different pitch are sounded simultaneously, beats are heard. This phenomenon plays an important role in music and in the behavior of musical instruments. In many pipe organs, for example, two pipes tuned to slightly different frequencies are used to produce the familiar tremolo effect.

The phenomenon of beats may be demonstrated with two tuning forks as shown in Figure 6–1. One fork is made slightly out of tune with the other by looping rubber bands tightly around the prongs. If the two forks are sounded simultaneously, the *resultant sound intensity* rises and falls periodically. This is illustrated by the vibration graphs shown in Figure 6–8. The upper curve represents the sound vibrations arriving at the ear from one fork, and the second curve represents the vibrations from the other fork. Both waves arriving at the ear are first *out of phase* (out of step) with each other, then *in phase* (in step), then out of phase, then in phase, and so on. The combined action of these two waves on the eardrum is represented by the third line in the figure. When the waves are in phase, as at (a), (c), and (e), the resultant has a large amplitude equal to the sum of the two amplitudes. When they are out of

FIGURE 6-8
Time graphs for the vibrations of two pure tones of different frequency, which, if sounded simultaneously, are heard to produce beats.

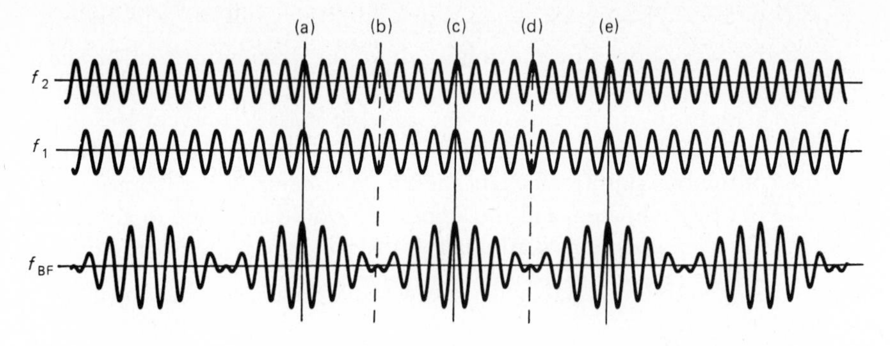

phase, as at (b) and (d), the resultant amplitude becomes zero.

Beats are defined as the pulsations of sound intensity caused by the periodic coincidence of the amplitudes of two sound waves of slightly different frequencies. The frequency of the pulsations in sound intensity caused by beats is called the **beat frequency**, f_{BF}, and is determined by the difference between the respective frequencies of the two sources of sound:

$$\text{beat frequency } f_{BF} = f_2 - f_1 \qquad [6b]$$

where f_2 is the higher frequency. The perceived pitch of the resulting fused tone is given by the average of the two frequencies:

$$f_{av} = \frac{f_1 + f_2}{2} \qquad [6c]$$

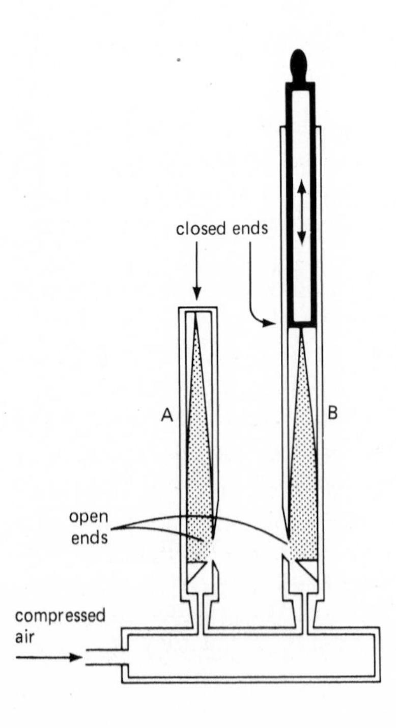

FIGURE 6-9
Two organ pipes with closed ends, one of which is adjustable in pitch, for demonstrating beat notes of different frequencies.

A good demonstration of this relation for beat frequency is obtained with two organ pipes, at least one of which is adjustable so as to produce different frequencies. See Figure 6–9. When the beat frequency lies between about 1 to 6 Hz, the ear perceives a single intertone halfway between the two sounded but periodically waxing and waning in intensity.

In summary, when two tones of slightly different frequency and comparable loudness are sounded simultaneously, one hears a single tone corresponding to the average frequency, which varies in loudness at the difference frequency. If the difference frequency exceeds about 15 Hz, the beats are no longer heard as periodic pulsing and new sensations occur, as we will see in Section 10.3.

6.6 The Doppler effect

Everyone has at some time or another heard the Doppler effect. The pitch of the horn on a car moving at high speed along the highway exhibits this phenomenon. As the car passes by, the pitch of the horn drops as much as several notes on the musical scale. The same change in pitch of a racing car motor is heard as the car approaches and recedes from the grandstand at a race track. The motor seems to slow down as the car passes by. The pitch of the exhaust from a propeller-driven airplane drops an octave or more in pitch as it approaches and then recedes from a ground observer. The whistle on a fast-moving train sounds considerably higher in pitch as it approaches an observer than it does after it has passed by. The transition from high to low pitch is usually quite fast but smooth and continuous. The pitch of a bell ringing at a railroad crossing, as heard by a passenger on the passing train, appears likewise to drop in pitch. In this case, the source is at rest while the observer is moving.

The apparent change in frequency due to the Doppler effect is attributed to the relative motions of the source and observer.[1] When the horn on a stationary automobile is blown, it sends out waves with the same velocity in all directions. The true pitch of the horn is heard by all stationary observers, in whatever direction they are located, since just as many waves arrive at the ear per second as there are waves leaving the source. This follows from the fact that waves do not accumulate anywhere.

When the car is moving and the observers are stationary, the horn also sends out waves with the same velocity in all directions. As shown in Figure 6–10, the horn is moving toward the waves moving to the left and away from the waves traveling to the right. With each new wave sent out by the source, the car is closer to the preceding wave sent to the left and farther from the preceding wave sent to the right. The result is that consecutive waves traveling to the left are shorter while those traveling to the right are longer. Since the velocity of sound is always the same in all directions, an observer at O_2 hears fewer waves per second, and an observer at O_1 hears more waves per second. The wavelength of the waves moving to the left is shorter than normal, while the wavelength of the waves moving to the right is longer. For observers at O_3 and O_4, at right angles and at some distance from the car, the pitch remains unchanged. At these side locations, the source is neither approaching nor receding from the observer, so approximately the same number of waves are received per second as there are waves leaving the source.

The mathematical relationship between the measurable quantities involved in the Doppler effect are given by the symmetrical equation

$$\frac{f_0}{V - v_0} = \frac{f_S}{V - v_S} \qquad [6d]$$

where f_0 is the frequency heard by the observer, in hertz, f_S is the frequency of the sound source, in hertz, v_0 is the velocity of the observer, v_S the velocity of the source, both in meters/second, and V is the velocity of sound. The direction of the velocity of sound V at the observer is positive, and its direction is taken as the positive direction for all velocities. Either v_S or v_0 is positive if it is directed along the positive direction, and it is negative if it is in the opposite direction.

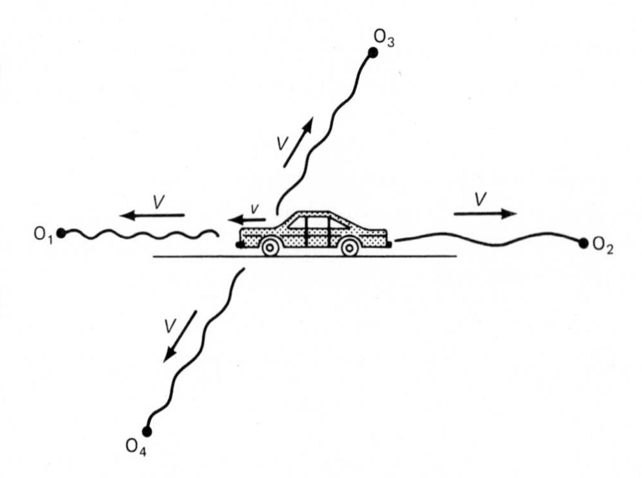

FIGURE 6–10
Doppler effect. The horn on a moving car sounds higher in pitch as it approaches a stationary observer in front of the car, lower in pitch behind the car, and unchanged to observers off at right angles to the car.

Example

A racing car passes the grandstand at a speed of 55.0 m/s, as shown in Figure 6–11. If the sound from the motor has a frequency of 800 Hz, what frequency is heard by each person in the grandstand as the car (a) approaches and (b) recedes? Assume the speed of sound to be 348 m/s.

Solution

The given quantities are $V = +348$ m/s, $v_S = +55$ m/s, $v_0 = 0$, and $f_S = 800$ Hz.
(a) In order to use the Doppler equation, Equation (6d), we first solve for the unknown, f_0.

$$f_0 = \frac{f_s(V - v_0)}{(V - v_s)} \quad [6e]$$

With the source moving in the same direction as the sound waves, V and v_s are both positive. Upon substituting the known quantities, we obtain

$$f_0 = \frac{800(348 - 0)}{348 - 55} = 950 \text{ Hz}$$

(b) For this part, V is positive, v_s is negative, and $v_0 = 0$. Substitution into Equation (6e) gives

$$f_0 = \frac{800(348 - 0)}{348 + 55} = 691 \text{ Hz}$$

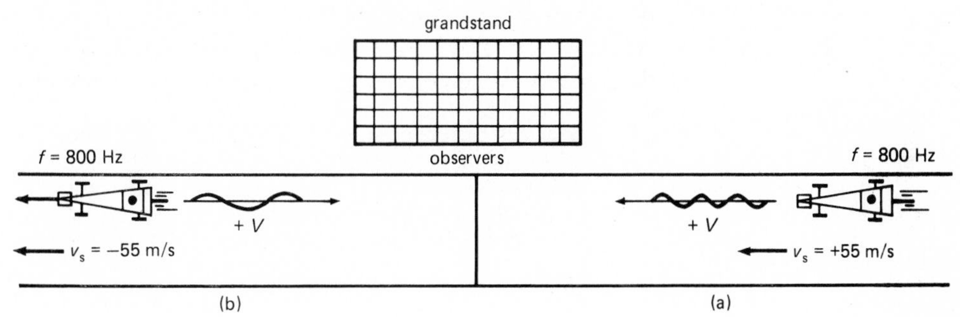

FIGURE 6–11
Doppler effect. The pitch of the motor of a racing car sounds (a) higher as it approaches observers in the grandstand and (b) lower in pitch as it recedes on down the track.

6.7 Laboratory experiment

Using two strips of wood, make a letter T, and clamp the lower end in a vise. Suspend a piece of metal from the end of a piece of string about 75 cm long and a small block of wood from another piece of string of the same length. Fasten their upper ends to opposite ends of the cross arm. Set the metal bob swinging and measure the maximum amplitude of the responding wooden bob. Leaving the length of the metal pendulum unchanged, vary the length of the wooden pendulum and record the maximum amplitude of its response. Plot a graph of amplitude versus difference in lengths of the two pendulums. What can you conclude from this graph?

QUESTIONS

1. Define or briefly explain in your own words each of the following: (a) resonance, (b) beats, (c) beat frequency, and (d) Doppler effect.

2. Briefly describe an experiment which can be performed to demonstrate the phenomenon of resonance.

3. Describe an experiment that can be performed in the classroom to demonstrate the Doppler effect.

PROBLEMS

1. Three simple pendulums have lengths of (a) 9.80 cm, (b) 98.0 cm, and (c) 980 cm. Find their periods and frequencies.

2. Three strings $A_3 = 220$ Hz, $E_4 = 330$ Hz, and $E_5 = 660$ Hz, are sounded simultaneously. (a) What are the frequencies of all the beat notes? (b) Give the ratios of all the frequencies heard by the ear.

3. An airplane traveling at 80.0 m/s (180 mi/h) "buzzes" a house on the ground. If the plane's exhaust produces a sound with a frequency of 325 Hz, find the frequency heard by a stationary ground observer as the plane (a) approaches and (b) recedes from the house. Assume the speed of sound to be 338 m/s.

4. Consider the case of the Doppler effect in which the observer is at rest and a source of musical sound is approaching a stationary observer at twice the speed of sound, or 700 m/s. What does the observer hear?

PROJECT

Assemble a set of eight to a dozen different-sized bottles and arrange them in an order of decreasing volumes to the right. Blow across each bottle opening, one after the other, and see if they sing out with anything like a musical scale. The pitch of any bottle can be raised by pouring in water to change its volume. The resonant frequencies of bottles is given (approximately) by the formula

$$f = k\frac{A}{\sqrt{vV}} \qquad [6f]$$

where A is the area of the opening at the top in square centimeters, v is the neck volume, V is the body volume, both in cubic centimeters, and $k = 5380$. By pouring water in some bottles, and removing some from others, see if you can establish a reasonably good musical scale.

NOTES

[1]R. E. Kelly, "Musical Pitch Variations Caused by the Doppler Effect," *American Journal of Physics* 42 (1974):452.

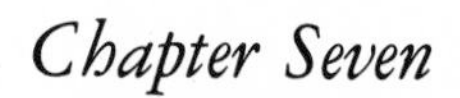

Chapter Seven

DIFFRACTION AND INTERFERENCE

In open air, sound waves travel outward from the source in straight lines. However, barriers between the source and observer do not seem to eliminate the transmission of sound. Sound waves are found to bend around corners and to spread out after passing through openings. Sound emitted from the clarinet spreads almost uniformly in all directions, whereas sound from a trumpet is strongly directed forward. In this chapter, we will see how this comes about.

7.1 Diffraction of sound waves

Consider sound waves arriving at a screen with an opening that is smaller than one wavelength. See Figure 7−1. Approaching the screen, the waves are shown traveling in straight lines. Parts of the waves are reflected or absorbed by the screen, while those that pass through the opening spread out as if S were a new source of waves. This bending of waves is called **diffraction.**[1] Diffraction may be defined as the phenomenon exhibited by waves passing through small apertures or by the bending of waves around the edges of obstacles, causing a redistribution of the energy. Diffraction is not to be confused with refraction, which was treated in Section 5.4.

Diffracted waves have their greatest amplitude in the forward direction, SC. At an angle to SC, the amplitude falls off rather rapidly. At $\theta = 90°$, the intensity drops to 25 percent of that in the forward direction, $\theta = 0°$.

If the small opening is broadened to widths larger than one wavelength, the transmitted waves still spread out but are strongly peaked in the forward direction. See Figure 7 – 2(b). For a slit opening of width b that is much larger than one wavelength, the waves are found to continue forward, with only minor amounts of the wave energy being diffracted to the side as in diagram (b). See Figure 7 – 2(c). In Figures 7 – 1 and 7 – 2(a), nearly all the energy is widely diffracted.

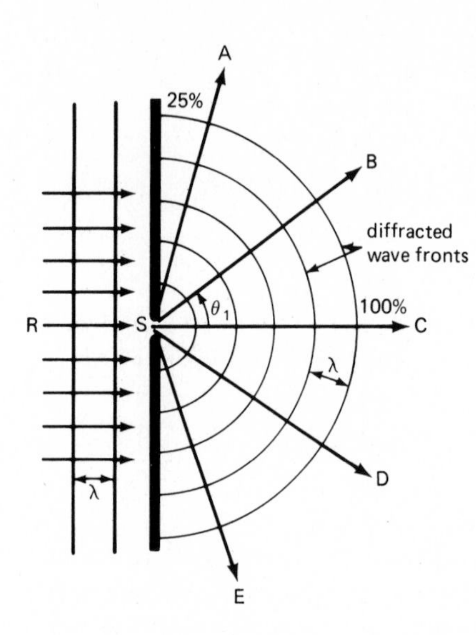

FIGURE 7 – 1
Sound waves passing through a narrow aperture in a screen spread out as if the aperture were a new source of waves.

The theory of diffraction shows that for the slit width b larger than the wavelength λ, the intensity is a maximum in the forward direction, and at some particular angle θ_1, it drops to zero (although a small amount of intensity will still be present for angles beyond θ_1). This **diffraction angle for slit aperature** depends upon the ratio between the wavelength λ and the aperture width b, and for $\lambda \ll b$, it is given by[1]

$$\theta_1 = \frac{\lambda}{b} \qquad [7a]$$

where λ and b are measured in the same units and θ_1 is measured in radians (1 radian = 57.3°).

For a given aperture b, sounds with $\lambda \ll b$ will be expected to pass through with little diffraction, as in Figure 7 – 2(c), whereas sounds with $\lambda \gg b$ will be strongly diffracted, as in diagram (a). For example, a normal doorway has a width of about 0.88 m. Sounds of this wavelength, $\lambda = 0.88$ m, have a frequency given by Equation (3a) of

$$f = \frac{V}{\lambda} = \frac{345 \text{ m/s}}{0.88 \text{ m}}$$
$$f = 392 \text{ Hz}$$

This frequency corresponds to about middle G on the musical scale. (See Table 14-6.) One would therefore expect notes below middle G to diffract widely beyond an open doorway and notes of considerably higher frequencies to pass through with little bending into the shadow.

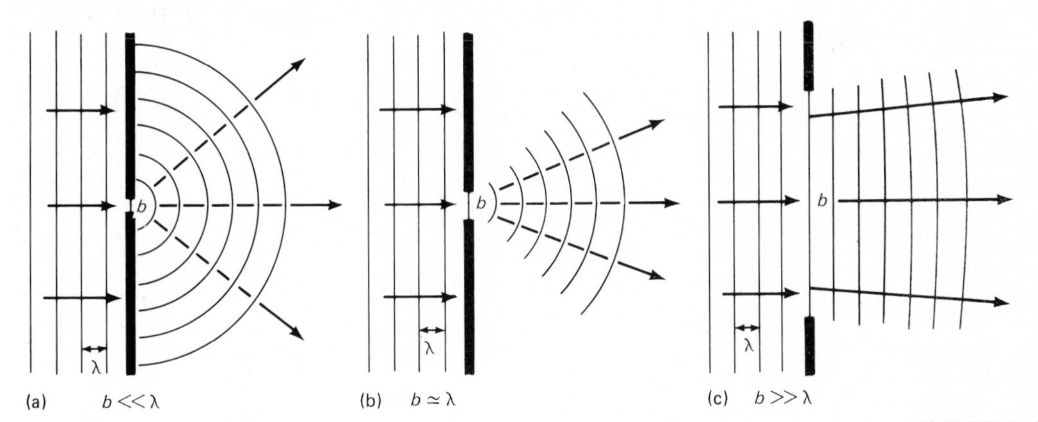

FIGURE 7-2 Diffraction of sound waves of the same wavelength but different sized apertures. (a) $b \ll \lambda$, wide diffraction, (b) $b \simeq \lambda$, medium-angled diffraction, and (c) $b \gg \lambda$, narrow-angled diffraction.

7.2 Directionality of loudspeakers

Most loudspeakers are round in shape and emit sound waves that emerge from the speaker in much the same way as waves emerge from a circular aperture. To illustrate the diffraction pattern of such sound waves, it is customary to plot the intensity I against the angle θ in **polar coordinates**. See Figure 7-3. The length of any arrow at any angle θ is drawn proportional to the relative intensity in that direction from the source. This is called a **lobe pattern.** At angles greater than θ_1, there are additional lobes, but these are relatively weak.

Since Figures 7-1 and 7-2 were assumed to represent long uniform slit apertures of width b, it should not be surprising to find that for a circular aperture of diameter b, the zero intensity angles P and Q occur at slightly larger angles than θ_1. Theory shows, and laboratory experiments confirm the results, that for parallel waves emerging from a circular aperture, the **diffraction angle for a circular aperture** is

$$\theta_1 = 1.22 \frac{\lambda}{b} \qquad [7b]$$

where λ and b are measured in the same units and θ_1 is the angle of zero intensity in radians, shown in Figure 7-3. Equation (7b) holds for small angles only and deviates appreciably for angles greater than 20°. The shorter the wavelength λ, and the larger the aperture b, the narrower is the lobe pattern. The longer the wavelength λ, and the smaller the aperture b, the wider is the lobe pattern. For circular apertures, the diffraction patterns are three dimensional, and the diagrams of Figure 7–3 should be rotated around their central axis SC. The pattern on a distant screen at right angles to SC would be a circular patch, growing weaker at the edges and becoming zero at θ_1. This area is called the **Airy disk**.

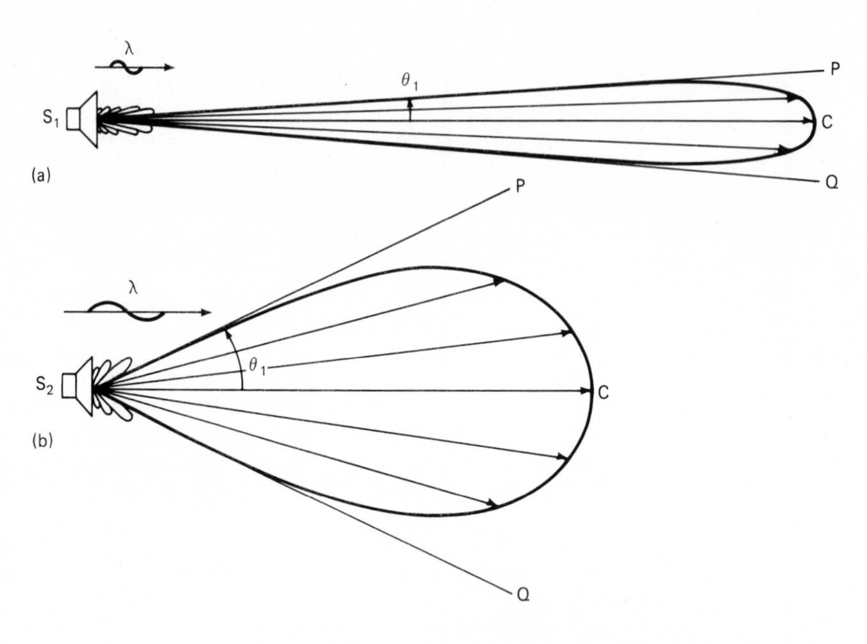

FIGURE 7–3
Two lobe diagrams showing the angular distribution of sound intensity *I*, radiated by diffraction from loudspeakers S_1 and S_2 of the same aperture but different wavelengths. (Wavelengths shown are greatly exaggerated). (a) Aperture width is much larger than the wavelength λ, and (b) aperture width is comparable in size to the wavelength λ. The small secondary lobes shown near the loudspeakers are negligible.

7.3 Public-address speaker arrays

Arrays of loudspeakers are used to direct the sound waves directly to an audience. One or more arrays like that shown in Figure 7–4 are arranged at appropriate positions high on the front or side walls and tilted at an angle to the vertical. As a rule, six to eight speakers are placed one above the other and connected electrically so that they vibrate in phase. In this way, they act as though the entire rectangular area of length l and width b were generating waves. The three-dimensional diffraction pattern from such an array may be represented by a lobe pattern that is wide horizontally [Figure 7–3(b)] and relatively narrow vertically [Figure 7–3(a)], thereby directing maximum sound energy at the audience. In an auditorium, this greatly reduces undesirable reflections from the walls and ceiling areas of the room.

7.4 Acoustical interference

If a stone is dropped into a still pond of water, the waves spread out in ever-widening circles. These waves can be represented by concentric circles, as

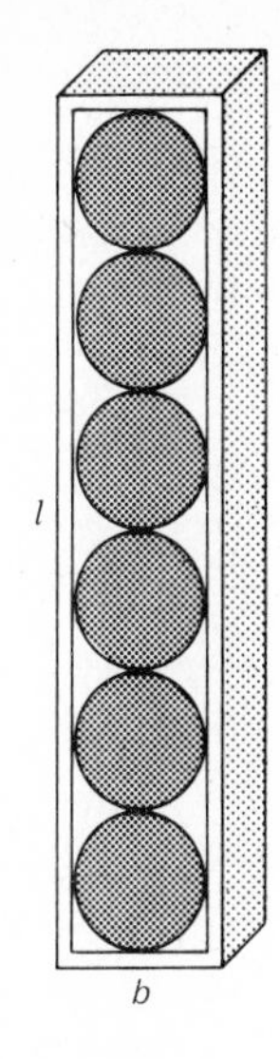

FIGURE 7–4 Rectangular cabinet using six circular loudspeakers to direct sound energy in a horizontal fan pattern toward an audience.

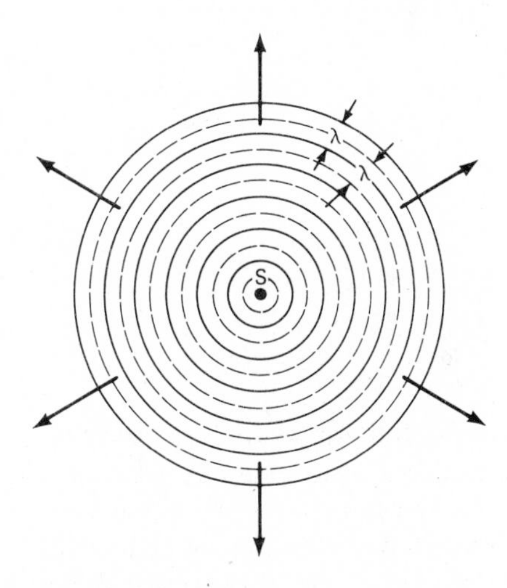

FIGURE 7–5 Concentric waves traveling outward with constant speed, caused by a stone dropped in a still pond of water.

shown in Figure 7–5. The crests of the waves are represented by solid lines, while the troughs are represented by dashed lines.

When two stones are dropped simultaneoulsy into the water, two sets of concentric wave patterns develop, as shown in Figure 7–6. In crossing each other, these waves add and subtract from each other, producing an **interference pattern.** Where the troughs come together with crests, along the lines u, v, w, x, y, and z, each wave is out of phase with the other. The amplitudes are opposite in direction and cancel each other. This is called **destructive interference.** Where two crests come together, at the solid lines a, b, c, d, and e, the waves are in phase, and the resultant amplitude is doubled. This is called **constructive interference.**[1] An instantaneous photograph of the pattern on a

FIGURE 7-6
Concentric waves traveling outward from two sources of waves, producing what is called an *interference pattern.*

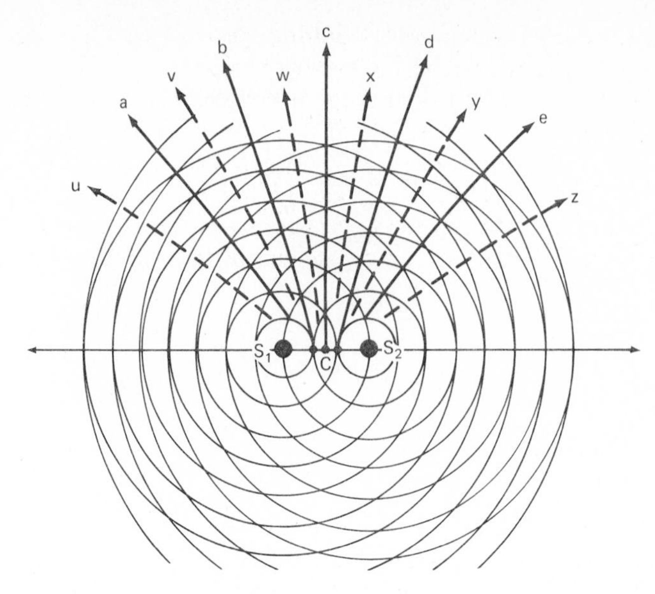

water surface, made with a ripple tank, is shown in Figure 7-7. Observe how clearly the destructive regions of null (zero) amplitude stand out.

FIGURE 7-7
Ripple tank photograph of the interference of water waves from two vibrator sources. (Courtesy of Physical Sciences Study Committee Project)

Similar interference patterns can be produced with sound waves, as shown in Figure 7-8. Two small radio loudspeakers S_1 and S_2 are mounted about 75 cm apart in a hollow box. The box is pivoted and free to turn about a vertical axis halfway between the two speakers. Both speakers are connected to the same audio generator G, so that each vibrating speaker cone sends out waves of equal amplitude with the same frequency and phase.

One or more listeners E are located in front of the box, where the sound waves from each source can be heard. As the box is turned slowly about its axis, the resultant sound intensity received by every ear or microphone pick-up will rise and fall periodically, indicating regions in the surrounding space of constructive and destructive interference. At positions of high intensity, the two waves arriving at the ear are in phase, and at positions of low intensity, the waves arrive out of phase.

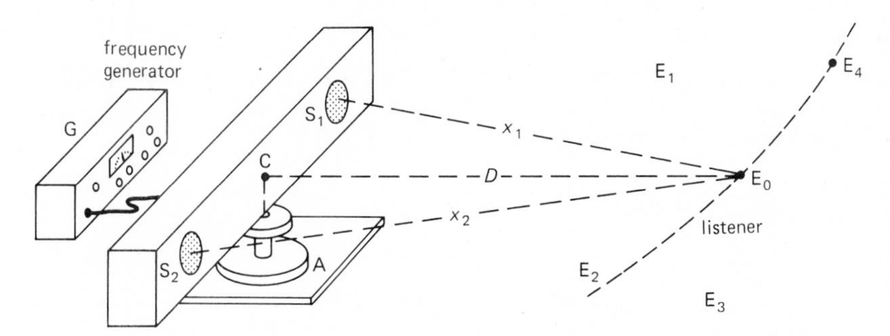

FIGURE 7–8 Experimental arrangement for determining the interference pattern for sound waves from two sources S_1 and S_2.

A diagram showing how the waves from the two sound sources S_1 and S_2 come together, in phase and out of phase, is given in Figure 7–9. Points b, c, and d correspond to those shown in Figure 7–6 where the waves arrive in phase, and points w and x correspond to those where the two waves arrive out of phase. At point b in Figures 7–6 and 7–9, for example, waves from S_2 have traveled one wavelength farther than the waves from S_1 and are again in phase. At point w, waves from S_2 have traveled one-half wavelength farther than from S_1 and are out of phase. At c, where waves 1 and 2 are in phase, the resultant amplitude is $A = 2a$. At equidistant points p and q between these two extremes, the resultant amplitudes are $A = 1.7a$ and $A = 1.0a$, respectively, and the corresponding path differences are $\frac{1}{6}$ and $\frac{1}{3}$ wavelength, respectively. A diagram of the waves themselves arriving at c, p, q, and w is shown in Figure 7–10.

If the path difference from S_2 and S_1 in Figure 7–9 to any given point P on the screen JK is equal to 0, λ, 2λ, 3λ, 4λ, and so on, waves arriving at that point will produce a maximum. These maxima slowly decrease in amplitude and intensity from the central maximum outward. An intensity graph for the interference pattern out to a path difference of $\frac{7}{2}\lambda$ is shown in Figure 7–11.

It can be seen in Figure 7–6 that from six to seven wavelengths out from the sources, the directions of constructive and destructive interference become essentially straight lines, and that these straight lines extended backward intersect at the center C. Assuming, therefore, that beyond ten wavelengths they are straight lines, as shown in Figure 7–9, a very simple approximate

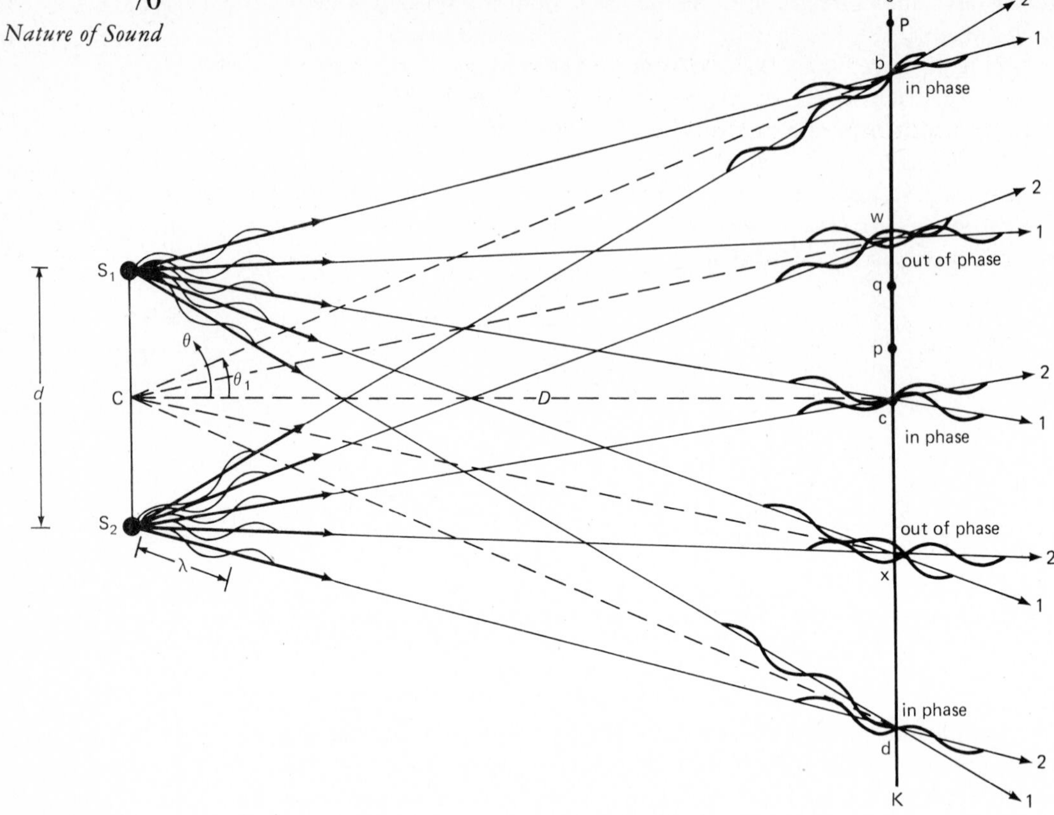

FIGURE 7–9
Diagram showing how sound waves from two sources come together at points b, c, and d and constructively interfere; at w and x they destructively interfere.

relation can be derived among θ, λ, and d. This relation, which is a valid approximation if θ is small, is

$$\theta = n \frac{\lambda}{d} \qquad [7c]$$

where θ is measured in radians and λ and d are measured in the same units. For interference maxima, n is given by whole numbers (both positive and negative values),

$$n = 0, 1, 2, 3, 4, \ldots \qquad [7d]$$

The zero positions (interference minima) u, v, w, x, y, and z, are given by half integers (both positive and negative values),

$$n = \frac{1}{2}, \frac{3}{2}, \frac{5}{2}, \frac{7}{2}, \frac{9}{2}, \ldots \qquad [7e]$$

For point c in Figure 7–9, $n = 0$ where $\theta = 0$. For point b, $n = +1$ where $\theta = \theta_1 = \lambda/d$. For point w, $n = +\frac{1}{2}$ and $\theta = +\frac{1}{2}(\lambda/d)$. For point d, $n = -1$ and $\theta = -\theta_1 = -\lambda/d$.

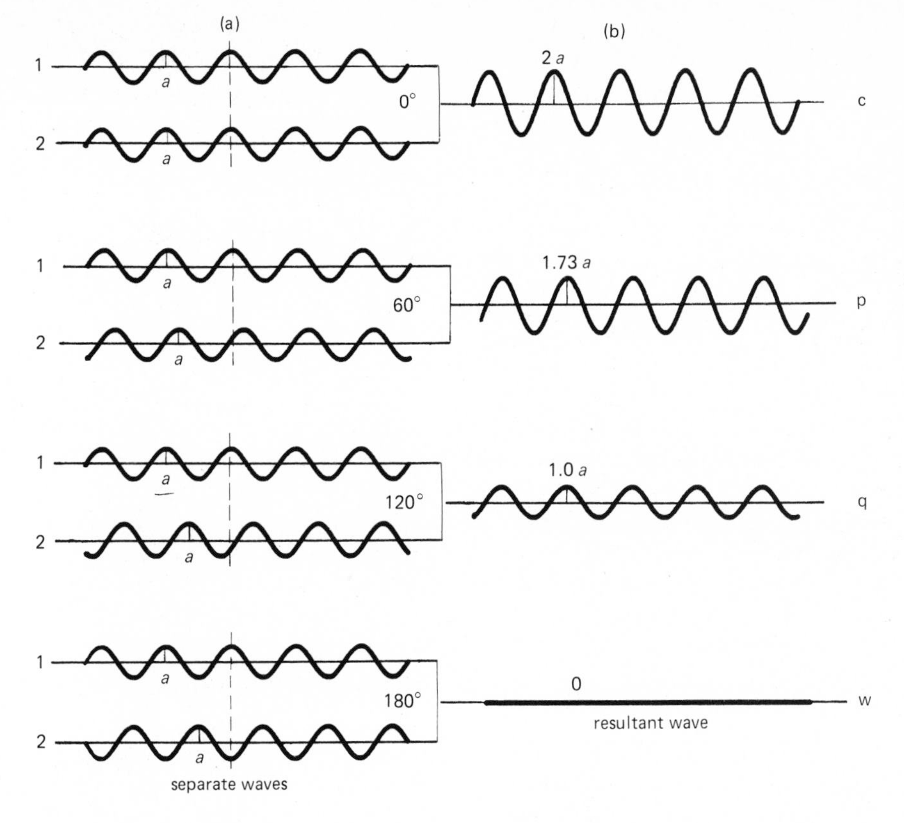

FIGURE 7–10 Time graphs of (a) pairs of waves from two sources S_1 and S_2 (see Figure 7–9) as they arrive at points c, p, q, and w and (b) the resultant waves at these same points.

Example

Two identical speakers, 1.40 m apart, emit sound waves in phase with a frequency of 1000 Hz. (a) Find the wavelength. At what angle beyond $\theta = 0°$ do the waves first interfere (b) destructively and (c) constructively? Assume the speed of sound to be 350 m/s.

Solution

(a) From the general wave equation $V = f\lambda$, we obtain, for the wavelength,

$$\lambda = \frac{V}{f} = \frac{350 \text{ m/s}}{1000 \text{ 1/s}} = 0.350 \text{ m} = 35.0 \text{ cm}$$

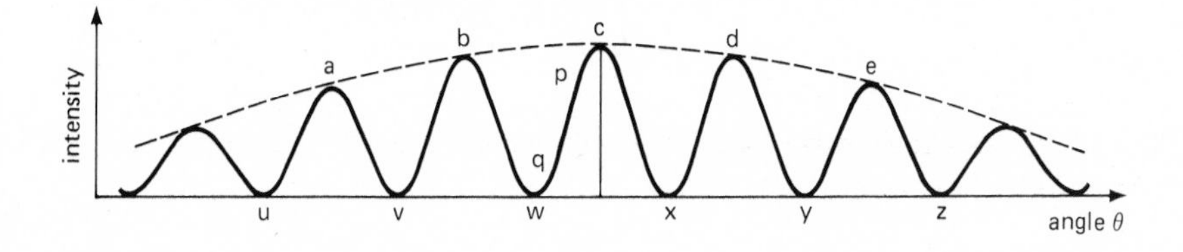

FIGURE 7–11 Graph of the resultant sound intensity bands along the plane JK in Figure 7–9.

(b) By direct substitution of the known quantities into Equation (7c), we obtain, for the angle between the central maximum and the first-side minimum,

$$\theta_{1/2} = \frac{1}{2}\frac{\lambda}{d} = \frac{35.0 \text{ cm}}{2 \times 140 \text{ cm}} = 0.125 \text{ rad} = 7.2°$$

(rad is the abbreviation for radian).

(c) The angle to the first maximum is found by the same formula to be approximately

$$\theta_1 = \frac{\lambda}{d} = \frac{35 \text{ cm}}{140 \text{ cm}} = 0.25 \text{ rad} = 14°$$

(1 rad = 57.3°.)

If the distance d between the two sound sources is increased, the maxima and minima of this pattern will be closer together, and there will be more of them within any given angle. If the frequency of the sound is increased, the wavelength will be shorter, the maxima will again be closer together, and there will be more of them.

7.5 Interference in stereophonic sound

When two loudspeakers are supplied with the same signal, positions can usually be found in the room where certain frequencies destructively interfere and cannot be heard, or are greatly reduced in intensity. However, the reflection of sound waves from the floor, walls, and ceiling may also contribute significantly to this effect.[2] In fact, room reflections usually degrade most acoustical experiments such as those discussed in this chapter. Room acoustics and construction is taken up in Chapters 26 through 28.

7.6 Laboratory experiment

An excellent laboratory experiment employing a sound generator and two small loudspeakers is set up as shown at the left in Figure 7–8. In place of the ear, a microphone M, connected to a sensitive ammeter A, is used to detect the resultant sound, as shown at the right in Figure 7–12. As the speaker box is turned slowly from side to side, interference maxima and minima can be accurately located by observing the meter readings on the ammeter.*

Using a meter stick or a tape measure, the distances x_1 and x_2 for each

*If the box in this experiment is left stationary and the microphone is moved to different positions, difficulty will be found in accurately locating maxima and minima due to reflections of sound waves from obstacles and the walls of the room.

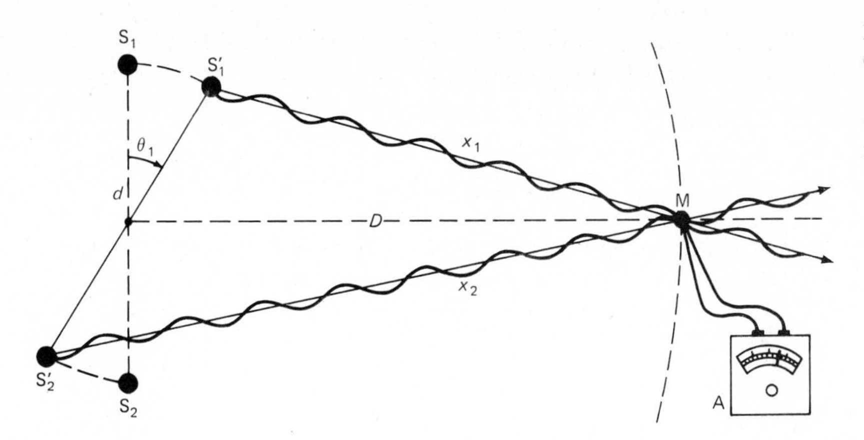

FIGURE 7–12
Geometry for the laboratory experiment shown in Figure 7–8.

maximum and minimum position should be recorded. When the box is perpendicular to D, the two sources S_1 and S_2 are equidistant from M, and the waves will arrive in phase. When the speaker arms are turned to position $S_1'S_2'$, the line between the two speakers has turned through the angle θ_1 such that the waves at M are again in phase. At angle θ_1 the path x_2 is exactly one wavelength λ longer than x_1. See Figure 7–9.

Turning the box slowly to greater angles than θ_1, another position can be found where the response is a maximum. This will occur when the path difference $x_2 - x_1$ is equal to exactly 2λ. If the wavelength selected for this experiment is short (10 to 20 cm), third and fourth positions can be located where the path difference $x_2 - x_1$ is equal to 3λ and 4λ, respectively.

Symmetrical positions on the other side of the center position E_0 will be found to be obeying the same relations between distances $x_2 - x_1$. In general, therefore, we can write

$$x_2 - x_1 = n\lambda \quad [7f]$$

where x_2 and x_1 are measured in meters and n is an integer or half integer given by Equations (7d) and (7e) for maxima and minima, respectively. Measurements can be tabulated under column headings of x_1, x_2, n, and λ.

QUESTIONS

1. Define or briefly explain in your own words each of the following: (a) constructive interference, (b) destructive interference, (c) diffraction, (d) Airy disk, and (e) lobe pattern.

2. Briefly describe an experiment for demonstrating the interference produced by two small sound sources, if they have the same frequency and are a distance b apart.

3. Why are low notes more audible than high notes when heard in a neighboring room connected by an open doorway?

4. What properties of wave combination can be used to describe beats and interference phenomena?

5. Suppose you were seated behind a post in an outdoor theater so that the band you wished to hear was not visible. What would you expect to hear?

PROBLEMS

1. Two identical sources of sound, both emitting waves of equal amplitude and a frequency of 2000 Hz, are 80 cm apart. See Figures 7–8 and 7–9. A microphone is located on the center line a distance of 2.0 m away. What is (a) the wavelength of the sound waves? If the microphone is moved along the line JK, at what angle θ should one observe the first maximum on each side of the center line (b) in radians and (c) in degrees? Assume the speed of sound to be 350 m/s.

2. Two similar loudspeakers are located 5.0 m apart at the front of an open-air theater. A musical note of frequency 264 Hz is sounded simultaneously from both speakers. The sources are exactly in phase. What are (a) the wavelengths of the two sound waves? Assume the speed of sound to be 348 m/s. At what minimum angle in degrees will the two waves (b) destructively interfere and (c) constructively interfere?

3. A radio loudspeaker is round and has an aperture of 30 cm. If it sounds a note with a frequency that produces a lobe pattern having a total angle of $2\theta_1 = 20°$, find (a) the total angle in radians, (b) the wavelength of the sound waves in centimeters, and (c) the frequency of the sound. Assume the speed of sound to be 349 m/s.

4. An auditorium speaker cabinet contains six round dynamic loudspeakers, each with a diameter of 20 cm. Arranged as a group, as shown in Figure 7–4, they give rise to waves emerging from a rectangular aperture 20 cm wide and 120 cm high. Assuming Equation (7a) is valid, and the speed of sound is 350 m/s, find (a) the wavelength emitted if a frequency of 1200 Hz is sounded. At what lobe angle θ_1, in degrees, will the waves destructively interfere (b) in the horizontal plane and (c) in the vertical plane?

PROJECTS

1. Make a study of the diffraction of sound as it occurs in various musical situations. (a) Estimate the angular width of the central diffraction lobe from a trumpet. Try several different notes. (b) Listen to recorded symphonic selection out of the direct line of sight with the loudspeaker. Which instruments do you still hear well?

2. Study diffraction of waves other than sound, such as these: (a) Diffraction of water waves through a gap or around an obstacle. (b) Diffraction of a laser beam by a human hair. What happens as the relation between object size and wavelength varies?

NOTES

[1]The interference and diffraction of sound waves parallels the treatment of the same subject for light waves, which is considered in detail in the textbook by F. A. Jenkins and H. E. White, *Fundamentals of Optics,* 4th ed. (New York: McGraw-Hill, 1976), chaps. 13–16.

[2]T. D. Rossing, "Acoustics Demonstrations in Lecture Halls; A Note of Caution," *American Journal of Physics* 44 (1976):1220.

PART TWO

Hearing and Harmony

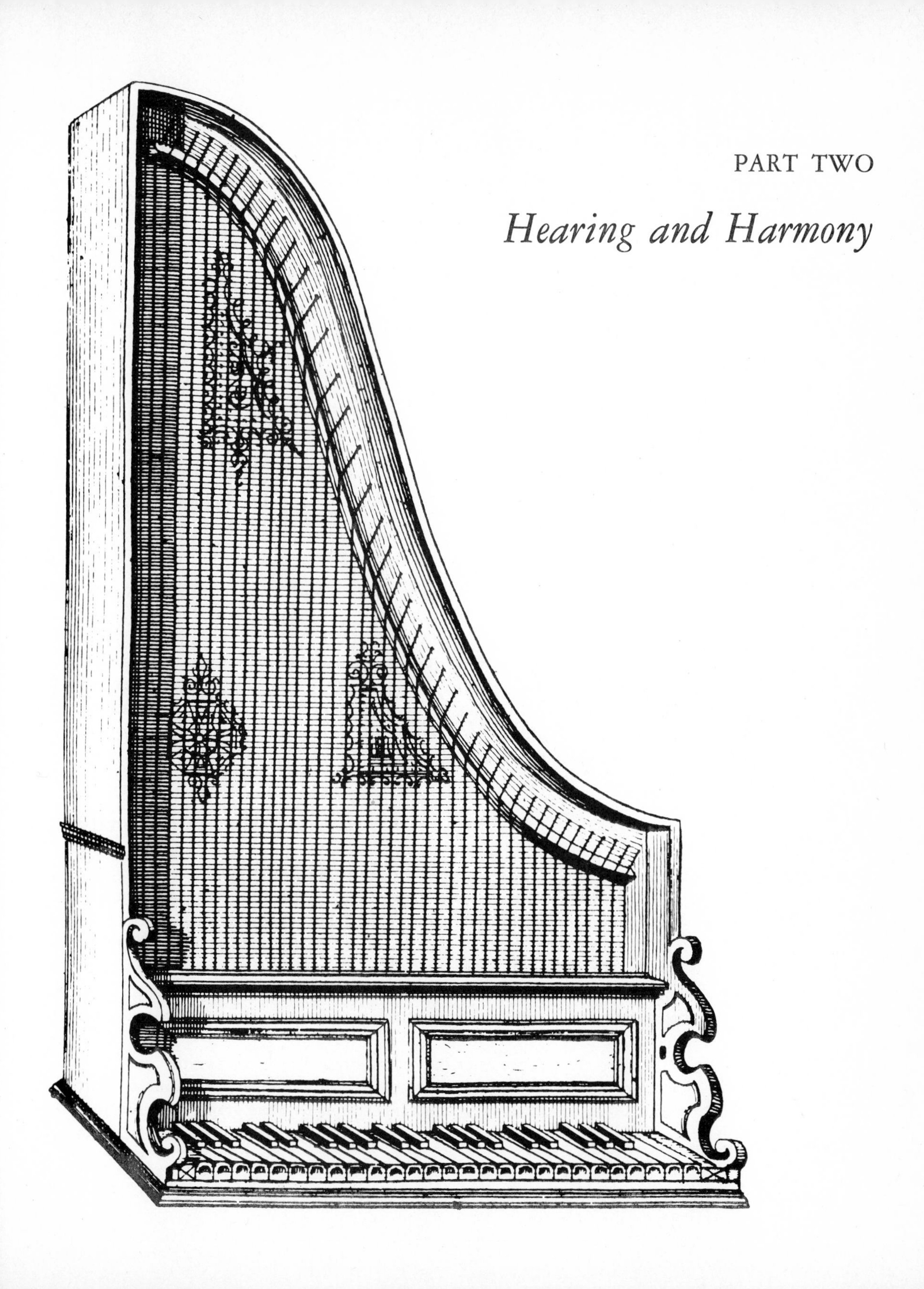

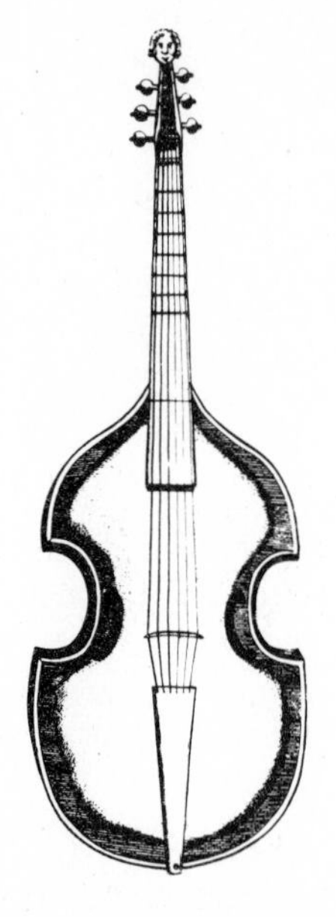

Chapter Eight

HARMONICS AND WAVE COMBINATIONS

There are an infinite number of musical **tones** that can be produced with instruments and the human voice, and yet each one can be described by specifying the intensity or loudness, the frequency or pitch, and the waveform or timbre. From the standpoint of music, the waveform or timbre of any complex tone is all-important and can be described by specifying the relative amplitudes and phases of all the different frequencies of which it is composed. We will study these concepts in this chapter.

8.1 Wave analysis

If a wave is generated by simple harmonic motion, it will be a *sinusoidal* or a *sine wave.* See Section 3.2 and Figure 3–3. A sine wave is indicative of one well-defined and definite frequency. The analysis of most musical tones shows that they are composed of a number of such components of various frequencies called **partials.** The process of adding these components to produce any complex vibration or wave is called **synthesis.** The converse of this process, breaking down any complex vibration or wave into its components, is called **analysis.**

Figure 8–1 represents two common graph forms for the same sound. Diagram (a) is a time graph representing the vibrations of a source emitting sound waves. Diagram (b) is a distance graph, or wave graph, representing the contour of the waves traveling to the right through the air with a velocity V.

Diagram (a) also represents a time graph of the vibrations of the eardrum, or a microphone diaphragm, detecting the sound.

If a wave graph (b) were drawn traveling to the left instead of to the right, it would look exactly like graph (a). Graph (a) is just the mirror image of graph (b), and vice versa. Since all three graphical representations look alike, it makes little difference which one is drawn to represent a given sound.

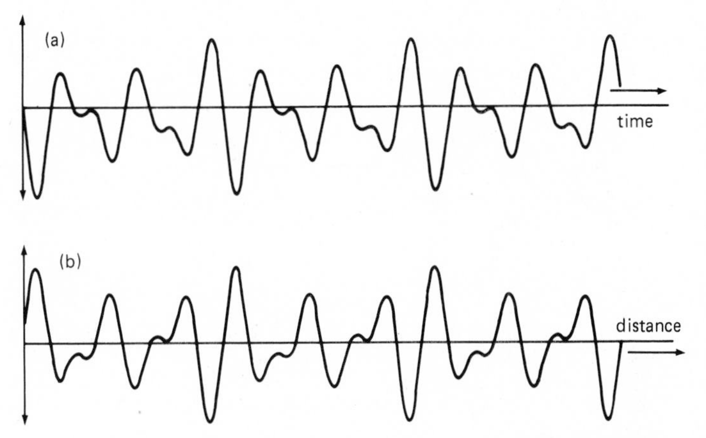

FIGURE 8–1
(a) Time graph of the vibrations of a musical source of sound or of the vibrations imposed on the eardrum by incident sound waves. (b) A wave graph of the same sound as the waves travel with a speed *V* to the right.

8.2 Partials and harmonics

The simplest waveform is a sine wave, usually drawn as a time graph of simple harmonic motion. See Figures 2–4, 2–5, and 2–6. A time graph of one of the prongs of a tuning fork, the waves transmitted through the air to an observer, and the vibration the waves impose upon the eardrum serve as good examples of this. See Figure 3–5. Any vibrating body that rapidly executes simple harmonic motion in air emits a sinusoidal sound wave. This sound wave is referred to as a **pure tone**, although the aural perception of even a pure tone is impure (see Section 15.2). Actually, nearly all tones produced by musical instruments, and other sources in general, are not pure tones but mixtures of pure-tone frequencies called **partials**. The lowest such frequency is called the **fundamental**. All partials higher in frequency than the fundamental are referred to as **upper partials**, or **overtones**.

In special cases, the frequencies of these overtones are exact multiples of the fundamental and are called **harmonics**. If we designate the frequency of any fundamental by f, all higher harmonics are designated by $2f$, $3f$, $4f$, $5f$, and so on. If, for example, we select a fundamental frequency of 200 Hz and call it the **first harmonic**, it and its higher harmonics are given by

First harmonic: $1f = 200$ Hz
Second harmonic: $2f = 400$ Hz

Third harmonic: $3f = 600$ Hz
Fourth harmonic: $4f = 800$ Hz

and so forth.

If singing voices, or different musical instruments, sound notes of the same pitch and loudness, we recognize the pitch as that of the fundamental, but the timbre or quality of each note differs from the others by virtue of the relative amplitudes of its partials. In most cases, particularly with the percussion instruments, the upper partials (overtones) are not exact multiples of the fundamental frequency. Such an overtone is called an **inharmonic partial,** and the combined tone is often unpleasant.

Most musical tones are composed of harmonics. In fact, the entire musical scale, as played by most musicians today, is based on a scale of harmonics. (See Chapter 14.) With these principles in mind, we begin our study with the combination of two pure tones, combine them, and find their *resultant waveform.*

8.3 Two pure tones in unison

If two pure tones of the same frequency are sounded simultaneously, and both waves arrive at the listener's ears, the resultant vibrations will have the same frequency. Such sources are said to be vibrating in **unison.** This is illustrated in Figure 8-2 by the combination of two SHMs, each having a frequency of 833 Hz and a period of 12×10^{-4} s but with different amplitudes, $a_1 = 8 \times 10^{-7}$ m and $a_2 = 6 \times 10^{-7}$ m, respectively. Vibration (a) has an initial phase angle $\phi_0 = 0°$, and vibration (b) has an initial phase angle $\phi_0 = +90°$. See Section 2.5 and Figure 2-6.

Since the frequencies are equal, the graph points p_1 and p_2 move around their circles of reference in the same time, always keeping the same phase angle difference of 90° between them. As a consequence, their resultant amplitude A always has the same magnitude of 10×10^{-7} m and an initial phase angle of $\phi_0 = 37°$. The amplitudes a_1 and a_2 are *added vectorially* in the left-hand side of diagram (c).

Each of the graph points p_1 and p_2, as well as the *resultant* graph point P, is seen to move once around its respective circle in the same time, and the SHMs along the y-axis trace out sinusoids with the period T. It will be observed that the vertical lines from 0 to 12 show that, at all points in time, the vertical displacements of curve (c) are always equal to the sum of the displacements of curves (a) and (b). The three time graphs are superposed in Figure 8-3. We conclude from this result that the combination of two SHMs of the same frequency will always give rise to a resultant SHM of the same frequency, but with a resultant amplitude that depends upon the two amplitudes and their phase angle difference.

This same principle is illustrated for two vibrations of the same frequency and amplitude but different phase angles in Figure 7–10. There, in diagram (a), the two sine waves are in phase, and the resultant amplitude at the right has a maximum value of $2a$. When the two components are 180° out of phase, as shown at the bottom, the resultant amplitude is zero. For all other phase angle differences, the resultant amplitude is between these two extremes.

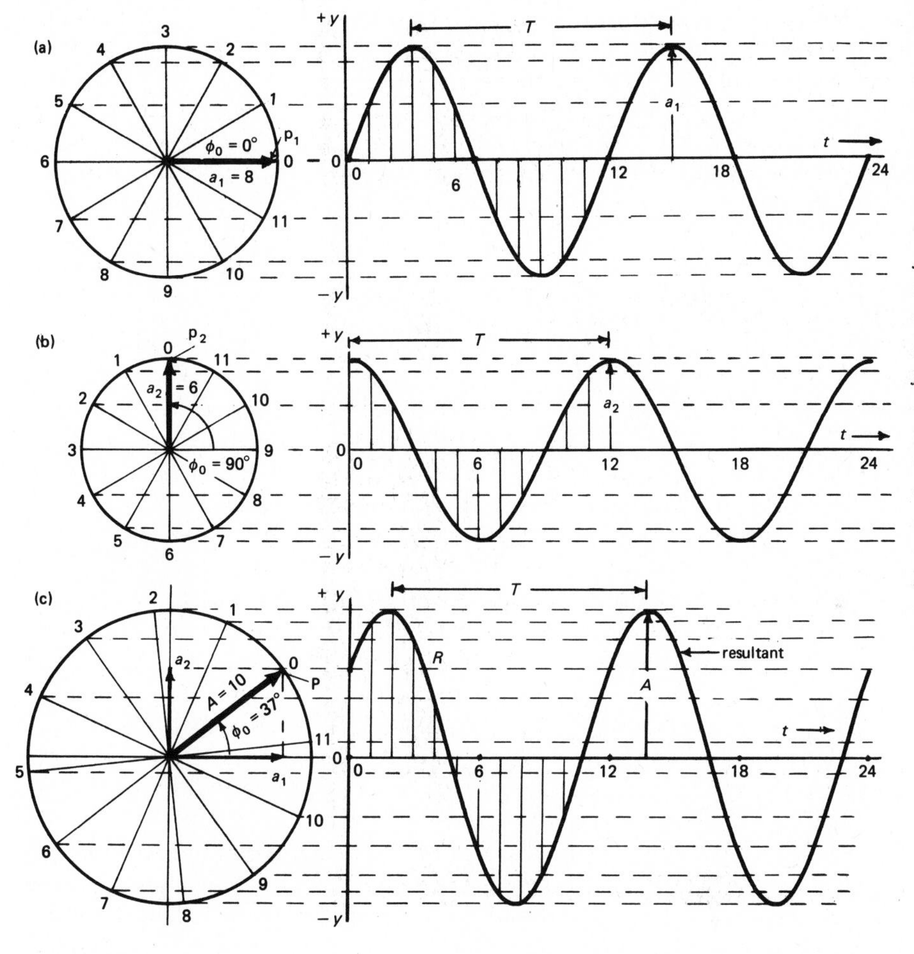

FIGURE 8–2 Circles of reference and time graphs for two tones of the same frequency and for their resultant. (a) First graph of frequency f, amplitude a_1, and initial phase angle $\phi_0 = 0°$. (b) Second graph of frequency f, amplitude a_2, and initial phase angle $\phi_0 = 90°$. (c) Resultant graph of the same frequency f, a different amplitude A, and initial phase angle $\phi_0 = 37°$.

If the two amplitudes are not equal, as in Figure 8–2, the resultant vibration will still have the same frequency. The maximum amplitude (waves in phase) will be equal to the arithmetic sum $a_1 + a_2$, and the minimum amplitude (180° out of phase) will be equal to the arithmetic difference $a_1 - a_2$. For all other phase differences, the resultant will be between these two extremes and found by the methods of vector addition (Al. 11).

The theory of vibrating bodies and the waves they emit shows that both contain energy and that their respective energies are given by the simple relation

$$E = cf^2a^2 \qquad [8a]$$

where f is the frequency, in hertz, a is the amplitude, in meters, E is the energy, in joules, and c is a *proportionality constant.* Since the intensity is a direct measure of energy flow (Section 9.2), this means that the intensity is directly proportional to *the square of the amplitude.*

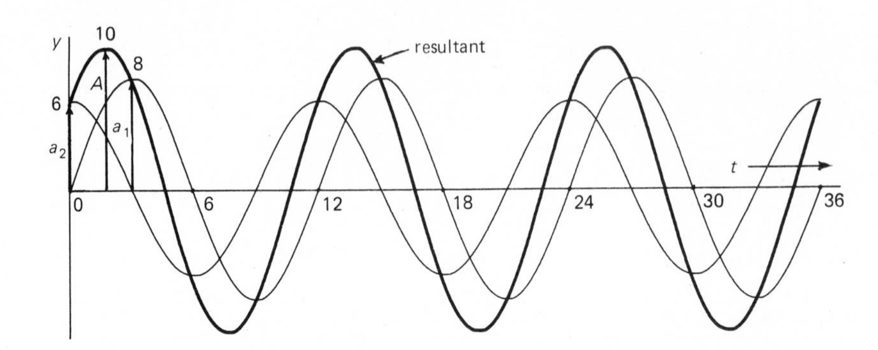

FIGURE 8-3 Superposed time graphs from Figure 8–2.

If two or more waves of the same frequency are combined, the intensity I is proportional to the square of the resultant amplitude A:

$$I = kA^2 \qquad [8b]$$

where k is a proportionality constant, which depends upon the medium through which the waves are traveling, and I is in watts/meter^2. See Chapter 9.

8.4 Many vibrations with random phases

In a symphony orchestra, there are a number of instruments of the same kind, all playing the same note. Since the sound intensity is given by the square of the amplitude, and amplitudes add vectorially to give a resultant amplitude A, it might be expected that a number of instruments, like violins, sounding the same note, might give rise to very high intensities. (See A1.11.) Let us consider, therefore, a large number of wave trains of the same amplitude and frequency arriving at a listener's ear. From what has been said in the preceding section, we can be sure that the resultant vibration will be another tone of the same frequency. Let there be n vibrations, each with an amplitude a. If these vibrations were all in phase, the resultant amplitude would be na, and the intensity by Equation (8b) would be $(na)^2$. If, for example, there is one violin, $n = 1$, the amplitude at the listener's ear is $1a$, and the intensity would be $(1a)^2$. If there are 10 violins, $n = 10$, the resultant amplitude would be $10a$, and the intensity would be $(10a)^2$, or $100a^2$. This, we know, cannot be correct, for the sound would be far to intense. We must assume, therefore, that the phase angles are a matter of pure chance, both at the sources and at the ears.

If we use the graphical method of combining amplitudes with random phases, we obtain a typical diagram as represented in Figure 8–4. The initial phase angles ϕ_0 take random values between 0° and 360°, the resultant amplitude is A, and the resultant intensity is proportional to A^2. A mathematical treatment for adding these equal amplitudes with random phases gives

$$I = na^2 \quad [8c]$$

and not n^2a^2, which means that the average intensity is just n times the intensity of one instrument. This also shows that, instead of averaging to zero, a large number of equal amplitudes a, added in random directions, must increase in magnitude as n increases.

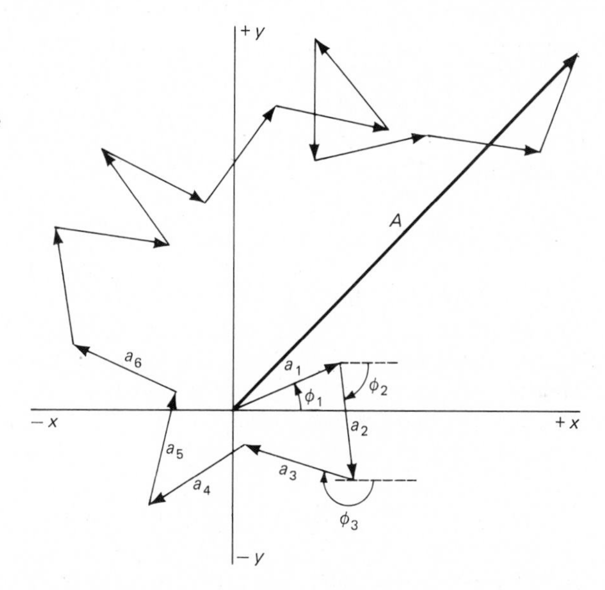

FIGURE 8–4
A vector diagram showing the resultant amplitude A of 17 equal amplitudes having random phases.

In general, any given number of sound sources emitting waves at the same time produce an intensity at a listener's ears that is just the arithmetic sum of the separate intensities.[1] This will be true whether the sources are all alike or different, and the intensity is independent of the frequencies the sources are emitting.

8.5 The chorus effect

When two or more musicians with the same kind of musical instruments are playing in unison, the quality of the combined sounds they produce is not the same as the sound from a single instrument greatly amplified. The harmonic structure of the same note played by two violinists, for example, will not be identical, for various reasons. No two instruments are exactly alike structurally, and no two musicians will bow their strings in exactly the same way. While the harmonic structures will all be slightly different, each one will, of course, sound like a violin. The sound spectrum of each note will have a fundamental, as well as the appropriate harmonics, but will vary slightly from one instrument to another.

We have seen in the previous section that the fundamentals of a group of violins will not have the same phase angles and that in general they will be random. It is also reasonable to assume that all musicians will not produce exactly the same frequency. This means that beat notes of different frequencies will be produced between fundamentals, between second harmonics, between third harmonics, and so on, and these will make the overall waveform from the group of violins more complex. The sound quality produced by the combined frequencies from a number of instruments of the same kind, playing the same note, is called the **chorus effect**. Although the primary purpose of using a number of violins in the string section of a symphony orchestra, for example, is to obtain a loudness balance with the other orchestral instruments, the chorus effect contributes to the overall richness of the musical sound.

8.6 Composition of first and second harmonics

Let us assume that a musical instrument sounds a tone in which the first and second harmonics, and no others, are present. The same resultant vibrations at the ear can be produced by sounding of one of the pure tones by one in-

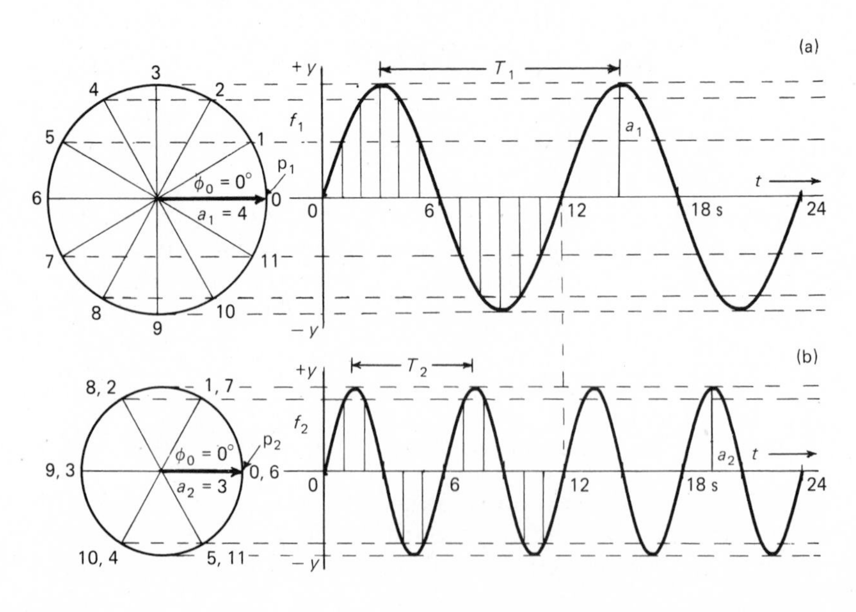

FIGURE 8–5 Time graphs for the generation of two SHMs with initial phase angles $\phi_0 = 0°$: (a) frequency f and amplitude 4, (b) frequency $2f$ and amplitude 3.

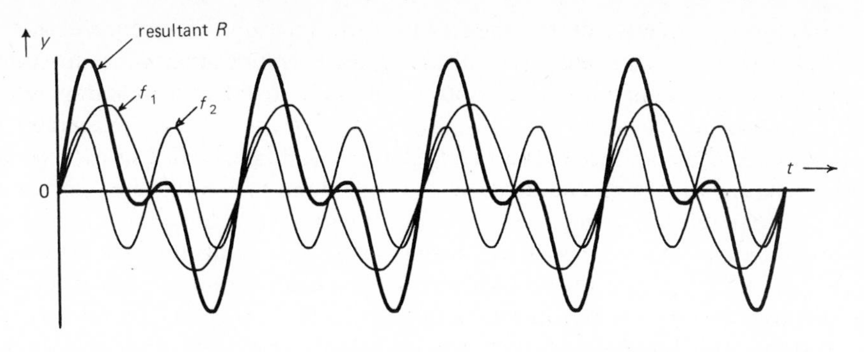

FIGURE 8–6
Composition of the two SHMs in Figure 8–5 showing the resultant **R** in relation to the amplitudes a_1 and a_2 of the separate components.

strument and the other pure tone by a separate instrument. These two SHMs, with frequencies in the ratio of 1 to 2 and initial phase angles both zero, are given graphically in Figure 8–5. The sum of the two displacements f_1 and f_2 (light lines) is shown by the resultant vibration R (heavy line) in Figure 8–6. Since f_2 has twice the frequency of f_1, the graph points p_2 and p_1 (on the circles

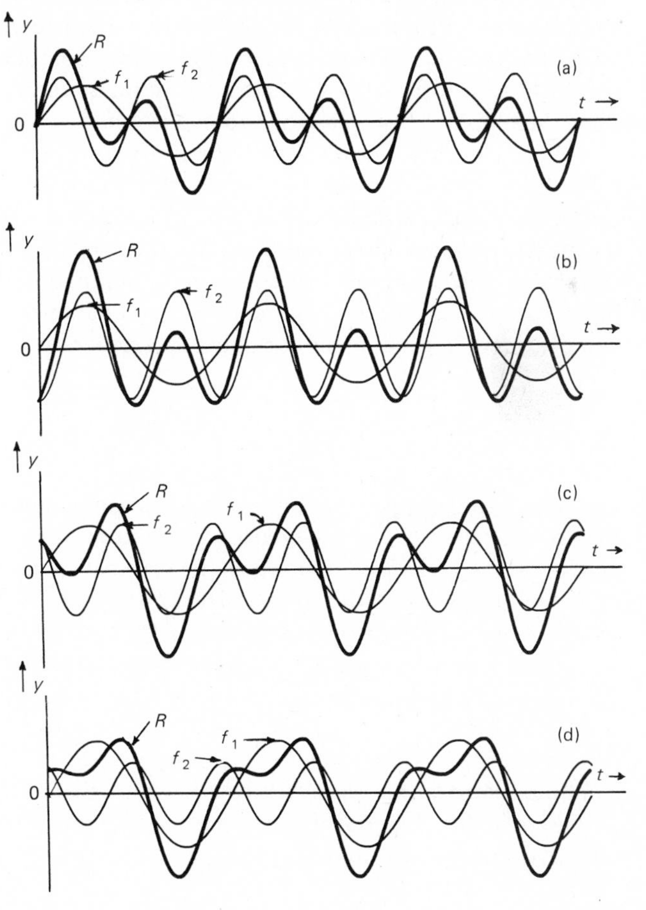

FIGURE 8–7
Time graphs combining the first and second harmonics of a fundamental frequency f_1 to form a resultant. Both frequencies f_1 and f_2 have (a) the same initial phase angles but different amplitudes, (b) different initial phase angles and different amplitudes, (c) different initial phase angles but equal amplitudes, and (d) different initial phase angles and different amplitudes. All four *resultants* (heavy lines) have different shapes but reveal the same two frequencies f_1 and f_2.

of reference) rotate with frequencies in the ratio 2 to 1. The second harmonic makes two vibrations for every one of the first harmonic. For this example, the periods are assumed to be 12×10^{-4} s and 6×10^{-4} s, corresponding to frequencies of 833 Hz and 1666 Hz, respectively. The resultant R in Figure 8–6 is obtained by adding the vertical displacements of f_1 and f_2 at each instant of time and drawing a smooth curve through them.

If the relative amplitudes are changed without changing the initial phase angles, we obtain curves of a different shape. Changing the relative amplitudes and the initial phase angles also changes the resultant curve. Typical graphs with such changes are shown in Figure 8–7. It should be pointed out that these are but a few of the infinite number of resultant vibration patterns that can be drawn. See Figure 15–5 for others.

8.7 Two, three, and four harmonics

Suppose we sound a pure tone of any given frequency, and then, one after another, we add the second, third, and fourth harmonics. The quality of each combination will depend upon the relative amplitudes, while the resultant vibration pattern becomes progressively more complex and, in many cases, more pleasant to hear. (Two consonant notes sounded together are called a *dyad,* three notes a *triad,* and four notes a *tetrad.*)

As an example, let us choose a first harmonic, or fundamental, of 833 Hz, followed by the second, third, and fourth harmonics. Let the relative amplitudes of the four harmonics be $a_1 = 8$, $a_2 = 6$, $a_3 = 4$, and $a_4 = 6 \times 10^{-7}$ m, and let the initial phase angles be $\phi_1 = 90°$, $\phi_2 = 45°$, $\phi_3 = -90°$, and $\phi_4 = -45°$. Graphs of these combinations are given in Figure 8–8. It can be seen that, as harmonics are added, the resultant vibration curve becomes more and more complex, and, in general, the tone becomes richer in quality.

8.8 Wave generation

The separation of any sound into its various components can be accomplished by mechanical or electronic devices called **analyzers**, and any set of components can be recombined to produce the original sound by similar mechanical or electrical devices called **synthesizers.**[2] In 1622 the French mathematician Fourier showed that it was possible to break down any complex periodic curve into a series of sinusoids whose frequencies are harmonically related. Stated another way, any periodic waveform can be constructed by combining a sufficient number of sine waves. This is called **Fourier's theorem.** This means that any periodic sound wave of arbitrary waveform will act acoustically as a combination of pure tones. While we will not go into the mathematics, we will graphically add, or synthesize, a number of SHMs to form several special vibration forms used by electronic engineers in the development of oscilloscopes and television receivers, by audio engineers in their development of electronic music devices for special effects, and by manufacturers in producing musical instruments of various kinds.

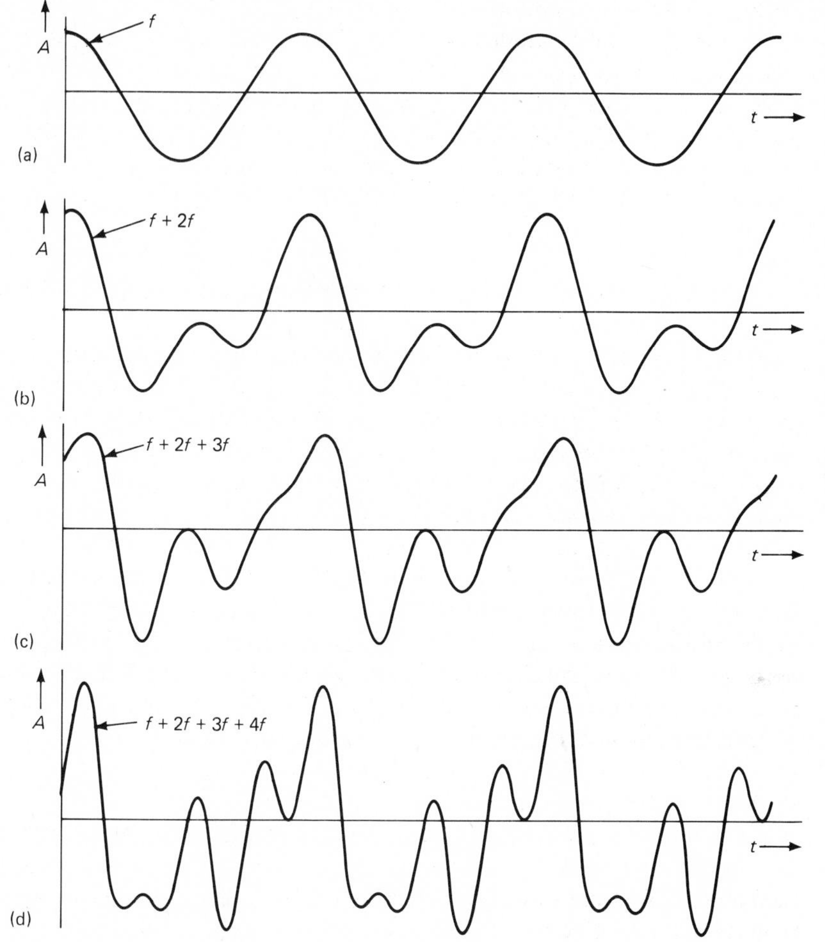

FIGURE 8–8
Time graphs for the addition of the first four harmonics of a given fundamental frequency *f*. Vibration modes for (a) the fundamental alone, (b) the first and second harmonics together, (c) the sum of the first, second, and third harmonics together, and (d) all four harmonics together.

The four simplest waveforms, or vibration forms, in common use in synthesizers today are called (a) **sine waves**, (b) **sawtooth waves**, (c) **square waves**, and (d) **triangular waves**. See Figure 8–9. Diagrams (b) and (d) belong to a family of straight-line forms called **ramp waves**. All four of these waveforms can be produced with relatively simple electronic circuits. Since the analysis of complex waves can be broken down into sine waves, and the synthesis of a number of sine waves (harmonics) can be compounded to produce complex waveforms, we can apply Fourier analysis—that is, a series of sine waves as harmonics—to reproduce sawtooth waves, square waves, and triangular waves.

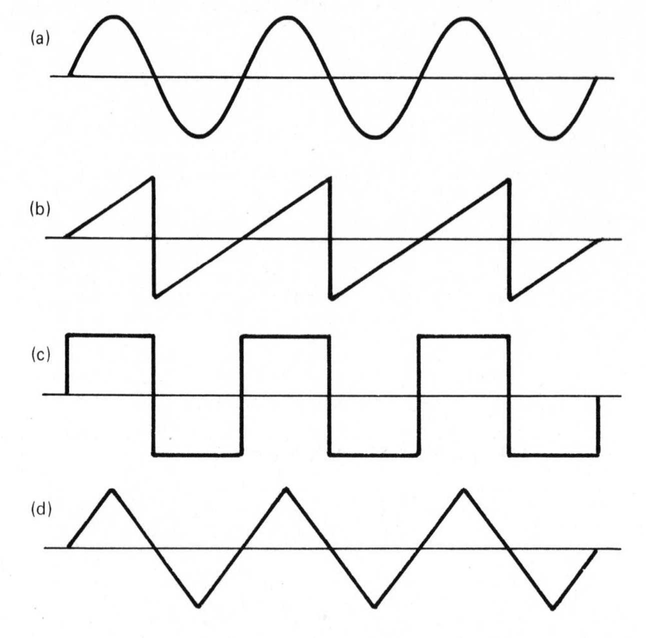

FIGURE 8–9
Four common types of waves produced by electronic sound or wave generators: (a) sine wave, (b) sawtooth wave, (c) square wave, and (d) triangle wave.

Sawtooth waves may be developed by starting with the repetition rate of the saw teeth as the fundamental frequency $1f$ with amplitude $1a$. This is followed by the addition of the second, third, fourth, fifth, and succeeding harmonics. The amplitudes of this series have values of $\frac{1}{1}a, \frac{1}{2}a, \frac{1}{3}a, \frac{1}{4}a, \frac{1}{5}a,$ and so on, and initial phase angles of 0°, 180°, 0°, 180°, 0°, . . ., respectively. The first four harmonics are shown in Figure 8–10, with three progressively synthesized curves below. Starting with the first harmonic in diagram (a), and adding the second harmonic with a phase change of 180°, that is, with a minus sign, in diagram (b) produces diagram (e). Adding to this the third, fourth, and fifth harmonics, with +, −, and + signs, respectively, produces diagram (f). Adding to this the sixth, seventh, and eighth harmonics, with their appropriate amplitudes and signs, produces diagram (g). It can be seen that by adding more and more of the higher harmonics in turn, with alternating initial phase angles, the synthesized wave progressively approaches the sawtooth wave.

The synthesis of square waves, shown in Figure 8–11, is produced by adding all the odd harmonics, $1f, 3f, 5f, 7f,$ and so on, with relative amplitudes of $1a, \frac{1}{3}a, \frac{1}{5}a, \frac{1}{7}a,$ and so on, and initial phase angles all zero, respectively. Starting with the repetition rate as the frequency of the first harmonic at the top and adding the third harmonic $3f$ produces diagram (e). Adding to this the fifth harmonic of diagram (c) produces diagram (f), and adding the seventh harmonic of diagram (d) produces diagram (g). It can also be seen that by adding in succession more and more of the odd harmonics, the resultant waveform approaches the square wave shown in Figure 8–9(c).

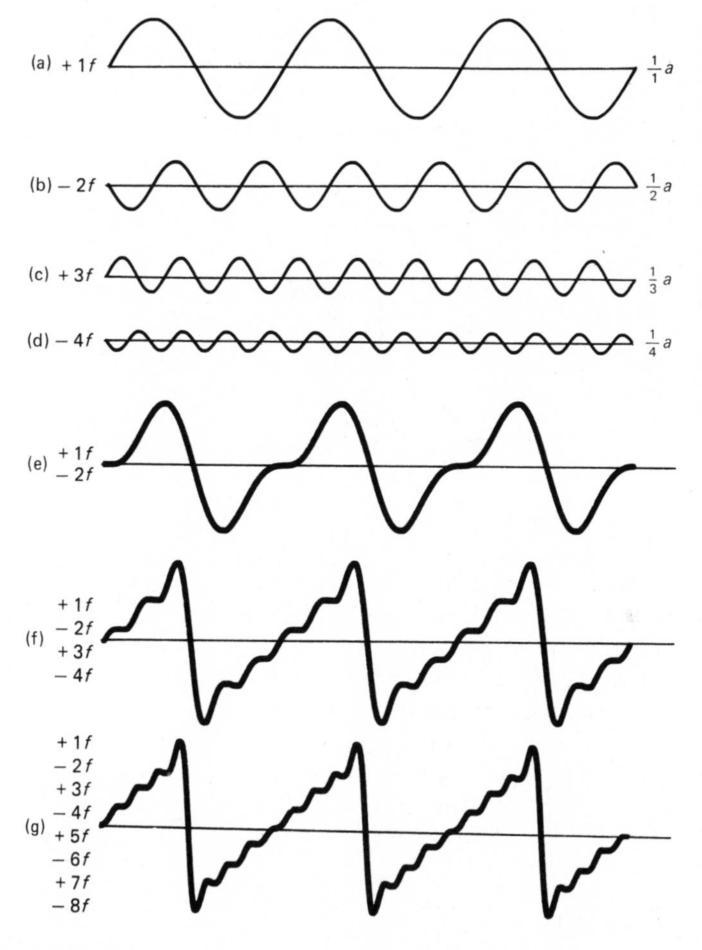

FIGURE 8–10
Sawtooth waves can be analyzed into a series of harmonics, with appropriate amplitudes and initial phase angles. They can be reproduced by synthesizing these same harmonics with their appropriate amplitudes and relative phases.

The synthesis of triangle waves is shown in Figure 8–12. Here again, only the odd harmonics $1f$, $3f$, $5f$, $7f$, and so forth, are added, with amplitudes $(1/1^2)a$, $(1/3^2)a$, $(1/5^2)a$, $(1/7^2)a$, and so on, and initial phase angles of alternate sign, +, –, +, –, and so on. Observe how quickly the addition of harmonics, with very small amplitudes, quickly approaches the waveform in Figure 8–9(d).

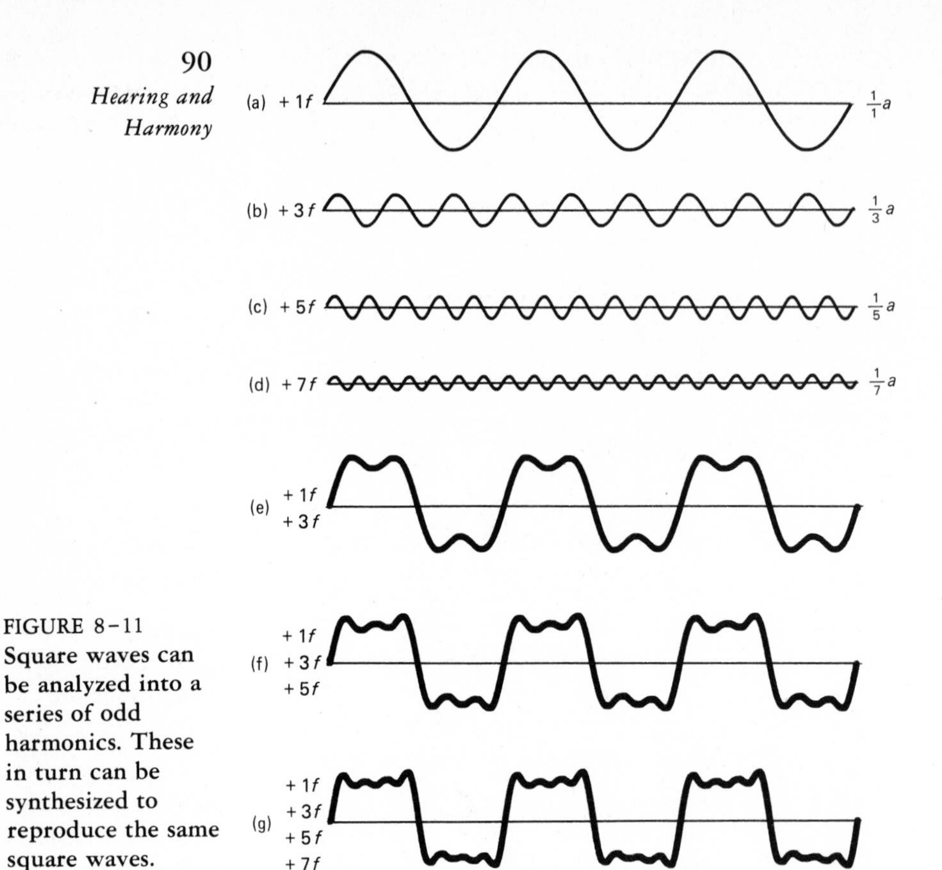

FIGURE 8–11
Square waves can be analyzed into a series of odd harmonics. These in turn can be synthesized to reproduce the same square waves.

8.9 Sound spectra

Since the quality or timbre of musical sounds depends on the relative amplitudes of harmonics, and very little upon the relative phase angles between them, we may characterize them by a diagram called a **sound spectrum.** (The term *spectrum* is derived from the optical analog of displaying light intensity as a function of wavelength or frequency.) As an example, the harmonics synthesized in Figure 8–8(d) can be represented by the diagram in Figure 8–13(a). The length of each vertical line is drawn proportional to the amplitude of the harmonic it represents. Harmonic numbers are plotted horizontally to a linear scale. The sound spectra for all three of the waveforms shown in Figures 8–10, 8–11, and 8–12 are given in diagrams (b), (c), and (d) of Figure 8–13. It will be seen in later chapters that sound spectra are not always as regular as the ones represented here, nor are they the same for all notes sounded by the same musical instrument.

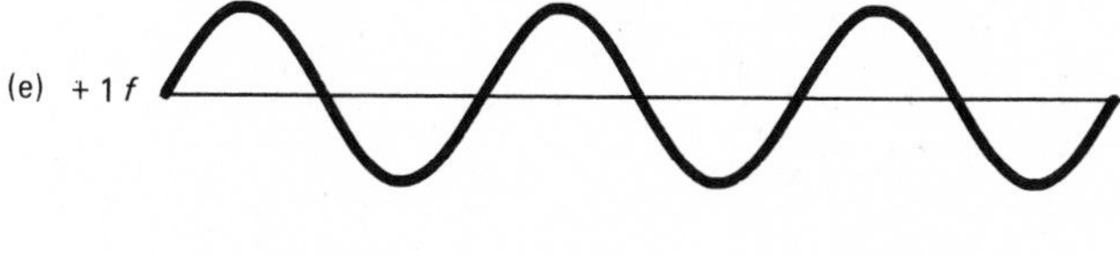

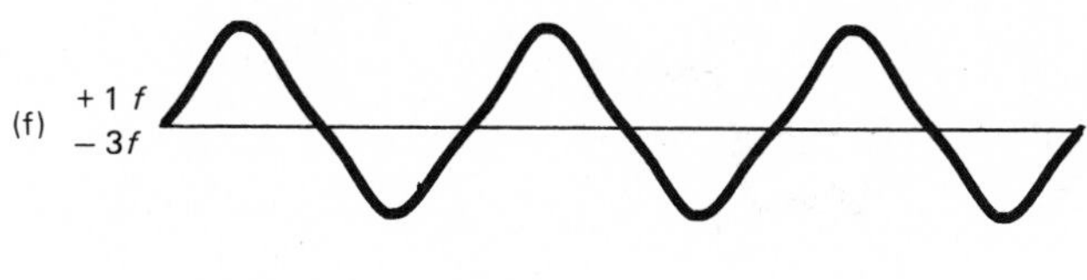

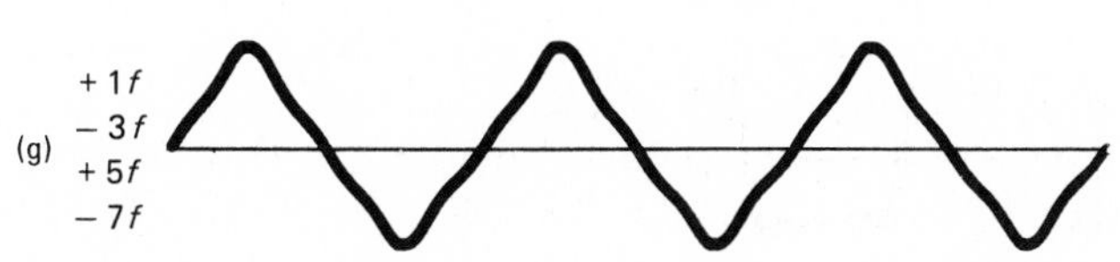

FIGURE 8–12 **Symmetrical triangle waves can be analyzed into a series of odd harmonics of rapidly decreasing amplitudes. These in turn can be synthesized to produce the same triangle waves.**

8.10 Formants

Since the tones from different musical instruments have different tone spectra, one might expect to correlate the two and find that each instrument has one and only one typical sound spectrum. Such is not the case. The first extensive research program performed to find the vibration patterns for different musical instruments was carried out years ago by Miller.[2] His mechanical method of recording the vibration curves for different musical sounds did not lend itself to the measurement of higher partials, however. More recently, high-fidelity microphones with oscilloscopes have been used to study vibration patterns. A study of recorded sound spectra clearly shows that each instrument produces a number of quite different patterns. For example, while the clarinet largely produces odd harmonics, not all tones composed of odd harmonics sound like the clarinet.

There are several good reasons why a sound spectrum alone is not used to

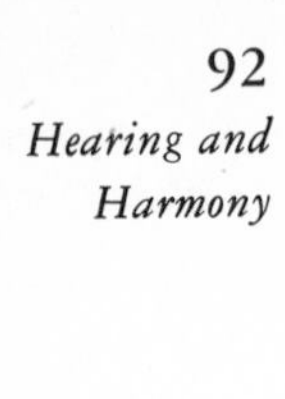

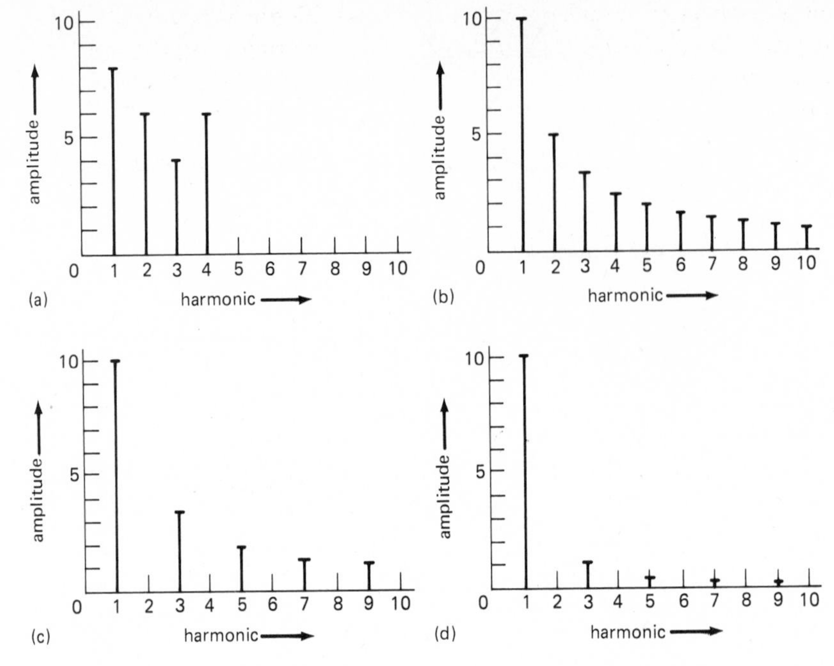

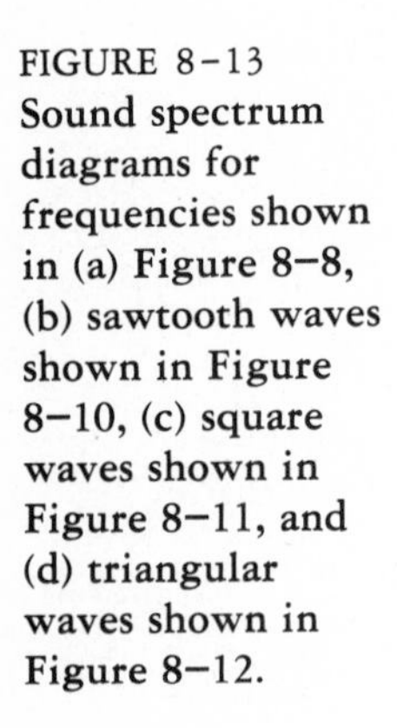
FIGURE 8–13
Sound spectrum diagrams for frequencies shown in (a) Figure 8–8, (b) sawtooth waves shown in Figure 8–10, (c) square waves shown in Figure 8–11, and (d) triangular waves shown in Figure 8–12.

classify musical instruments. First of all, the harmonic structure of any given note on an instrument depends upon its loudness.[3] Loud notes usually contain many more of the higher harmonics covering a greater frequency range than do softer notes. Furthermore, the tone quality depends upon how the musician plays the instrument, and the sound spectrum varies over the entire frequency range. Notes in the low frequency range vary widely from those in the high frequency range. (It is still one of nature's secrets that, even though the sound spectra of each instrument vary widely over its playing range, the trained ear can overlook the differences and identify the instrument.)

As another aid to instrument identification, the number and amplitude of various partials that are heard during the initial buildup of a note are of great importance.[4] The plucked strings of a violin, for example, sound quite different from the same string bowed to sound the same notes. However, a more successful approach to the establishment of a correlation between a musical instrument and its sound spectrum is to use the displays from many notes to establish the formant of the instrument. A **formant** of a musical tone is a frequency band in its sound spectrum where sound energy is largely concentrated. Two hypothetical sound spectra of two different fundamental frequencies from the same musical instrument are shown in Figure 8–14. Diagram (a) represents the sound spectrum for a fundamental note of 100 Hz, and diagram (b) represents the sound spectrum for a fundamental note of twice the frequency. Both show a formant region in the frequency range between 400 and 800 Hz.

One of the astonishing results of studying many sound spectra is the fact that the fundamental frequency that determines the pitch of the note is frequently not the loudest of the harmonics, and it is often relatively weak. As one might suspect according to the principle of formants, the high-intensity, upper partials often determine the timbre of the instrument.

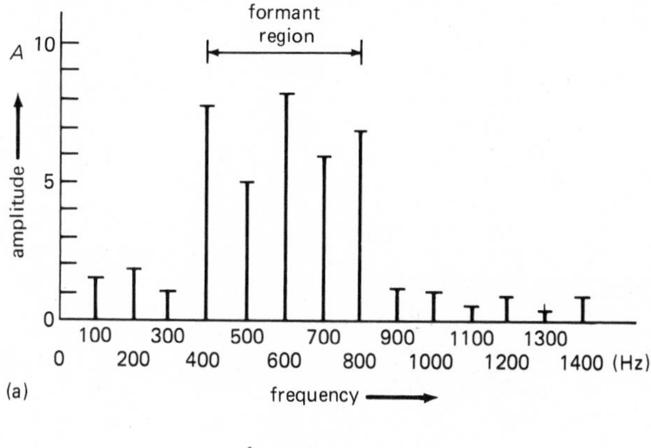

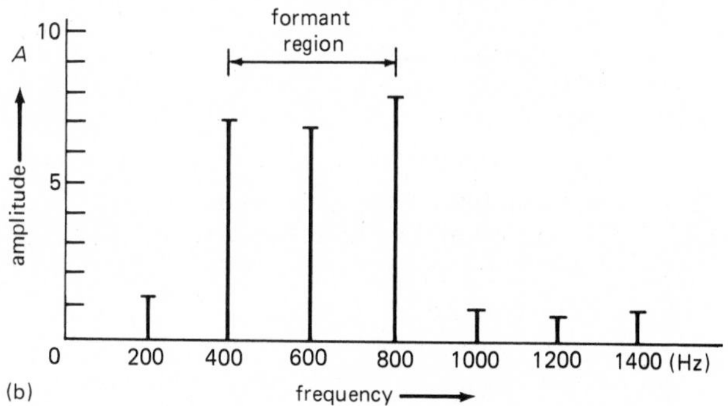

FIGURE 8–14 **Hypothetical sound spectra of the same musical instrument sounding (a) one frequency, *f*, and (b) twice the frequency, 2*f*.**

8.11 Vibrato

The character of a musical note is usually enriched by producing periodic variations in pitch, loudness, or timbre at the rate of 4–8 Hz. The word *vibrato* has been used in the past to describe all of these. More recently, the term **vibrato** has come to refer specifically to pitch modulation and **tremolo** to loudness modulation.* Modulation in timbre is simply called **timbre vibrato.** In practice, especially with electronic systems, vibrato is produced by **frequency modulation** and tremolo (also called tremulant) by **amplitude modulation.** See Figure 23–11(c) and (d). This is done by a variety of mechanical methods as well as by electronics.

*Some confusion exists since vocalists often use *tremolo* to describe widely modulated *vibrato.* Actually, the tone warbling of vocalists usually involves both tremolo and vibrato.

The correlation of pitch to frequency, loudness to amplitude, and so on is never perfect, as we have seen. This is illustrated by examples such as the vibraphone (see Section 20.4), for which variations in timbre and loudness are accompanied by slight variations in pitch. Vibrato and tremolo sound quite similar.[5] Vibrato is usually more pleasing to the ear, but it is often more difficult to produce.

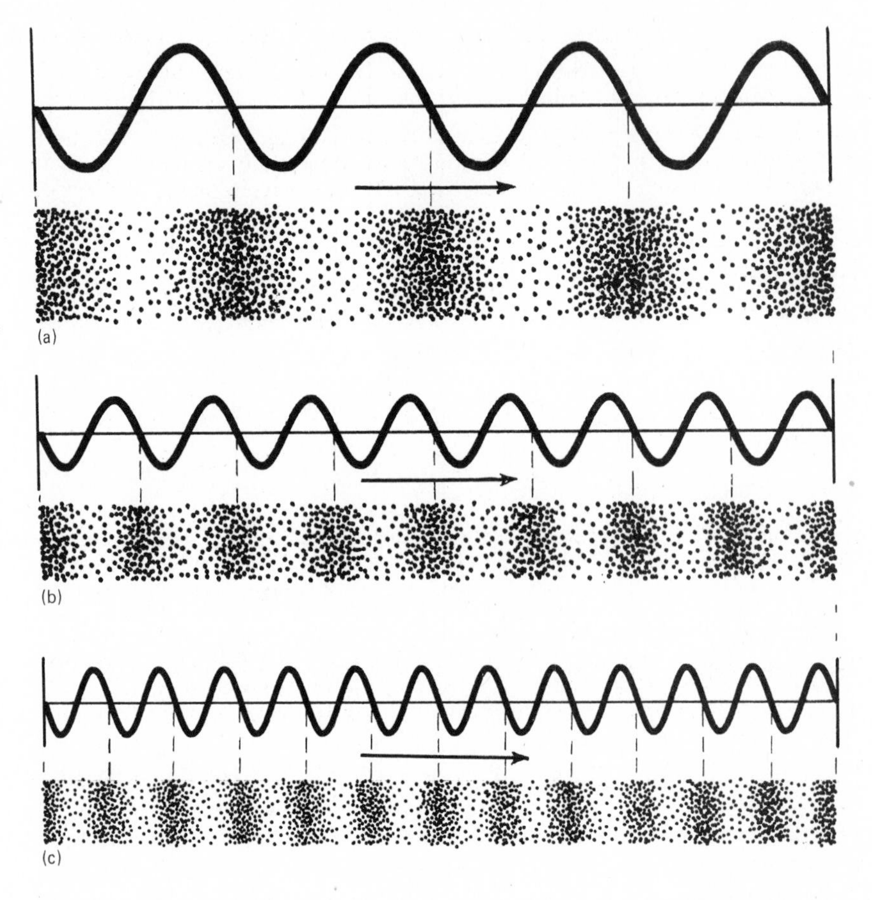

FIGURE 8-15
Comparison of sine waves with their corresponding sound waves, showing the air molecules: (a) the first harmonic, (b) the second harmonic, and (c) the third harmonic.

Before we end our discussion of time graphs of harmonics and combinations of waveforms, remember that sound waves are longitudinal in character. Their representation by sinusoids or combinations of sinusoids, instead of relative positions of molecules as shown in Figure 8-11, is one of great graphical convenience and is much more informative. Not only is it time-consuming to draw dot patterns instead of sine curves, but imagine drawing dot patterns for waveforms like Figure 8-8d.

QUESTIONS

1. Define or briefly explain in your own words each of the following: (a) harmonics, (b) overtones, (c) partials, (d) unison, and (e) inharmonics.

2. Define or briefly explain each of the following: (a) random phases, (b) synthesis, (c) Fourier analysis, (d) sawtooth wave, and (e) sound spectrum.

3. Draw a hypothetical sound spectrum showing its formant. Briefly explain the region in which it is located.

4. Draw a circle of reference and a time graph for the following SHM: amplitude $a = 15 \times 10^{-7}$ m, period $T = 24 \times 10^{-4}$ s, and initial phase angle $\phi_0 = 30°$. Draw the graphs for two complete vibrations.

5. A wave that looks the same forward and backward (e.g., triangular) is called *symmetric,* whereas a wave that looks different forward and backward (e.g., sawtooth) is called *asymmetric.* How is the symmetry of a wave related to the type of harmonics necessary to synthesize it?

6. If a string is plucked in the very center, which harmonics do you think might be absent? Why?

7. If a string is plucked at one-sixth the length of the string from either end, which partials are probably absent? Explain.

PROBLEMS

1. Construct a sound spectrum graph for the following pure tones; $f_1 = 60$ Hz, $a_1 = 1.8 \times 10^{-7}$ m; $f_2 = 120$ Hz, $a_2 = 0.60 \times 10^{-7}$ m; $f_3 = 180$ Hz, $a_3 = 1.20 \times 10^{-7}$ m; $f_4 = 240$ Hz, $a_4 = 1.5 \times 10^{-7}$ m; $f_5 = 300$ Hz, $a_5 = 0.30 \times 10^{-7}$ m; $f_6 = 360$ Hz, $a_6 = 0.80 \times 10^{-7}$ m.

2. The following pure tone is sounded: $f = 50$ Hz, $a = 2.0 \times 10^{-6}$ m, and $\phi_0 = 90°$. Using a compass, a ruler, and a protractor, draw to scale a circle of reference and divide it into 16 equal parts. Using this circle, draw a waveform diagram for one wavelength.

3. The following pure tone is sounded: $f = 100$ Hz, $a = 1.6 \times 10^{-6}$ m, and $\phi_0 = 120°$. Using a compass, a ruler, and a protractor, draw a circle of reference to scale and divide it into 12 equal parts. Using this circle, draw a time graph for two periods.

4. Two sine waves of equal frequency are to be compounded, (1) $a_1 = 3.0 \times 10^{-6}$ m and $\phi_0 = 60°$ and (2) $a_2 = 4.0 \times 10^{-6}$ m and $\phi_0 = 150°$. (a) Calculate their resultant amplitude. (b) Using the protractor, measure the resultant's initial phase angle.

5. Two pure tones with amplitudes $a_1 = 25 \times 10^{-7}$ m and $a_2 = 25 \times 10^{-7}$ m, and the same frequency $f = 100$ Hz, are sounded with initial phase angles $\phi_0 = 30°$ and $120°$, respectively. Using a compass, protractor and ruler, (a) draw a circle of reference; (b) find their resultant amplitude and initial phase angle; (c) divide the circle into 12 equal parts; (d) draw a vibration graph for one complete vibration. (e) What is the period of vibration?

6. Using a compass and ruler, draw an amplitude diagram for nine pure tones sounded simultaneously with equal amplitudes and random phases. Make the resultant amplitude conform to Equation (8a) and have an initial phase angle of 30°.

PROJECT

Bring a stringed instrument to class, or use a laboratory sonometer, and find answers to the following questions: (a) If one of the strings is bowed, where should it be touched to raise the fundamental pitch one octave? (b) Where should the string be touched to raise the pitch two octaves? (c) Where should it be touched to raise the pitch three octaves?

NOTES

[1]F. A. Jenkins and H. E. White, *Fundamentals of Optics,* 4th ed. (New York: McGraw-Hill, 1976), p. 248.

[2]D. C. Miller, *The Science of Musical Sounds* (New York: Macmillan, 1926), p. 78.

[3]M. Clark, Jr., and P. Milner, "Dependence on Timbre of Tonal Loudness Produced by Musical Instruments," *Journal of the Acoustical Society of America* 12 (1964):28.

[4]E. L. Saldanha and J. F. Corso, "Timbre Cues and Identification of Musical Instruments," *Journal of the Acoustical Society of America* 36 (1964):36.

[5]N. H. Crowhurst, *Electronic Musical Instruments* (Blue Ridge Summit, Pa.: Tab Books, 1976), p. 49.

Chapter Nine

SOUND INTENSITY AND HEARING

Our sense of hearing is one of nature's most marvelous gifts. The human ear is an extremely sensitive mechanism, capable of detecting the slightest changes in sound frequency, intensity, and timbre. The average person, for example, can hear sound frequencies over the enormous range of 20 to 16,000 Hz, sound energies from 1 watt (W) to one-trillionth of a watt per square meter, and musical sounds over a very wide range in quality. Furthermore, trained musicians can analyze musical sounds into their stronger partials and concentrate on them one at a time.

9.1 The human ear

The complex process we call hearing has long been studied by research investigators in a number of branches of science. Although experts agree on the general structure and mechanical processes taking place within the ear, controversies still exist about some of the physiological and psychological processes concerned with hearing.

The intricate mechanical actions taking place in our hearing process are generally understood. The parts in which these actions occur are shown in Figure 9 – 1.[1] These are

outer ear middle ear inner ear

The outer ear consists of the **pinna**, F, which helps to collect sound waves from outside and transmit them through the air-filled auditory canal M (**meatus**) to the **eardrum** D (**membrana tympani**). The eardrum is not flat but conical in shape, and it has its maximum vibration near the center.[2]

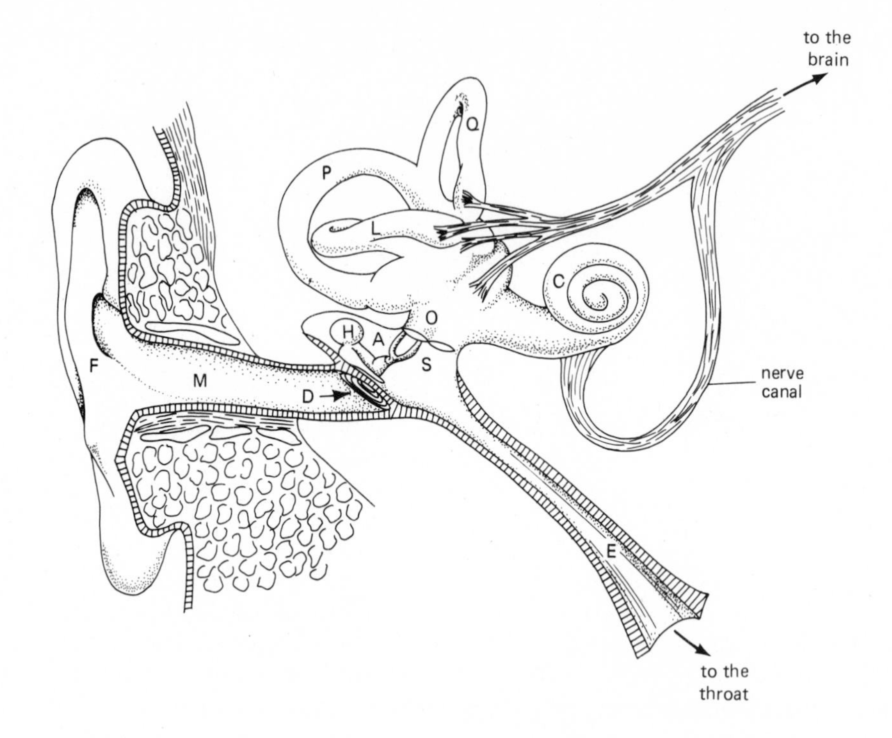

FIGURE 9–1
A cross-sectional diagram of the human ear. (With permission of D. Van Nostrand Company)

The middle ear is a small cavity within the most dense of all the bones in the human body, and it is filled with air at atmospheric pressure. This cavity contains three small bones called **ossicles**, and, because of their shape, they are designated the **hammer** (mallus), **anvil** (incus), and **stirrup** (stapes), respectively. In the normal ear, the hammer, anvil, and stirrup are elastically connected one to the other, and they serve as a lever system to transmit the vibrations of the outer eardrum to the inner ear. The end of the hammer handle, called the **manubrium**, is fastened to the center of the eardrum, while the footplate of the stirrup is fastened to an elastic membrane called the **oval window**. This window separates the middle and inner ear. Atmospheric pressure is maintained between the outer and inner ear through the **Estachian tube** E, which opens into the **nasal pharynx**. (See Figure 19–16.) As the hammer handle moves back and forth, the footplate rocks back and forth.

The inner ear is confined to a small cavity in the solid bony structure, called the **bony labyrinth**. This space is composed of two major parts: the **cochlea**, C, and the **semicircular canals**, P, L, and Q. The entire labyrinth is filled with a watery liquid called the **endolymph**, through which the sound vibrations from the oval window are transmitted to the sensitive detecting membranes of the cochlea. In the cochlea are thousands of tiny hair cells, which are stimulated by sound vibrations and give rise to our sense of hearing. In the semicircular canals are to be found the nerve endings that give rise to our sense of balance. As far as is known, the semicircular canals have little to do with normal hearing and will not be discussed here.

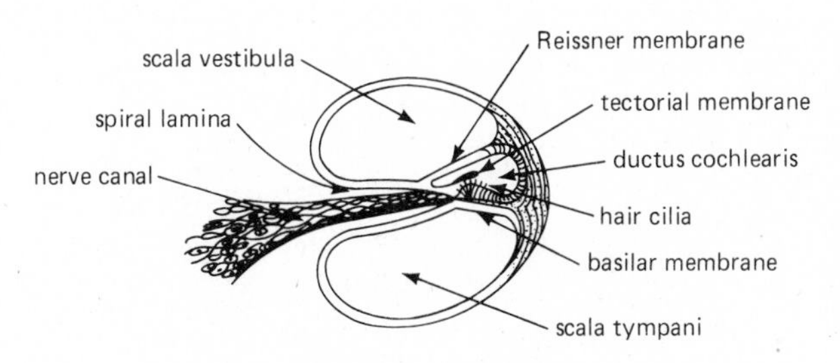

FIGURE 9–2
A cross-sectional diagram of the inner ear showing three chambers of the cochlea: the scala vestibula, the ductus cochlearlis, and the scala tympani. (With permission of D. Van Nostrand Company)

The cochlea is composed of two and one-half turns of a spiral cavity shaped like a snail shell. This cavity is shown by a cross section at the middle of the spiral in Figure 9–2 and by a lengthwise cross section as it would appear if the entire cavity were straightened out in Figure 9–3. Separate chambers in the cochlea are formed by a bony structure called the **spiral lamina**, extending about two-thirds of the way across the cavity, and by two flexible structures called the **Reissner membrane** and the **basilar membrane**. These partitions divide the cochlea into three long cavities: the **scala vestibula**, the **ductus cochlearis**, and the **scala tympani**. A small opening called the **helicotrema** at the far end forms a liquid passage between the three chambers.

The total length of the basilar membrane, from one end to the other, is

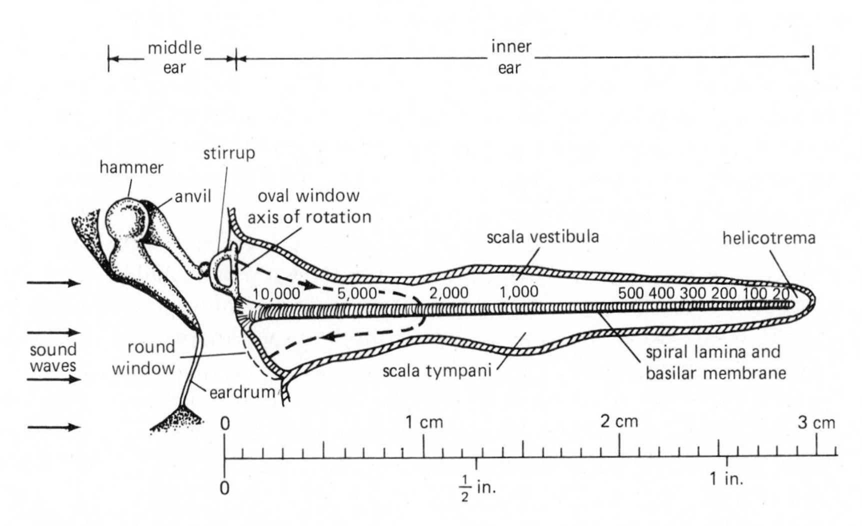

FIGURE 9–3
Cross-sectional diagram of the middle and inner ear showing the action of the hammer, anvil, and stirrup and the paths of sound waves of different frequencies crossing the basilar membrane in the cochlea. (With permission of D. Van Nostrand Company)

approximately 3.0 cm, and it contains approximately 30,000 hair cells arranged in rows. See Figure 9–4. Protruding from each hair cell into the liquid above are 12 to 40 **hair cilia** (four each are shown in the diagram). Over these cilia, and touching them, is a soft pad called the **tectoral membrane.**[3] The auditory nerve canal (about 1 mm in diameter) entering the spiral lamina consists of a bundle of several thousand nerve fibers. Each nerve fiber, or neuron, is connected at one end to about five hair cells. At the other end, after passing through the temporal bone, it terminates in the brain.

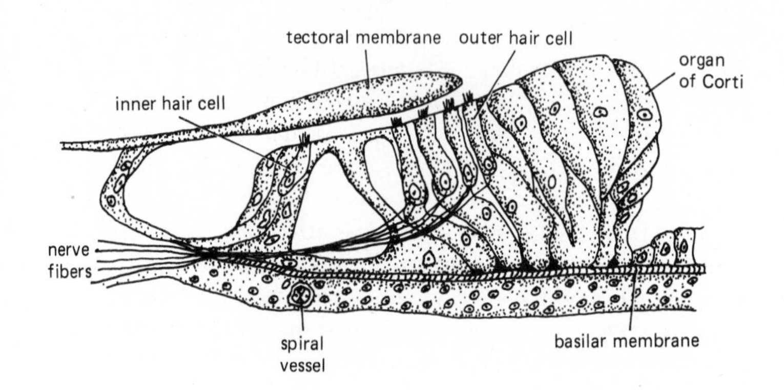

FIGURE 9–4
Detailed diagram of the organ of Corti, the tectoral membrane, hair cells, and related parts of the inner ear.

The eardrum, the chain of three bones of the middle ear, and the oval window entrance to the inner ear form a coupling mechanism between the air vibrations in the ear canal and the liquid vibrations of the inner ear. This lever system acts as an *impedance-matching transformer.* (See Section 5.6.) The lever system itself has a mechanical advantage of only 1.3 to 1. However, the effective area of the eardrum is much larger than the effective area of the oval window. As a result, the ratio of the sound pressure acting on the oval window to the sound pressure acting on the eardrum amounts to a magnification of about 10 at a frequency of 100 Hz to about 20 at 2200 Hz. This large pressure change improves the impedance match between the pressure vibrations on the outside of the eardrum and the volume displacement of the endolymph inside the oval window. This permits a larger amount of vibration energy to be transmitted to the inner ear.

In the transmission of energy to the inner ear, as the stirrup rocks back and forth in response to a relatively high frequency, such as 2000 Hz, longitudinal waves travel to the right through the scala vestibula, across the basilar membrane about one-third of the way from the oval window, and back to the round window. See Figure 9–3. Since the enclosed liquid is practically incompressible, the round window moves out when the oval window moves in, and vice versa. When the stirrup responds to a low frequency of 100 Hz, the longitudinal waves travel the full length of the scala vestibula, across the basilar membrane near the helicotrema, and back to the round window. The selectivity of crossover regions for the entire audio range is indicated in the diagram; high-frequency regions are close to the windows and low-frequency regions are farther and farther away.

An extensive amount of research on the mechanism of the crossover points along the cochlear partition has been carried out. Three different theories have been proposed over the years, and it would appear that all three mechanisms are most probably involved in the complete description.[4] As vibrations pass through the Reissner membrane and across the edge of the spiral lamina, there is a relative motion between the tectoral membrane and the hair cilia, thereby creating tiny electrical impulses that are sent through the appropriate nerve fibers to the brain. One of the most difficult subjects yet to be solved is the question of how the mechanical vibrations of the endolymph fluid in the inner ear give rise to electrical impulses that travel to the brain. Experiments show that the electrical signals generated are proportional to the displacement of the basilar membrane, not the speed. It appears that the vibrations serve to trigger or release a pool of electrical potential, about 80 millivolts (mV), already present in the endolymph.[5] Cochlear potentials seem to be due to electrolytic differences and to chemical activity.

9.2 Sound intensity

We have seen in previous chapters that one of the three major characteristics of all sounds is their intensity or loudness. Intensity is a physical quantity that, in principle, can be measured with instruments, whereas loudness is a subjective aural response, a sensory magnitude that depends upon the condition of each individual's hearing acuity.

Sound intensity is defined as the average time rate at which sound energy passes through a unit area normal to the direction of the waves.* That sound waves constitute a flow of energy may be demonstrated in a number of different ways. See Figure 9–5. A vibrating fork is placed near the larger of the two openings of a Helmholtz resonator and a very lightweight pinwheel near the other. A good pinwheel for this demonstration is a solar radiometer (which can be purchased at almost any novelty shop) with the glass envelope removed. The pinwheel contains four lightweight mica vanes, mounted in the form of a cross and free to turn on a needle point. Pulses of air from the prongs of a tuning fork, traveling through the hollow cavity, vibrate in resonance to the fork and come out as reinforced waves striking the vanes of the pinwheel. If the vibrating fork is removed, the pinwheel stops rotating, a direct indication that sound waves constitute a flow of energy (see Question 8).

*Musicians often use the word *intensity* to refer to musical passages intended to elicit strong emotional response. To prevent confusion, such usage will be avoided here.

FIGURE 9–5
A demonstration showing that sound waves have energy. Sound energy from a tuning fork, passing through a Helmholtz resonator, sets a pinwheel rotating.

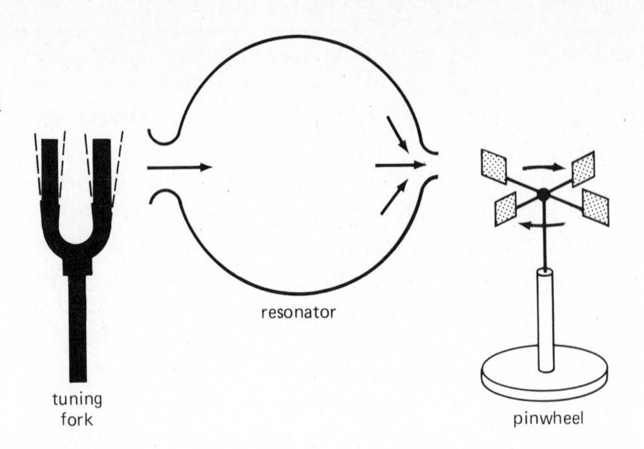

Sound, heat, light, atomic, electrical, and mechanical energy are all measured in joules (J). (See Appendix I.) **Power** is defined as the time rate of flow of energy and is given by

$$\frac{1 \text{ joule}}{1 \text{ second}} = 1 \text{ watt} \qquad [9a]$$

where the joule is the SI unit of energy in kilograms × meters²/second².

The delicate construction of the pinwheel in Figure 9–5 demonstrates how small the energy flow in sounds of normal intensity really is. For this reason, it has become convenient in practice to use the picowatt (pW) as a unit of sound power, a unit that is one-trillion times smaller than the watt (see Appendix II).

$$\begin{aligned} 1 \text{ pW} &= 10^{-12} \text{ W} \\ 10^{+12} \text{ pW} &= 1 \text{ W} \end{aligned} \qquad [9b]$$

We therefore usually express sound intensity as the power in picowatts flowing through 1 square meter of area normal to the direction of sound propagation. See Figure 9–6.

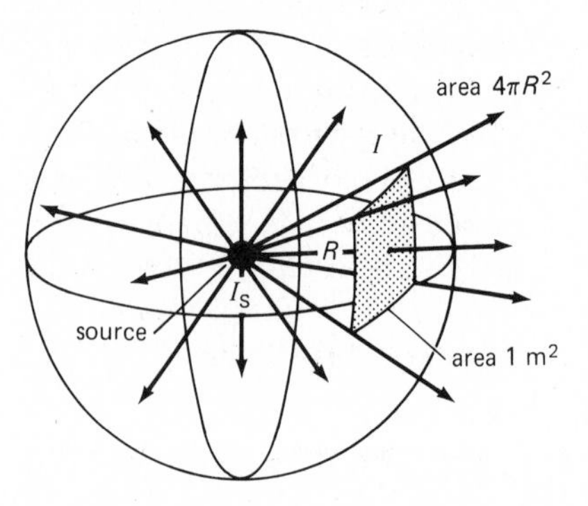

FIGURE 9–6
Sound intensity is defined as the power, in watts or picowatts, through 1 square meter of area, the area taken normal to the direction of energy flow.

9.3 Intensity versus loudness

We have seen that sound intensity is an objective physical quantity based upon energy flow per second and can be related to instrumental measurements, whereas loudness is a sensory quantity based on a subjective response that is estimated by each individual. Musicians with trained aural acuity have set up a rough scale of loudness and designated the various levels as given in Table 9–1.[6] Observe how the exponents of the physical scale of intensity in the right-hand column of the table provide a quantitative measure of linearity and strongly suggest a logarithmic scale of intensities.

TABLE 9–1: Musicians' Scale of Loudness

LOUDNESS			INTENSITY
DESCRIPTION	TERMINOLOGY	DESIGNATION	(W/m^2)
Very very loud	Fortississimo	fff	10^{-3}
Very loud	Fortissimo	ff	10^{-4}
Loud	Forte	f	10^{-5}
Moderately loud	Mezzoforte	mf	10^{-6}
Soft	Piano	p	10^{-7}
Very soft	Pianissimo	pp	10^{-8}
Very very soft	Pianississimo	ppp	10^{-9}

9.4 Sound intensity levels

Since the human ear responds to sound energy over the tremendous range of 1 to 1 trillion (1 to 1,000,000,000,000), musicians, sound engineers, audiologists, and physicians (ear specialists) have adopted a logarithmic intensity scale. This follows directly from the last sentence in the previous section. The general rule is known as the **Weber-Fechner law.** According to this law, the response of any sense organ is approximately proportional to the logarithm of the magnitude of the stimulus.* It has long been known from laboratory experiments, for example, that in order to double the apparent brightness of a surface illuminated by visible light, or to double the apparent loudness of audible sounds, the measured energy or power of the radiation must be increased *approximately* tenfold. The actual figure is more nearly ninefold.

The normal ear is most sensitive to frequencies in the range of 1000 to

*Real sensory responses are actually somewhat more complex than that given by the Weber-Fechner law.

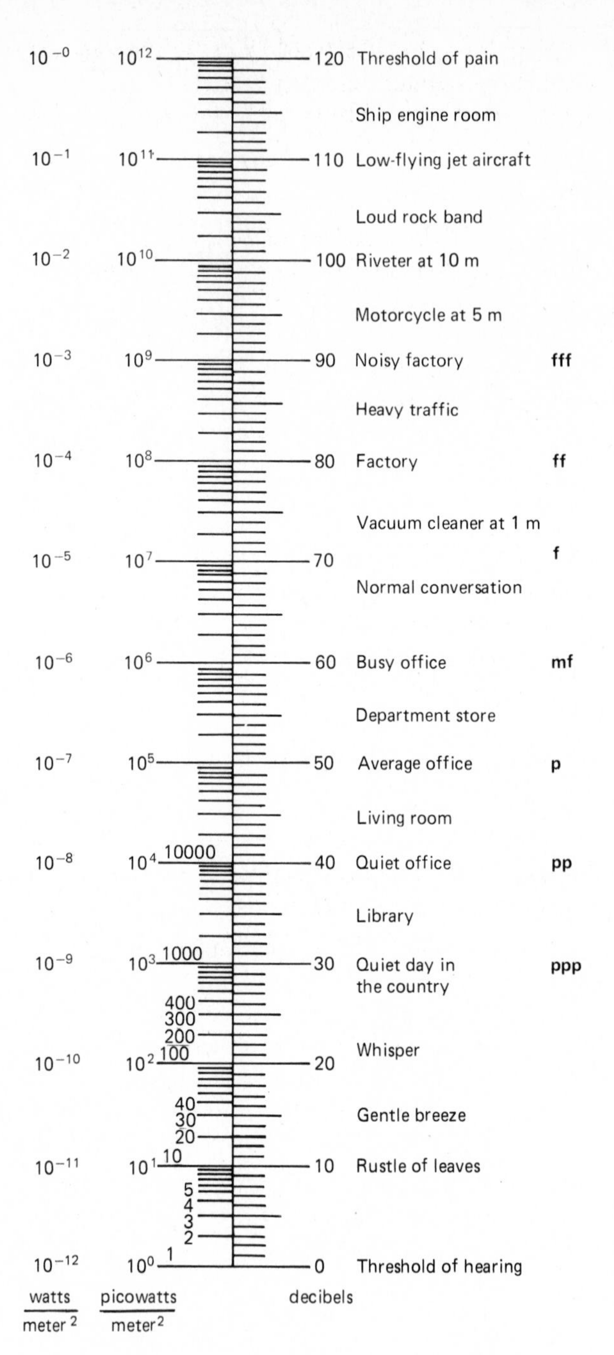
10^−0
10^12
120 Threshold of pain
Ship engine room
10^−1
10^11
110 Low-flying jet aircraft
Loud rock band
10^−2
10^10
100 Riveter at 10 m
Motorcycle at 5 m
10^−3
10^9
90 Noisy factory
fff
Heavy traffic
10^−4
10^8
80 Factory
ff
Vacuum cleaner at 1 m
10^−5
10^7
70
f
Normal conversation
10^−6
10^6
60 Busy office
mf
Department store
10^−7
10^5
50 Average office
p
Living room
10^−8
10^4 10000
40 Quiet office
pp
Library
10^−9
10^3 1000
30 Quiet day in the country
ppp
400
300
200
10^−10
10^2 100
20
Whisper
40
30
20
Gentle breeze
10^−11
10^1 10
10 Rustle of leaves
5
4
3
2
10^−12
10^0 1
0 Threshold of hearing
watts/meter^2
picowatts/meter^2
decibels

FIGURE 9–7
Sound intensity scales and levels.

7000 Hz. National technical committees have selected 1000 Hz as a standard frequency to be used in establishing a sound intensity scale.*

At 1000 Hz, the faintest pure tone that can be heard by the human ear has the intensity of approximately 1×10^{-12} W/m^2.

$$I_0 = 1 \times 10^{-12} \text{ W/m}^2 \qquad [9c]$$

While this value has been chosen as the standard of reference, the exact threshold values vary with each individual listener. Fairly recent evaluations indicate the average value, determined by using a large number of trained listeners, is a little higher than I_0. The value of I_0 in Equation (9c), however, is still used in all calculations. The measured lower limit (see Figure 9–7) is often referred to as the **threshold of hearing.** At very high sound levels, the upper limit is based upon what the ear can endure. Above approximately 1 W/m^2, the sound waves produce chest pains, and this value is frequently called the **threshold of pain.**

It is also common practice today to compare all sound intensities I, in watts/meter^2, with the standard I_0 given by Equation (9c) and to find the logarithm of the ratio of the two powers. For example, if the power of an unknown sound I is ten times the standard I_0, the ratio of $I/I_0 = 10$, the logarithm of 10 is 1.0, and the unknown is 1.0 bel (B). (The bel is a unit named in honor of Alexander Graham Bell, the inventor of the telephone.) The general relation is written as

$$B = \log_{10} \frac{I}{I_0} \qquad [9d]$$

where B is in bels. For values of this sound intensity scale in bels, see Table 9–2. In the table, the numbers in column 4 are just the logarithms of the numbers in column 3. To use most pocket calculators containing a *log key* and a *10^x key*, enter the number 1000 and press the log key. Read 3 as the answer. For the reverse process of finding the relative powers from log 3, enter 3, press the 10^x key, and read 1000.

The bel as a unit of sound intensity is divided into ten equal parts called decibels (dB). From this definition, we can write

$$S_{IL} = 10 \log_{10} \frac{I}{I_0} \qquad [9e]$$

where S_{IL} stands for **sound intensity level** and is expressed in **decibels.**

Table 9–3 illustrates specific values I/I_0 and the corresponding values of S_{IL} in decibels. Each power ratio in the left-hand column is 26 percent greater

*The United States government broadcasts daily, from Fort Collins, Colorado, over radio frequency bands of exactly 5, 10, 15, and 20 megahertz (MHz), a pure steady tone with a precise frequency of 1000 Hz.

TABLE 9-2: Sound Intensity Scale (B)

	SOUND INTENSITY I (W/m^2)	RELATIVE INTENSITY I/I_0	INTENSITY LEVEL B(B)
Threshold of pain	1	10^{12}	12
fffff	10^{-1}	10^{11}	11
ffff	10^{-2}	10^{10}	10
fff	10^{-3}	10^{9}	9
ff	10^{-4}	10^{8}	8
f	10^{-5}	10^{7}	7
mf	10^{-6}	10^{6}	6
p	10^{-7}	10^{5}	5
pp	10^{-8}	10^{4}	4
ppp	10^{-9}	10^{3}	3
pppp	10^{-10}	10^{2}	2
ppppp	10^{-11}	10^{1}	1
Threshold of hearing	10^{-12}	1	0

TABLE 9-3: Sound Intensity Scale (dB)

RELATIVE INTENSITY (I/I_0)	SOUND INTENSITY LEVEL (S_{IL}, in dB)
1.00	0
1.26	1
1.58	2
2.00	3
2.51	4
3.16	5
3.98	6
5.01	7
6.31	8
7.94	9
10.00	10

than the preceding value. Since a 26 percent *change* in power is just detectable by the normal human ear, the decibel is a convenient and practical unit of intensity.[7]

Example

The intensity of a given sound is measured and found to be 1.250×10^{-9} W/m^2. What is the sound intensity level in bels and decibels?

Solution

The given quantities are $I = 1.250 \times 10^{-9}$ W/m² and $I_0 = 1 \times 10^{-12}$ W/m². Using Equation (9d), the ratio of I to I_0 is given by

$$\frac{I}{I_0} = \frac{1.250 \times 10^{-9} \text{ W/m}^2}{1.0 \times 10^{-12} \text{ W/m}^2} = 1250$$

The logarithm of 1250 can be found by using a pocket calculator or a logarithm table; it is 3.10 B. In decibels, the answer is 31.0 dB.

A chart comparing sound intensities, in watts/meter² and in picowatts/meter², with sound intensity levels, in bels and decibels, is given in Figure 9–7.

9.5 The inverse square law

Let us imagine a source of sound waves I_S that sends out waves of equal intensity in all directions. Being propagated in straight lines, as shown in Figure 9–6, the acoustical power radiated by the source must all pass through the surface of every sphere of radius R and area $4\pi R^2$. Since the sound intensity I is defined as the energy per second passing through each square meter of area, we can write

$$I = \frac{I_S}{4\pi R^2} \qquad [9f]$$

where I_S is the power of the source, in watts, R is the radius of a sphere, in meters, and $4\pi R^2$ is the area of the surface of the sphere. We conclude, therefore, that in the absence of reflections from the walls of a room, or the absorption of energy by the air, the sound intensity is inversely proportional to the square of the distance from the source. This is called the **inverse square law.**

The inverse square law is often applied to the relative intensities of a source at different distances from an observer. See Figure 9–8. Whatever acoustical energy flows through an area A at a distance of 1 m must also pass through an area B at 2 m, an area C at 3 m, an area D at 4 m, and so forth. Since these areas, normal to the flow of energy, have the ratios of 1 to 4 to 9 to 16 and so on, the energy flow per second through unit area at each distance will be $I, \frac{1}{4}I, \frac{1}{9}I, \frac{1}{16}I$, respectively.

To find the ratio of the intensities at any two different distances R_1 and R_2 from the same source, we can write

$$\frac{I_2}{I_1} = \frac{R_1^2}{R_2^2} \qquad [9g]$$

This is another way of writing the **inverse square law.**

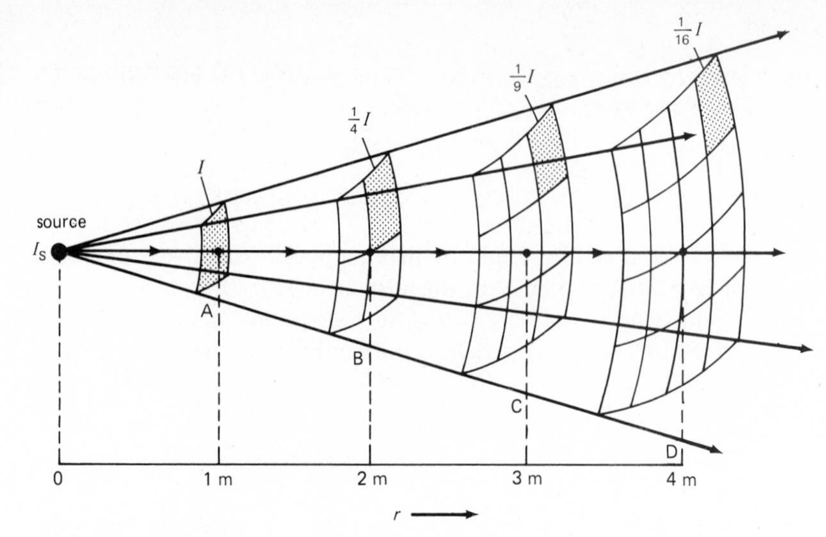

FIGURE 9–8 Diagram illustrating the inverse square law.

Example

The noise from a jet plane flying only 25 m overhead is measured with a sound level meter and found to be 100 dB. What will be the measured sound intensity level of the plane, in decibels, when it flies overhead at an altitude of 800 m? See Figure 9–9.

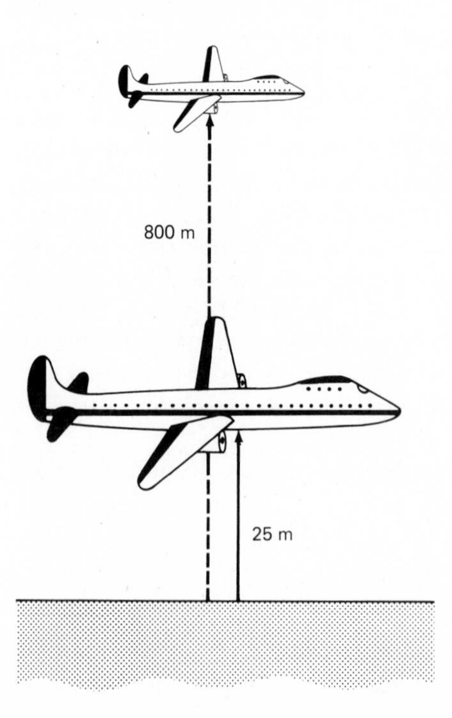

FIGURE 9–9 Diagram illustrating the inverse square law applied to the sound intensity level at ground level from a plane flying overhead at two different altitudes.

Solution

The given quantities are $I_1 = 100$ dB, $R_1 = 25$ m, and $R_2 = 800$ m, and the unknown is I_2. Using the chart of Figure 9–7, we find

$$I_1 = 100 \text{ dB} = 1 \times 10^{10} \text{ pW/m}^2$$

Using Equation (9g), solving for the unknown, and substituting known quantities, we obtain

$$I_2 = I_1 \frac{R_1^{\,2}}{R_2^{\,2}} = 10^{10} \text{ pW/m}^2 \times \frac{625 \text{ m}^2}{640{,}000 \text{ m}^2}$$

$$I_2 = 9.77 \times 10^6 \text{ pW/m}^2$$

Using Equation (9e), we find

$$S_{\text{IL}} = 10 \log_{10} 9.77 \times 10^6 \text{ pW/m}^2$$

giving

$$S_{\text{IL}} = 69.9 \text{ dB}$$

which means a drop from 100 dB to 69.9 dB, or a total of 30.1 dB.

9.6 Sound pressure levels

Sound engineers and acoustical laboratories measure sound pressure with an instrument called a **sound level meter.** See Figure 9–10. The principal components of such instruments consist of a microphone, an amplifier, a calibrated attenuator, and a milliammeter registering sound pressure levels in decibels. Using an oscilloscope, the rapidly changing pressure exerted by the waves themselves can be observed.

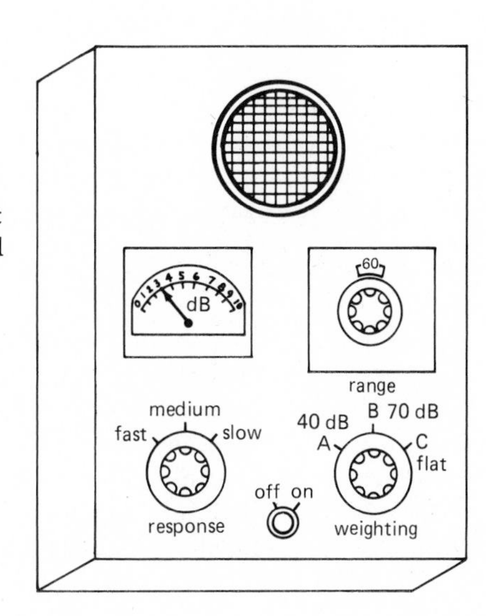

FIGURE 9–10
A sound level meter. The weighting switch positions A and B permit adjustment to simulate typical ear responses.

If we plot a graph of sound wave pressure against the distance x or the time t, we obtain a curve like that shown in Figure 9–11.[8] As the longitudinal sound waves travel to the right with a speed V, the pressure at each point along the wave oscillates about normal atmospheric pressure P. At point A, the air molecules are closest together (maximum pressure increase), and at B,

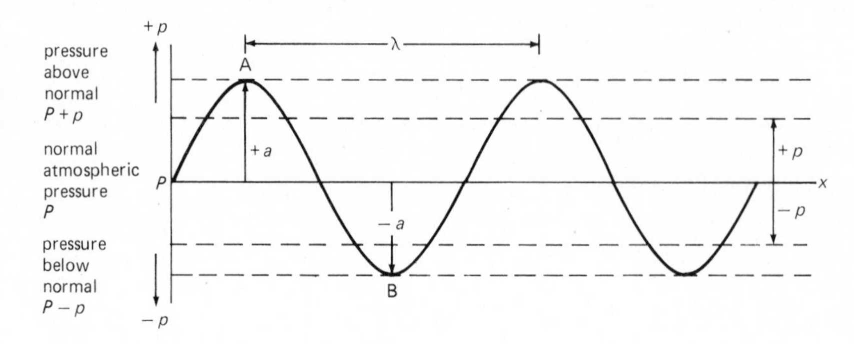

FIGURE 9–11
Graph of the pressure variations *p* in a longitudinal sound wave in air.

they have moved farthest apart (maximum pressure decrease). The effective pressure variation p is a constant value that would give the same intensity as the variable pressure. Its value is equal to the variation amplitude a divided by $\sqrt{2} = 1.414$.

$$p = \frac{a}{\sqrt{2}} = 0.7071a \text{ N/m}^2$$

It may be shown from basic principles in mechanics that the intensity I of a given sound is given by

$$I = \frac{p^2}{Vd} \text{ W/m}^2 \qquad [9h]$$

where I is in watts/meter^2, p is in newtons/meter^2, V is the speed of sound in meters/second, and d is the density of air in kilograms/meter^3. At the threshold of hearing I_0, the sound pressure variation $p_0 = 2 \times 10^{-5}$ N/m^2, so that, on the decibel scale given by Equation (9e),

$$S_{PL} = 20 \log_{10} \frac{p}{p_0} \qquad [9i]$$

where S_{PL} stands for **sound pressure level** expressed in decibels, p is the effective pressure variation, and p_0 is the effective pressure variation at threshold, both in newtons/meter^2 and both at 1000 Hz. For a beam of sound waves from a distant source, and in the absence of reflected waves from the walls of a room, the sound pressure levels given by Equation (9i) and the sound intensity levels given by Equation (9e) are equivalent. Both relations give approximately the same result in decibels, and either one can be used.

Using a sound level meter, measurements of sound levels from musical instruments have been made under somewhat standard conditions. At a distance of 10 m, and playing fortissimo, the strings and woodwinds produce a sound pressure level of about 60 dB, and the brasses produce a sound pressure level of about 75 dB.[9] Although the power output of musical instruments varies over a wide range, these values can be taken as representative, but not exact. Since 75 dB represents an intensity of 3.2×10^{-5} W/m^2, we can use the inverse square law, Equation (9g), to find the power of the source; it is 0.040 W.

QUESTIONS

1. Define or briefly explain in your own words each of the following: (a) sound intensity, (b) picowatt, (c) logarithm, (d) bel, (e) decibel.

2. Define or briefly explain each of the following: (a) threshold of hearing, (b) sound intensity level, (c) sound pressure level.

3. Make three diagrams (from memory) of the human ear, showing (a) the outer ear, (b) the middle ear, and (c) the inner ear. Label as many of the principal parts as you can.

4. State the inverse square law in your own words, and give an example of where and how it would apply.

5. At what approximate frequency is the normal human ear most sensitive? What is meant by the threshold of hearing? What is the threshold of pain?

6. What are the approximate values for the threshold of hearing and the threshold of pain (a) in watts/meter^2, (b) in picowatts/meter^2, (c) in decibels?

7. Tchaikovsky uses the dynamic (loudness) range from ppppp (Symphony *Pathetique*) to ffff *(1812 Overture)*. What is the equivalent sound intensity range?

8. It can be argued that the pinwheel demonstration (Section 9.2) can be accounted for by a flow of air from the tuning fork to the pinwheel. Is this what is happening? How could you find out?

PROBLEMS

1. The intensity level of the sound from a violin is measured with a sound level meter and found to be 47 dB. What is its sound intensity in (a) watts/meter^2 and (b) picowatts/meter^2?

2. The intensity level of a trumpet at a distance of 10 m is measured with a sound level meter and found to be 78 dB. Find (a) its sound intensity in picowatts/meter^2 and (b) the power of the source in watts.

3. The sound intensity in a room is measured and found to be 3.68 W/m^2. Find the sound intensity level in (a) bels and (b) decibels.

4. The sound intensity of a jet engine at a short distance away is found to be 7.25×10^{-2} W/m^2. Find its sound intensity level in (a) bels and (b) decibels.

5. The sound from a racing car 500 m away from a sound level meter is found to be 42 dB. Find (a) the intensity of the sound in watts/meter^2, (b) the intensity of the source in watts, and (c) the sound intensity level in decibels when the car is only 10 m away.

6. If five violins sound the same note of 1000 Hz, each at the same sound pressure level of 60 dB, what is their total intensity in decibels, compared with one violin alone?

7. Three sounds at standard frequency of 1000 Hz have sound intensity levels of 65 dB, 68 dB, and 76 dB, respectively. Find their resultant sound intensity level.

8. Twenty clarinets sound a note of 1000 Hz simultaneously and at the same sound pressure level of 60 dB. What is the resultant sound pressure level?

9. Three tuning forks at the same standard frequency of 1000 Hz produce sound intensity levels of 55 dB, 60 dB, and 65 dB, respectively. What will be their combined sound intensity level?

10. Seventy-six trombones are playing the same tune, each at a sound pressure level of 70 dB. What is the combined sound pressure level, as compared with one trombone alone?

PROJECT

Investigate the causes and forms of deafness and the measures that can be taken to aid partially or totally deaf persons.

NOTES

[1]H. Fletcher, *Speech and Hearing in Communication* (New York: Van Nostrand, 1953), p. 107. (Also published by Krieger Publishing Co., P.O. Box 542, Huntington, New York; 1971.)

[2]G. von Békésy, *Experiments in Hearing* (New York: McGraw-Hill, 1960), p. 101. A monumental book on the ear and the various theories of hearing.

[3]W. A. Yost and D. W. Nielson, *Fundamentals of Hearing* (New York: Holt, Rinehart and Winston, 1977), pp. 45–65.

[4]Von Békésy, p. 444.

[5]Von Békésy, p. 636, and Yost and Nielson, pp. 66–91.

[6]J. G. Roederer, *Introduction to the Physics and Psychophysics of Music,* 2d ed. (New York: Springer-Verlag, 1975), p. 78.

[7]Fletcher, p. 146.

[8]Roederer, p. 71.

[9]M. Clark, Jr., and D. Luce, "Intensities of Orchestral Instrument Scales," *Journal of the Acoustical Society of America* 13 (1965):151.

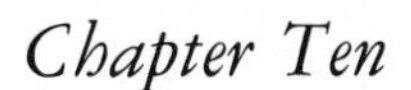

Chapter Ten

PERCEPTION OF LOUDNESS

In the preceding chapter, we were introduced to the physical scale of sound intensity I, in watts/meter^2, the measurement of sound pressure p, in newtons/meter^2, and the way in which both are expressed on the physical scale of decibels. In this chapter, we take up the concept of loudness as a *subjective evaluation* of sound intensity, and we consider the attempts that have been made to tie this loudness scale to the physical scale of decibels. With our present knowledge of hearing, we then apply these principles to the addition of two or more sounds to find the resultant loudness of the combination.

10.1 Sound loudness level

In this section, a scale of sound loudness is developed for the entire audible frequency range. It is based upon the physical measurements of sound pressure and then a comparison of these measurements with subjective estimates of loudness based upon a person's hearing acuity. To do this, a large number of trained subjects with normal hearing have been tested and their observations recorded. In such experiments, each subject hears a pure tone of the standard frequency of 1000 Hz and at a specified sound pressure level in decibels. A different specified frequency is then sounded, and its intensity is raised or lowered until the two sounds seem to have the same loudness. That level value is recorded in decibels. This procedure is carried out for many frequencies between 20 and 15,000 Hz and for many sound pressure levels between 0 and 120 dB. The averaged results of such experiments are plotted as a family of curves in Figure 10–1. Each curve represents an equal loudness level for all frequencies over the entire frequency range.[1] For example, a pure

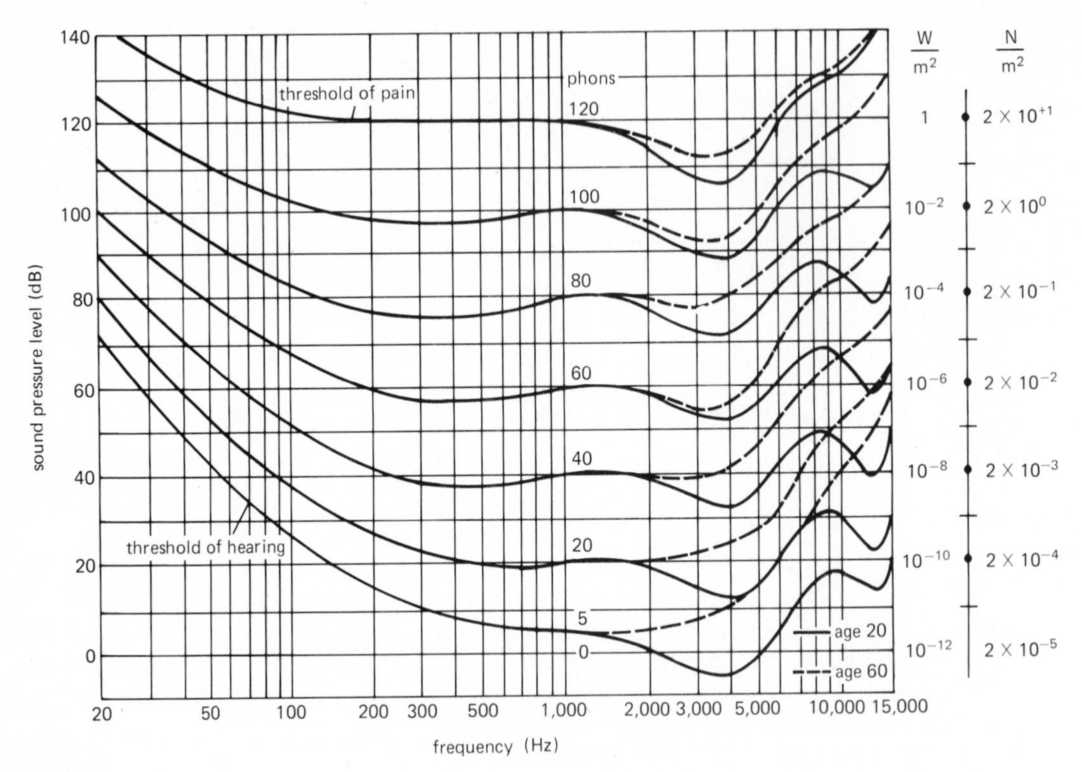

FIGURE 10–1
Curves of equal sound level loudness over the entire audio range of frequencies. Vertical scales of sound intensity levels, in decibels (far left); loudness levels, in phons (middle); energy levels, in watts (right); and sound pressure variations, in newtons/meter² (far right).

tone of 40 dB at 1000 Hz has the same loudness level as 65 dB at 50 Hz, 51 dB at 100 Hz, 42 dB at 200 Hz, 38 dB at 500 Hz, 38 dB at 2,000 Hz, 36 dB at 5,000 Hz, 48 dB at 10,000 Hz, and 50 dB at 15,000 Hz.

To relate physical measurements of sound pressure in decibels to the subjective determinations of loudness, the unit of the **phon** has been adopted. All points along any given loudness level curve in Figure 10–1 are assigned the same number of phons as the decibel value at the standard frequency of 1000 Hz. The curve labeled 80 phons, for example, has the same loudness level S_{LL} for all frequencies along that curve. The highest curve, labeled 120 phons, has the same loudness level, $S_{LL} = 120$ phons, at all frequencies, and the bottom curve, representing the threshold of hearing at 1000 Hz, has a sound loudness level of approximately 5 phons at all frequencies. The threshold of hearing at a frequency of 40 Hz has a sound loudness level of $S_{LL} = 5$ phons and a sound pressure level of 50 dB.

Beyond the standard frequency of 1000 Hz, all curves show one or two minima. All curves show a pronounced minimum between 1000 Hz and 5000 Hz. Here the ear reaches its maximum sensitivity, a feature attributed to resonance in the auditory canal. This air column, open at one end and closed at the other by the eardrum, acts as a resonant cavity. (See Figure 9–1 and Section 11.2.) At standard temperature and pressure and at a frequency of 3000 Hz, sound waves have a wavelength of 11.0 cm and will resonate with a maximum amplitude in a closed pipe of $\frac{1}{4}\lambda$, or 2.75 cm. This is the average length of the auditory canal. Because of the departure from an ideal tube, the resonance is broad and covers a range of several hundred hertz.

10.2 Adding sounds of the same frequency

Suppose we sound two violins with the same frequency f and the same intensity, and we ask the question, "Is the resultant sound twice as loud?" The answer is no. Adding two pure tones of 60 dB, and arriving at 120 dB, for example, is obviously wrong, since such a sound would be painful to hear.

From the standpoint of energy conservation, which is a universal law, two violins emitting sound energy at the same rate would result in a total rate of energy emission double that for one violin alone. Starting with this premise, we carry out the following procedure to find the correct answer: To find the sound intensity level of two sounds together, as compared with one of them alone, we use Equation (9e). Take the difference between two sound levels. Using superscripts l and s to distinguish the *louder* combination (two sounds of 60 dB each) from the *softer* sound (one sound of 60 dB), we obtain

$$S_{IL}^{l} - S_{IL}^{s} = 10 \log \frac{I^{l}}{I_0} - 10 \log \frac{I^{s}}{I_0} \qquad [10a]$$

which can be written as

$$S_{IL}^{l} - S_{IL}^{s} = 10 \log \frac{I^{l}}{I^{s}} \qquad [10b]$$

Using Figure 9–7, we obtain

$$S_{IL}^{l} - S_{IL}^{s} = 10 \log \frac{(10^{-6} + 10^{-6})\,\mathrm{W/m^2}}{10^{-6}\ \mathrm{W/m^2}} = 10 \log 2 \qquad [10c]$$

Since the log of 2 is 0.301, we have

$$S_{IL}^{l} - S_{IL}^{s} = 3.01\ \mathrm{dB} \qquad [10d]$$

Doubling the intensity of a given sound of 60 dB, then, increases the intensity to 63 dB. This is true for all frequencies and intensities as long as they are both equal. For example, if each sound is 40 dB, the two together will be 43 dB, and if each sound is at 75 dB, the two together will be 78 dB.

If 10 violins are playing the same pure tone of 1000 Hz and at the same intensity of 60 dB, we can find the rise in intensity on the decibel scale by using the procedure above. With Equation (10b), we can write

$$S_{IL}{}^{l} - S_{IL}{}^{s} = 10 \log \frac{10 \times 10^{-6}\ \mathrm{W/m^2}}{1 \times 10^{-6}\ \mathrm{W/m^2}} = 10 \log 10$$

Since log 10 = 1, we have

$$S_{IL}{}^{l} - S_{IL}{}^{s} = 10\ \mathrm{dB} \qquad [10e]$$

The intensity increases from 60 dB for one violin to 70 dB for ten violins. Note this increase of 10 dB is independent of the frequency and the intensity. If ten pure tones of 50 dB at 200 Hz are sounded simultaneously, their total intensity I of 10 times one pure tone alone gives 60 dB. It is reasonable to assume that the phase angles of a number of instruments are random. (See Section 8.4.)

If we now add two sounds at *the same frequency* but *different intensities,* the procedure above may still be followed. For example, if we add two sounds of 60 dB and 65 dB, respectively, the softer sound has an intensity of 1×10^{-6} $\mathrm{W/m^2}$ (see Figure 9–7), and the louder sound an intensity of $3 \times 10^{-6}\ \mathrm{W/m^2}$. The louder sound, therefore, has three times the intensity of the softer sound. The sum of the two together is four times that of the softer. Following Equation (9e), we can write

$$S_{IL}{}^{l} - S_{IL}{}^{s} = 10 \log \frac{I^{l}}{I^{s}} = 10 \log 4$$

which gives

6.0 dB

The louder sound raises the softer sound by 6 dB, or from 60 dB to 66 dB.

The *loudness level increase* in the example above (in phons) is precisely equal to the intensity increase (in decibels) only at the standard frequency of 1000 Hz. At other frequencies, it may be more accurate to first determine the intensities involved and use Figure 10–1 to determine the loudness levels.

The process described above of adding sounds to find a resultant intensity level, or loudness level, applies to *sounds of the same frequency.* If the frequencies to be added are *different,* another process must be used. See Section 10.4.

10.3 Frequency discrimination and the critical band

We have seen (in Section 6.5) that the superposition of two waves with the same amplitudes but slightly different frequencies f_1 and f_2 produce a beat frequency f_{BF} of

$$f_{BF} = f_2 - f_1 \qquad [10f]$$

Observe in Figure 10–2 that when two slightly different frequencies are combined (light lines, f_1 and f_2) to produce a resultant (heavy line R), there is a slowly changing phase difference between the two components. The resultant has its maximum amplitude at D where the two waves are in phase (phase difference is 0). At B and F, the two vibrations are out of phase (180° phase difference), and the two destructively interfere to give zero amplitude. At A, C, E, and G, they are 90° out of phase (one is a maximum, the other zero), and the resultant amplitude is $\sqrt{2}\,a$.

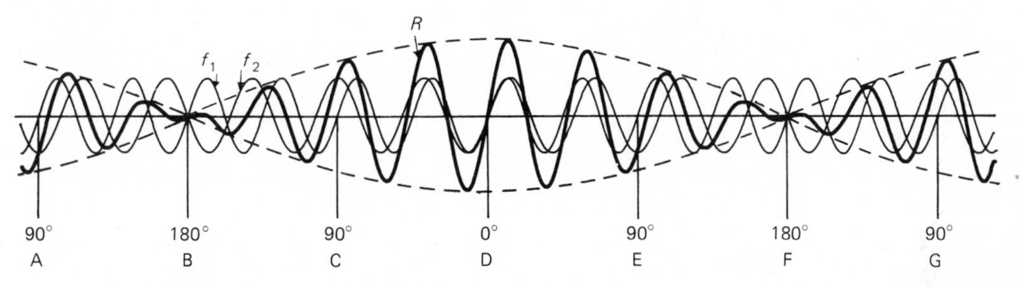

FIGURE 10–2
Beats are produced by two waves of slightly different frequencies as they periodically move in and out of phase.

Not knowing that the heavy line in the figure is the resultant of two tones, the eardrum follows this resultant vibration pattern of varying amplitude. At the oval window, this vibration gives rise to two wave trains traveling through the cochlea, one for each of the original frequencies f_1 and f_2. If the frequency difference f_{BF} is great enough, two resonance regions on the basilar membrane respond, and we hear the two original tones of constant loudness.[2] This ability of the cochlea to analyze a complex pattern into its original pure tones is called **frequency discrimination.** In order to identify the two separate tones, their difference in frequency must exceed an amount called the **limit of frequency discrimination, Δf_{FD}.** If the beat frequency f_{BF} is smaller than a certain amount, the resonance regions on the basilar membrane overlap, and we hear but one tone of intermediate frequency, but beating in loudness. These are the beats known to every musician and are used by them in tuning their instruments.

Suppose two electronic tone generators are used to produce the same pure tones of equal amplitude, and we slowly raise the pitch of one and listen for all sound effects. At first, we continue to hear one single tone of slightly higher pitch, corresponding to the **average frequency**

$$\bar{f}_{AF} = \frac{f_1 + f_2}{2} \qquad [10g]$$

and loudness, beating with a frequency $f_{BF} = f_1 - f_2$. When the beat frequency reaches about 10 to 15 Hz, the beating disappears and gives way to a roughness sensation. When f_{BF} surpasses the limit of frequency discrimination Δf_{FD}, we start to hear two separate tones with frequencies f_1 and f_2. What has happened on the basilar membrane is that the two resonance regions have barely separated. At that distance, roughness still persists, particularly with sounds in the low-frequency range. After passing a larger frequency difference Δf_{CB}, called the **critical band,** the roughness sensation slowly smooths out as it gives way to two clear tones. It is an experimental fact that the critical band Δf_{CB} is roughly independent of loudness and seems to be related to the structure of the basilar membrane. The spacial extension along the membrane is almost constant at 1.2 mm and involves a roughly constant number of 1,300 receptor cells out of a total of 30,000. A graph showing the frequency width of the critical band Δf_{CB} and the limit of frequency discrimination Δf_{FD} is shown in Figure 10–3.[3]

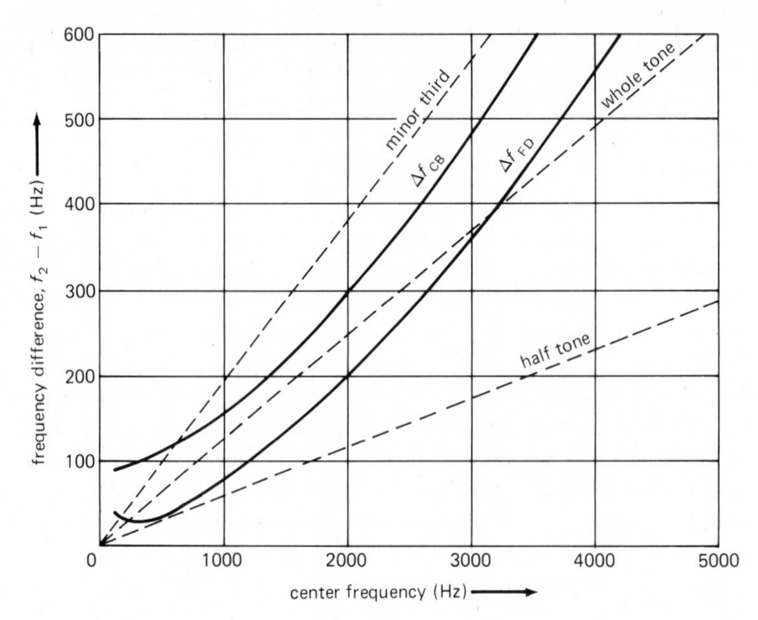

FIGURE 10–3 **Curves showing the critical band Δf_{CB} and the limit of frequency discrimination Δf_{FD} plotted against the center frequency of a two-tone stimulus.**

If the outputs of two electronic tone generators are fed separately into the two ears (using separate earphones), and the frequency differences discussed above, from zero on, are sounded, the beat frequency and roughness sensations are completely absent at all frequencies, and both tones can be discriminated even when the frequency difference Δf is far below Δf_{FD}. What happens is that only one region on each basilar membrane is activated, with no chance of overlapping of the resonance areas in the cochlea, and no beats are heard.

10.4 Subjective loudness

When we wish to add two sounds of different frequencies, we cannot employ the principles outlined in Section 10.2. Although we can apply the equal-loudness-level curves of Figure 10–1 to find the two sound intensities, and add them to get the correct sum of their powers in watts/meter^2, we cannot use this sum and work backward on the curves to find their combined loudness level. The reason for this is due to the experimental fact that the different frequencies activate different regions on the basilar membrane, while sounds of the same frequency activate the same region.

To combine different frequencies, a **subjective scale of loudness, S_{LL},** has been developed. Such a scale is designed to cover the range of musical instruments from 40 dB to 120 dB. Like the sound loudness level in phons, it is based on observations made with a large number of trained listeners. For this reason, both scales are actually subjective. In the experiments, subjects were asked to compare two sounds and to determine when one sound appeared to be twice as loud, or half as loud, as another. The average of many observations were used to compose the curve shown in Figure 10–4. The quantity S_{SL}, denoting **sound subjective loudness,** is given in units called **sones,** whereas the loudness level S_{LL} is given in *phons.*[4] The various scales introduced above are summarized in Table 10–1.

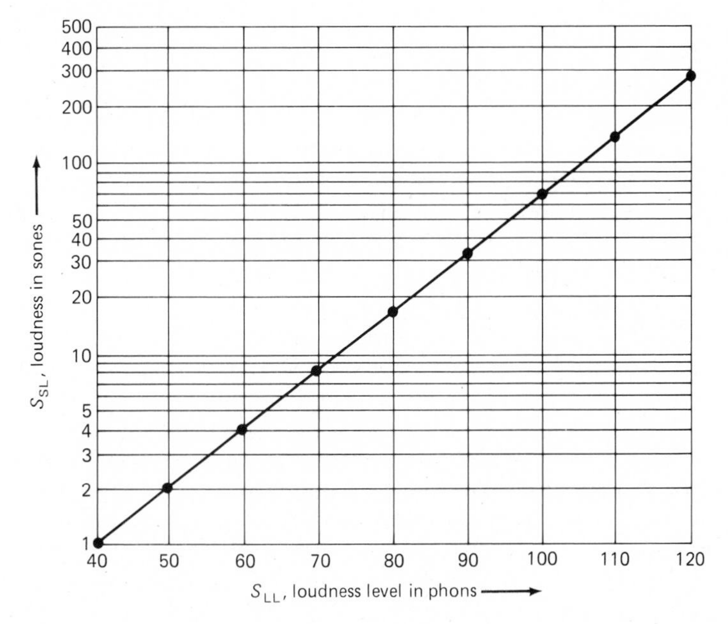

FIGURE 10–4
Musicians' scale of loudness, in sones, plotted on a logarithmic scale against different loudness levels measured in phons.

Observe that the S_{SL} scale is logarithmic, whereas the S_{LL} scale is linear. On the S_{LL} scale, the very soft sound of 40 phons has been set as the starting point, and it is arbitrarily set equal to 1 sone. A sound loudness of 2 sones is twice as loud as a sound of 1 sone; a loudness of 4 sones is twice as loud as one of 2 sones; 30 sones is twice as loud as 15 sones; and so forth. The loudnesses S_{SL} of individual sounds in sones are added directly to find their resultant loudness in sones. A sound of 60 phons corresponds to 4 sones; a sound of 70 phons corresponds to 8 sones; the sum $4 + 8 = 12$ sones corresponds to 75 phons. Adding 60 phons at 200 Hz to 60 phons at 700 Hz gives a loudness of $4+4=8$ sones, which is a loudness level of 70 phons. Adding a third sound of 60 phons at 1000 Hz gives a total loudness of $8+4=12$ sones and a total loudness level of 76 phons. Adding a fourth sound of 60 phons at 2000 Hz gives a total loudness of $12+4$ sones and a total loudness level of 80 phons.

TABLE 10-1: Intensity and Loudness Designations

SYMBOL	TERMINOLOGY	UNITS
I	Sound intensity	watts/meter^2
P	Sound pressure	newtons/meter^2
S_{IL}	Sound intensity level	decibels
S_{PL}	Sound pressure level	decibels
S_{LL}	Sound loudness level	phons
S_{SL}	Sound subjective loudness	sones

We have seen in Section 10.2 that adding two sounds of the same loudness produces an increase of about 3 phons, if the frequencies are the same. Here we see that two sounds of the same loudness produce a combined loudness increase of 10 phons, if the frequencies are quite different. The explanation involves the response of the basilar membrane and the width of the critical band. With one sound of any particular frequency, a critical band is activated on the basilar membrane, and doubling the sound pressure level at the same frequency increases the loudness level by only 3 phons. Exciting the basilar membrane at one point with one frequency does not change the sensitivity of other points outside the critical band. As a result, two frequencies far enough apart so that their critical bands do not overlap will produce a subjective impression of greater loudness than if they do overlap.

If a complex sound is composed of two sounds with frequencies close together, we would recognize a certain loudness level. If the frequencies of these sounds are slowly spread apart, the overall loudness would remain unchanged until the frequency spacing becomes greater than the critical bands, at which time the loudness increases. For the case of musical instruments or an entire orchestra, we have frequency combinations spanning all regions, resulting in interesting but complex situations.[5]

10.5 Masking

At one time or another we have all been in a crowded room where the sound level is so high that we have difficulty in hearing a nearby person speaking to us. A sound that is clearly audible under quiet conditions may become inaudible in the presence of a louder sound. Such an effect is called **masking.**

Suppose we have an 800-Hz tone that can barely be heard. Now a 400-Hz tone is sounded at constant intensity, making the 800-Hz tone inaudible. In this case, the 400-Hz tone is called the masking tone, and the 800-Hz tone is called the masked tone. We now raise the intensity of the 800-Hz tone until it can barely be heard again. This increase in intensity is called the **threshold increase of the masked tone,** expressed in decibels.

Extensive research experiments have been made at the Bell Telephone Laboratories on masking effects in which the rise in threshold of a given masked tone is determined in the presence of a louder masking tone at a specified level above threshold. Measurements were made and published in the forms of graphs, using masking tones of 200, 400, 800, 1200, 2400, and 3500 Hz.[6] Three of these six sets of curves are reproduced in Figures 10–5, 10–6, and 10–7.

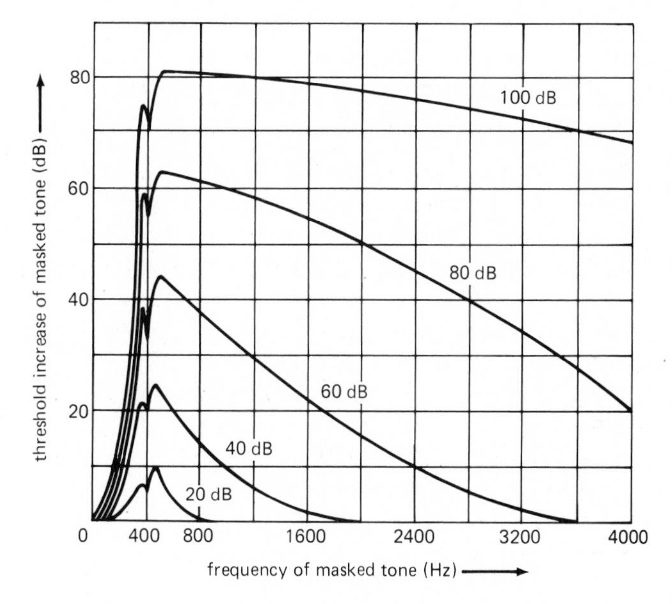

FIGURE 10–5 **Graphical representation of masking effects produced by a masking tone of 400 Hz. Intensities of the masking tones are labeled on the individual curves.**

When the frequencies of the two tones are close together, beat frequencies between the two can be heard, thus accounting for the sharp drop in the curves at 400 Hz in Figure 10–5, 800 Hz in Figure 10–6, and 1200 Hz in Figure 10–7. At high intensities of the masking tone of 1200 Hz, aural harmonics of 2400 and 3600 Hz are produced in the inner ear, and these form beat notes with the masked tones, making them easier to hear. (See Chapter 15.) This accounts for the dips in the curves at the frequencies shown.

Observe in all charts that at low loudness levels, up to about 20 dB, masking occurs only when the two frequencies are relatively close together. At 60 dB in Figure 10–5, the masked sound level at about 400 Hz, for example, must be raised to 44 dB to be heard. This is still 16 dB below the level of the masking tone.

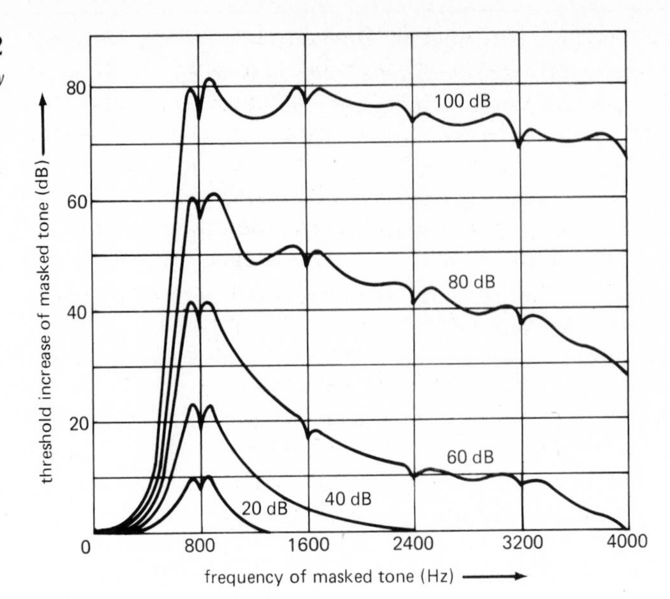

FIGURE 10–6 Graphical representation of masking effects produced by a masking tone of 800 Hz. Intensities of the masking tones are labeled on the individual curves.

For the masking tone of 400 Hz in Figure 10–5, and low intensities up to about 25 dB, masking is attributed to the relatively narrow region of the critical band, over which the basilar membrane is activated. Little masking occurs for tones of much lower or higher pitch. At high intensity levels of 80 to 100 dB for the masking tone in all diagrams, little masking occurs for the low frequencies, but strong masking occurs for all higher frequencies.

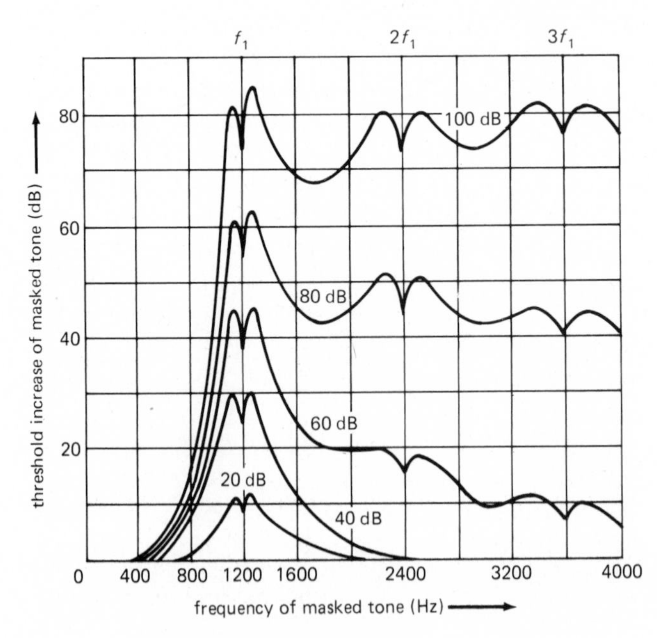

FIGURE 10–7 Graphical representation of masking effects produced by a masking tone of 1200 Hz. Intensities of the masking tones are labeled on the individual curves.

It is observed from all curves that as the intensity level of the masking tone is increased, all the higher frequencies must be increased to fairly high levels before they can be heard at all. For example, in Figure 10–5, the high frequencies must be raised about 75 dB above the threshold to be heard in the presence of a 400-Hz masking tone having an intensity level of 100 dB. Even for these high-intensity masking tones of 400 or 1200 Hz, frequencies a little below them are heard when they are only slightly above their threshold level.

An example of masking effects occurs in wireless telegraphy, where dot-dash-dot signals of many frequencies from as many different transmitters can be heard on the same receiving band. Due to masking, low-pitched signals are more easily heard over the higher-pitched background notes of other transmitters. Masking is also of importance to musicians. The masking of one instrument by the louder sounds from another instrument is of interest in polyphonic music, particularly in orchestration. When the brass section is playing fortissimo, the strings require concentrated effort by the listener to be heard. For this reason, there is little cause for them to play at the same time.

10.6 Hearing losses

Almost everyone is subject to a progressive hearing loss as he or she grows older, particularly in the high-frequency range. This ageing process is called **presbycusis.** Hearing losses of various kinds may occur at any age and may be brought about in many different ways. For example, if a person works all day, day after day, in a noisy place, such as a boiler factory, or if musicians play in a band using amplifiers turned up to very high intensity levels, they all run the risk of permanent hearing losses and should have periodic hearing tests.

To test a person's hearing, an instrument called an **audiometer** is generally used in medical clinics and hospitals. Such electronic devices generate sounds in a pair of earphones that are worn by the patient; these sounds are usually set at fixed frequencies of 125, 250, 500, 1000, 2000, 4000, and 8000 Hz. One after the other, frequencies are fed into the earphones at periodically decreasing and increasing sound levels to make repeated determinations of hearing thresholds. The average threshold for each frequency, and sometimes for each ear separately, is plotted on a chart called an **audiogram.** Many audiograms have the form shown in Figure 10–8. The frequencies commonly tested are printed across the top of the chart, and the thresholds are plotted on the vertical scale on the left. Straight lines are then drawn between the plotted points. The zero line near the top of the chart represents *normal hearing;* any deviations below this line, but within the limits labeled as normal hearing, are not considered to be significant.

In Figure 10–8, a person's audiogram with normal hearing is shown by curve (a). While most audiograms show the hearing record for but one patient, the records of four patients are shown here on the same chart for comparison. A pronounced decline below the normal hearing limit, in each ear separately or in both ears together, signifies a definite hearing loss in that frequency region. For example, a person who requires 100,000 times the normal intensity to hear a sound of 4000 Hz would have a hearing loss of 50 dB at that frequency. Such a drop would appear as a dip in the audiogram as shown in curve (b). This is a case in which the subject was exposed to the sound of gunfire.[7] Curve (c) is a case of congenital deafness in both ears, while curve (d) is a case of presbycusis. The latter is a typical case of the decline of hearing in both ears due to normal ageing of the subject. Such declines at the high-frequency end of the chart are not important in normal communication or in the appreciation of music.

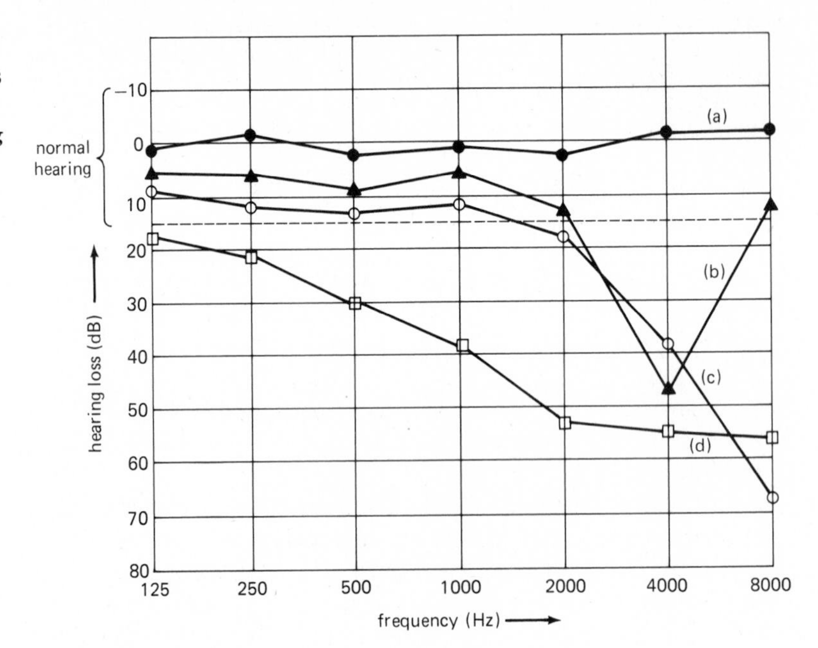

FIGURE 10–8
Human andiograms for (a) normal hearing, (b) hearing loss in the region of 4000 Hz, caused by exposure to loud gunfire, (c) hearing loss for a case of congenital deafness in both ears, and (d) impaired hearing due to normal ageing (presbycusis).

QUESTIONS

1. Briefly define or explain in your own words each of the following: (a) phons, (b) sound loudness level, (c) critical band, (d) audiogram, and (e) average frequency.

2. Define or briefly explain in your own words each of the following: (a) subjective loudness, (b) masking sound, (c) sones, (d) frequency discrimination, and (e) audiometer.

3. Describe briefly the steps used in finding the resultant loudness level in phons of two tones of the same frequency and the same sound pressure level of 70 dB.

4. Explain briefly how you would find the resultant subjective loudness in sones of two widely different frequencies if their loudness levels are both 70 phons.

1. The sound from a clarinet at a distance of 5 m from a sound level meter is found to be 52 dB. If the frequency is 1000 Hz, find (a) the sound pressure level in decibels, (b) the sound loudness level in phons, (c) the sound intensity in watts/meter^2, and (d) the power of the source in watts.

2. A tuning fork with a frequency of 100 Hz and a loudness level of 50 phons and another tuning fork with a frequency of 500 Hz and loudness level of 60 phons are sounded simultaneously. Find (a) their separate subjective loudness in sones, (b) their combined subjective loudness in sones, and (c) their combined loudness level in phons.

3. A tuning fork with a frequency of 300 Hz produces a sound pressure level of 55 dB. A second fork with a frequency of 400 Hz produces a sound pressure level of 60 dB. A third fork with a frequency of 200 Hz produces a sound pressure level of 65 dB. Find (a) the loudness level for each of the forks in phons, (b) the loudness for each fork in sones, (c) the loudness of all three forks combined, and (d) the loudness level of the combination.

Note: For additional problems, see W. Savage, *Problems for Musical Acoustics* (New York: Oxford University Press, 1977), chap. 13.

PROJECT

The legend about the breaking of a wine goblet by a singer producing a loud sustained note describes an extremely unlikely occurrence. However, the shattering of glass tumblers and laboratory beakers with a powerful audioamplifier has been accomplished.[8] This makes an interesting student project.

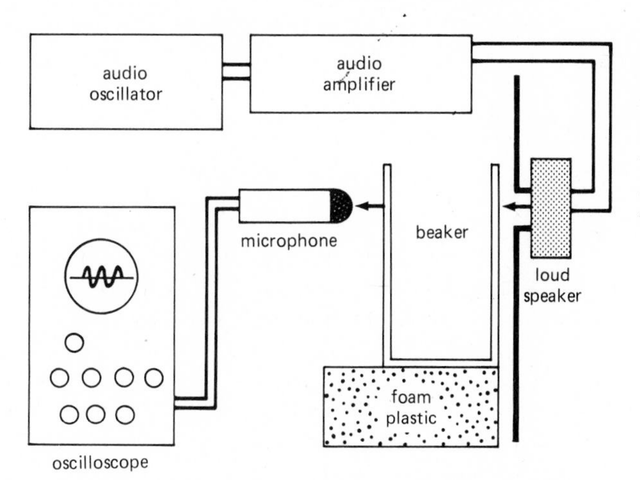

FIGURE 10–9
Block diagram of the apparatus for shattering glass vessels with high-intensity sound waves.

To do this experiment, you will need an audio oscillator, a powerful 45-watt audioamplifier, and a 30-watt loudspeaker. An oscilloscope and microphone are recommended and are almost necessary to determine the exact frequency to be used for each tumbler or laboratory beaker. See Figure 10–9.

In one such typical experiment, a glass tumbler with a diameter of 7.6 cm resonated to a frequency of 721 Hz and required an intensity of 135 to 140 dB. This is far louder than any singing voice.

Caution: Ear damage may result from this experiment unless proper precautions are taken, such as using ear protectors and keeping away from the sound source.

NOTES

[1]L. J. Sivian and S. D. White, "On Minimum Audible Sound Fields," *Journal of the Acoustical Society of America* 4 (1933):288. D. W. Robinson and R. S. Dadson, "Threshold of Hearing and Equal Loudness Relations for Pure Tones," *Journal of the Acoustical Society of America* 29 (1957):1284.

[2]E. Zwicker, G. Flottorp, and S. S. Stevens, "Critical Bandwidth in Loudness Summation," *Journal of the Acoustical Society of America* 29 (1957):548. R. Plomp, "The Ear as a Frequency Analyzer," *Journal of the Acoustical Society of America* 36 (1964):1628.

[3]J. G. Roederer, *Introduction to Physics and Psychophysics of Music,* 2d ed. (New York: Springer-Verlag, 1975), p. 30.

[4]S. S. Stevens, "Measurement of Loudness," *Journal of the Acoustical Society of America* 27 (1955):815. D. W. Robinson, "The Subjective Loudness Scale," *Acustica* 7 (1957): 217. J. C. Stevens and M. Guirao, "Individual Loudness Functions," *Journal of the Acoustical Society of America* 36(1964):2210.

[5]A. Benade, *Fundamentals of Musical Acoustics* (New York: Oxford University Press, 1976), chap. 13.

[6]H. Fletcher, *Speech and Hearing in Communication* (Huntington, N.Y.: Krieger Publishing, 1972), p. 155.

[7]J. Backus, *The Acoustical Foundations of Music,* 2d ed. (New York: Norton, 1977), p. 104.

[8]W. C. Walker, "Demonstrating Resonance by Shattering Glass with Sound," *Physics Teacher* 15 (May 1977):294.

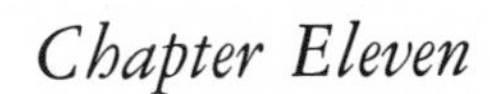

Chapter Eleven

VIBRATING STRINGS

Stringed instruments constitute a large part of the symphony orchestra and employ tightly stretched strings mounted on hollow boxlike resonating cavities. Six examples of stringed instruments in common use today are

violin
cello
harp
piano
guitar
banjo

We will consider the sound characteristics of vibrating strings in this chapter.

11.1 The fundamental frequency

There are two principal reasons that stringed instruments of different kinds do not have the same tone quality: first, the design of the instrument and second, the method by which the strings are set into vibration. For example, the strings of a violin and cello are bowed with long strands of tightly stretched horsehair; the strings of the harp, guitar, and banjo are plucked with the fingers or picks; and the strings of the piano are hammered with felt mallets.

Under very special conditions, a string may be made to vibrate with nodes at the two ends only, as shown in Figure 11 – 1. In this state of motion, every string gives rise to its lowest possible note, and the string is said to be vibrating in its **fundamental mode** at its **fundamental frequency.** We have seen in Chapter 8 that the fundamental frequency is called the first harmonic regardless of the method by which the instrument is made to sing out.

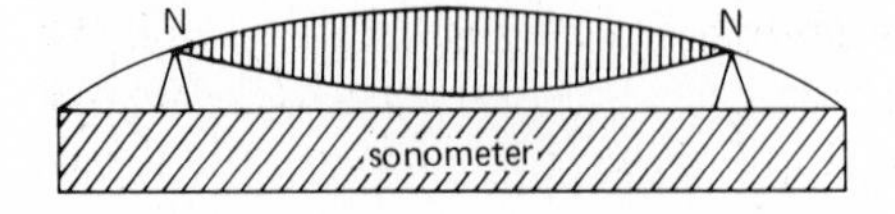

FIGURE 11-1
A string stretched across the top of a hollow box is called a *sonometer,* a device for studying the relations governing the frequency of vibrating strings.

11.2 Modes of string vibration

A violinist plays "in harmonics" by touching the strings lightly at various points while bowing. This sets the string vibrating in two or more segments, as shown in Figure 11-2. If a string is touched at its center, a node is formed at that point. In this case, the vibration frequency is double that of the funda-

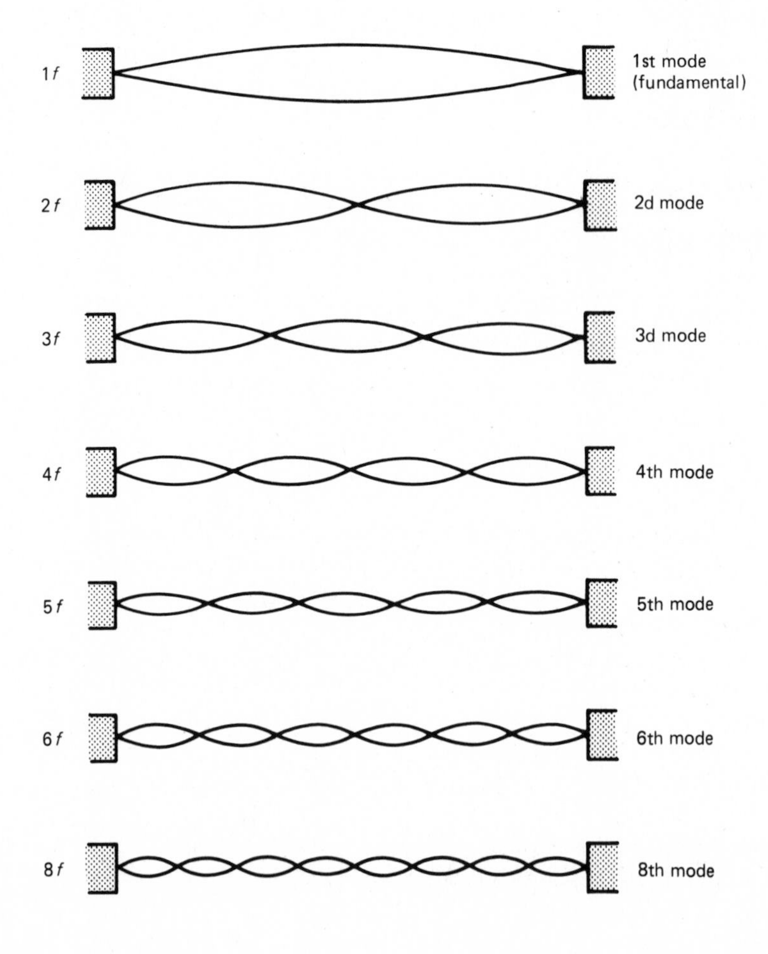

FIGURE 11-2
Vibration modes for strings of musical instruments.

mental. If the string is touched at a point one-third the distance from the end, it will vibrate in three sections and have a frequency three times that of the fundamental.

In elementary theory of vibrating strings, it is assumed that each string is thin, uniform in cross section, and highly flexible, and that it vibrates with a small amplitude between unyielding supports. For such an ideal string, the various vibration modes have frequencies exactly equal to whole-number multiples of the fundamental frequency f_1 and are true harmonics. (See Chapter 8.) The frequencies of all harmonics of such a vibrating string are given by the relation

$$f = nf_1 \qquad n = 1, 2, 3, 4, \ldots \qquad [11a]$$

To show how closely a real string meets these ideal conditions, the measured frequencies of a piano string whose fundamental frequency is 32.70 Hz are given in Table 11–1. This note C_1 is three octaves below middle C on the musical scale. (See Table 14–6.) The increasing inharmonicity with mode number is due largely to the stiffness of steel piano strings. Since most vibrating strings are very close to ideal, the partials are close to being exact harmonics. For convenience we will therefore treat them as harmonics in this chapter.

TABLE 11–1: Measured Harmonic Frequencies of a Piano String

MODE NUMBER	MEASURED FREQUENCY	HARMONIC FREQUENCY	RATIO
1	32.70	32.70	1.000
2	65.52	65.40	2.003
3	98.39	98.10	3.008
4	131.4	130.8	4.018
5	164.7	163.5	5.038
6	198.4	196.2	6.066
7	232.4	228.9	7.106
8	266.8	261.6	8.159

When a string is normally plucked or bowed, it will almost always be set vibrating in its fundamental and several of its high modes at the same time. As an illustration, a diagram of a string vibrating in two modes at the same time is shown in Figure 11–3(a). As the string vibrates in a single loop or segment with a frequency $1f$, the two halves move up and down out of phase as two loops with twice the frequency, or $2f$.

Again, if a string is made to vibrate in its first and third modes, it will divide into three sections, as shown in Figure 11–3(b), and as these sections vibrate with a frequency $3f$, the whole string moves up and down with the lower frequency $1f$ of the fundamental. The sound waves emerging from the string are a combination of the first and third harmonics, and in their combined form they have the shape given in Figure 11–4(c).

Of the many ways in which modes may be combined on a vibrating string, two more of the simpler ones are shown in Figures 11 – 3(c) and (d). Time graphs for all four of the vibrating string patterns shown in Figure 11 – 3 are shown in Figure 11 – 5.

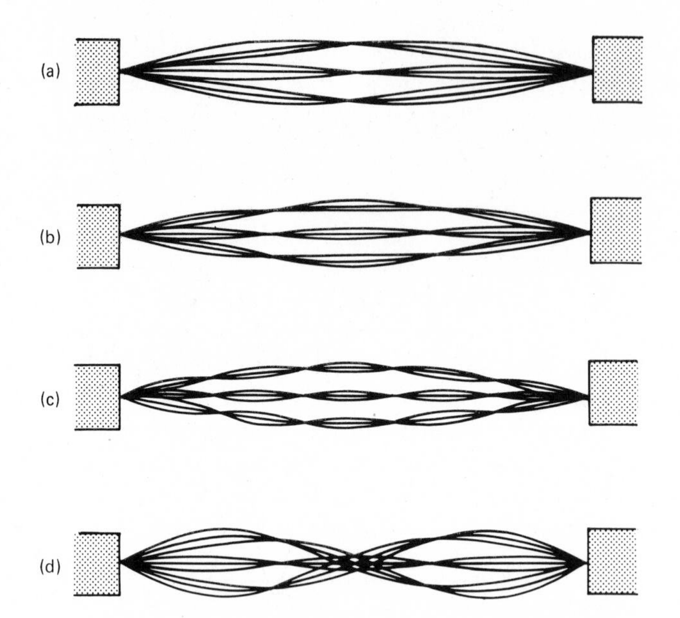

FIGURE 11 – 3
Four of the many ways in which a string can vibrate. The strings are vibrating with (a) the first and second harmonics, (b) the first and third harmonics, (c) the first and fifth harmonics, and (d) the second and third harmonics.

An interesting experiment that can be performed with a vibrating string is diagramed in Figure 11 – 6. Light from an arc lamp is focused on the central sections of a stretched string, which, except for a small vertical slot, is masked by a screen. An image of the slot and the string section seen through it is focused on another screen by a second lens, after reflection from a rotating mirror. As the string vibrates up and down, only a blurred image of the short section of the string is seen, but when the mirror is rotating, the string section image draws out a clearly visible curve W. If the string is plucked lightly at

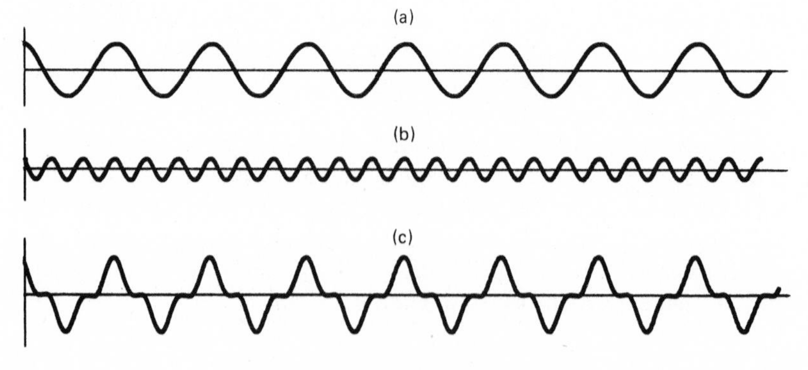

FIGURE 11 – 4
Time graphs for the sound waves from a string vibrating with two of its natural modes simultaneously. Graph for (a) the first harmonic alone, (b) the third harmonic alone, and (c) the first and third harmonics simultaneously. See Figure 11–3(b).

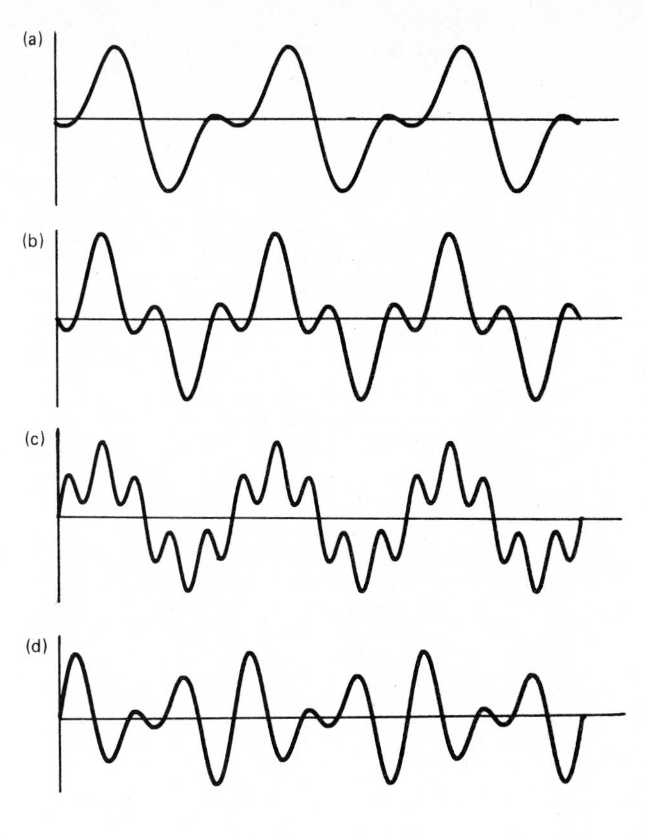

FIGURE 11–5
Time graphs for the sound waves emitted by a string vibrating with the four modes shown in Figure 11–3. Vibration for (a) the first and second harmonics, (b) the first and third harmonics, (c) the first and fifth harmonics, and (d) the second and third harmonics.

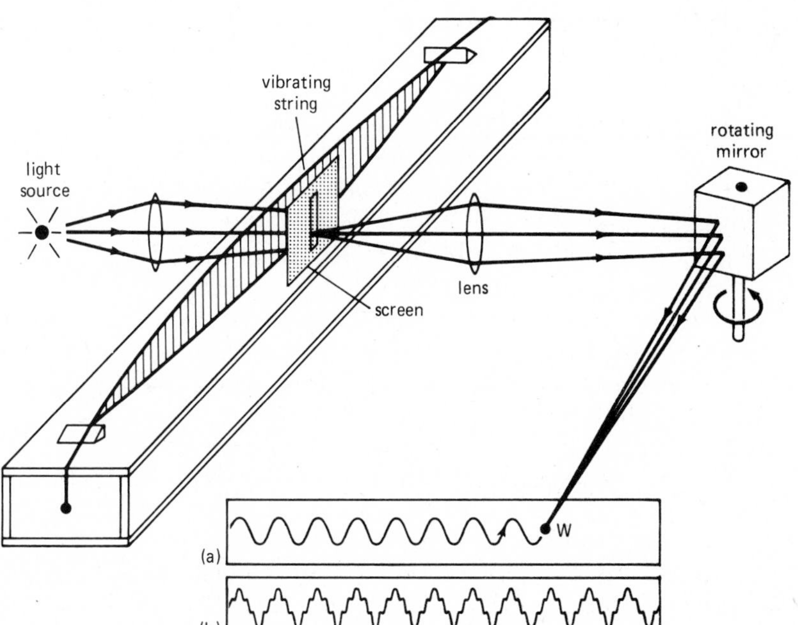

FIGURE 11–6
Demonstration experiment for showing an audience the actual vibrations of different string sections on a sonometer. The vibration patterns shown on the screen here are for (a) the first harmonic alone and (b) the first and fifth harmonics. See Figures 11–3 and 11–5.

the center, a smooth waveform (a) is drawn out on the screen. If it is plucked near the end to produce a harsh-sounding note, the waveform is more complex, as shown in trace (b). In the first case, the string is vibrating largely with its fundamental mode, while in the second case, higher modes are clearly present. (See Chapter 8.)

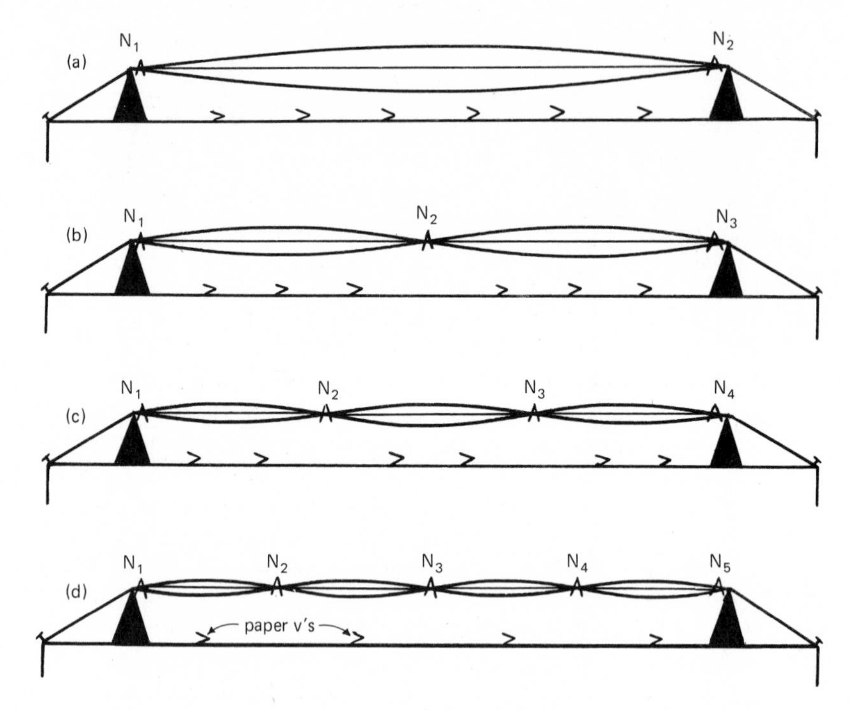

FIGURE 11-7 Demonstration showing the positions of nodes and antinodes on strings vibrating with the first four harmonics. Vibration modes shown are for (a) the first harmonic, or fundamental, (b) the second harmonic, (c) the third harmonic, and (d) the fourth harmonic.

The following experiments show that a string actually vibrates with definite nodes and loops. Starting with a **sonometer**, as shown in Figure 11-7, and a violin bow, the string can be set vibrating with any of its harmonics. Small strips of paper, bent in the form of an inverted v, are hung over the string at a number of points, as shown. Bowing the string lightly near one end will cause the fundamental to be sounded, whereupon all the bent paper strips, except the two at the very ends, bounce off the string. If the string is touched lightly at its center point—N_2 in diagram (b)—all paper v's, except those at the two ends and the very center, will bounce off. When the string is touched one-third of its length from one end—N_2 or N_3 in diagram (c)—all paper v's, except those at the four nodes, will bounce off. When the string is touched lightly one-quarter of its length from one end—N_2 or N_4 in diagram (d)—all v's except those at the five nodes will bounce off; and so on. These nodes can all be verified by using the arrangement shown in Figure 11-6, locating that section of the string being studied directly behind the slit in the screen.

11.3 Mersenne's rules

The first systematic study of vibrating strings was carried out by Pythagoras in the sixth century B.C. However, the definitive studies were carried out in the seventeenth century by Père Mersenne, a Franciscan friar, who summarized his findings, along with those of his predecessors, in four general rules. **Mersenne's rules** can be stated as follows:

The frequency of a vibrating string is

1. Inversely proportional to its length.
2. Proportional to the square root of the tension applied to the string.
3. Inversely proportional to its diameter.
4. Inversely proportional to the square root of its density.

Let us examine these rules in detail. First, it is found that increasing the length of a string for the same tension causes the pitch and frequency to decrease, and decreasing the length for the same tension causes the pitch and frequency to increase. The frequency is inversely proportional to the length:

$$f \propto \frac{1}{L} \qquad [11b]$$

Second, in tuning the strings of any musical instrument, increasing the tension causes the pitch and frequency to rise, and decreasing the tension causes the pitch and frequency to fall. Experimentally, one finds that the frequency is proportional to $\sqrt{F}$:

$$f \propto \sqrt{F} \qquad [11c]$$

Finally, for a constant tension, strings of increasing mass per unit length vibrate with lower pitch and frequency, and with decreasing mass per unit length, the pitch and frequency rise. Experimentally:

$$f \propto \sqrt{\frac{1}{m}} \qquad [11d]$$

where m is the mass per unit length of the string and is given in kilograms/meter. This third relation follows from rules 3 and 4 since m depends upon both the diameter and density of the string.

11.4 The theory of vibrating strings

Let us next investigate how Mersenne's rules follow from the behavior of a vibrating string. We have already seen that a string set into a natural mode of vibration, with nodes and antinodes, is an example of a standing transverse wave. Let us now view the process in terms of traveling waves.

Suppose a disturbance produced near one end, as in the plucking of a string, sends a wave along the string to be reflected from end to end. Since the fundamental frequency of vibration is equal to the number of times per second that the wave arrives at the same end, the pitch will depend upon the speed of the waves and the distance they have to travel. The speed V of transverse waves along a string under tension is given by

$$V = \sqrt{\frac{F}{m}} \tag{11e}$$

where F is the tension, in newtons, and m is the mass per unit length of the string in kilograms/meter. See Figure 11–8. The distance the wave must travel during one round trip is $2L$. The time for the wave to make one round trip is

$$t = \frac{2L}{V} \tag{11f}$$

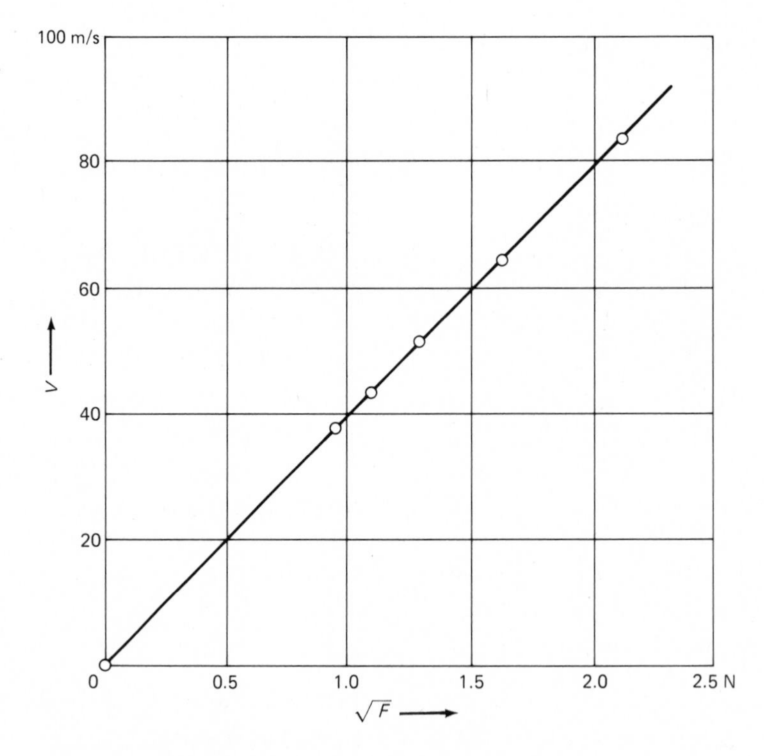

FIGURE 11–8
Graph showing the relation between the speed of waves along a stretched string and the square root of the string tension. Experimental arrangement for this is shown in Figure 3–7.

(See A1.13.) Since the frequency is the reciprocal of the period, we arrive at the expression for the frequency of the round trips:

$$f = \frac{V}{2L} \tag{11g}$$

or

$$f = \frac{1}{2L}\sqrt{\frac{F}{m}} \qquad [11h]$$

We now note that the mass per unit length, m, is given by the mass per unit volume (density) times the cross-sectional area:

$$m = \rho \times \frac{\pi}{4} d^2 \qquad [11i]$$

where ρ is the density of the material and d is the diameter. Combining with Equation (11h), we have

$$f = \frac{1}{2L}\sqrt{\frac{F}{\rho \times \frac{\pi}{4} d^2}} \qquad [11j]$$

or

$$f = \frac{1}{Ld}\sqrt{\frac{F}{\pi\rho}} \qquad [11k]$$

This formula summarizes all four factors that determine the fundamental frequency of a vibrating string, as found by Mersenne.

One last step is to determine the dependence of the mode on the frequency. We note that the number of loops on a standing wave is proportional to the mode number. See Figure 11–2. Viewing the standing wave now as reflecting traveling waves, we would expect the number of crests to also be proportional to mode number. Therefore, the number of crests arriving at any point per unit time will also be proportional to mode number. We can therefore infer that the frequency is proportional to mode number:

$$f \propto n$$

Example

A harp string that sounds a note two octaves below middle C on the musical scale is 75.0 cm long and has a total mass of 150 grams (g). (See Table 14–5.) If the string is under a tension of 1960 N, find the frequency of its fundamental mode.

Solution

The given quantities are $L = 0.75$ m, $F = 1960$ N, and $m = 0.150$ kg/0.750 m or 0.20 kg/m. Inserting these quantities directly into Equation (11h), we obtain

$$f = \frac{1}{2 \times 0.75 \text{ m}}\sqrt{\frac{1960 \text{ kg} \cdot \text{m/s}^2}{0.20 \text{ kg/m}}}$$

$$f = \frac{1}{1.50 \text{ m}} \times 99.0 \text{ m/s}$$

$$f = 66.0 \text{ Hz}$$

11.5 Laboratory experiment

This is an experiment for discovering the relationship between the tension F in a vibrating string, the length of a single loop L on the string or the wavelength λ, and the speed V of the waves along the string. The apparatus consists of an electrically driven tuning fork with a specified frequency, a piece of string or cord, and a set of weights. See Figure 3–7. By varying the weights added at one end, the string is made to vibrate in one after another of its natural modes. (See Figure 11–2.)

Typical recorded data from such an experiment is given in Table 11–2. The mass M is recorded in kilograms, the length of a number of equally spaced loops in meters, the number of loops selected n, and the known frequency of the fork f in hertz. From these data, values for the quantities listed under the column headings for Table 11–3 can be calculated. The wavelength $\lambda = 2L$, the string tension $F = Mg$, where $g = 9.80$ m/s^2, and the speed $V = f\lambda$ are the relations to be used.

TABLE 11–2: Recorded Measurements for a Vibrating String

TRIAL	M (kg)	nL (m)	n	f (Hz)
1	0.095	1.157	6	95.0
2	0.130	1.126	5	95.0
3	0.188	1.081	4	95.0
4	0.293	1.013	3	95.0
5	0.520	0.900	2	95.0

TABLE 11–3: Column Headings for Calculated Values from Table 11–2

TRIAL	L (m)	λ (m)	F (N)	$\sqrt{F}$ ($\sqrt{N}$)	V (m/s)

A study of these calculated values by the student will show that the greater the tension F, the greater is the wavelength λ. Upon plotting a graph of λ against F, a curve will be obtained that shows that λ is not proportional to F. If we plot λ against $1/F$, then λ against F^2, we will again obtain curves that are not straight lines, showing that λ is not proportional to $1/F$ or F^2. If, finally, we plot λ against $\sqrt{F}$, we will obtain a straight line, which shows that λ is proportional to $\sqrt{F}$. One can then write the proportionality as one of the steps in arriving at part of Equation (11h). This is left as an exercise for the student. By a similar procedure, the student should find how V and F are related. See Equation (11e).

QUESTIONS

1. What is the relationship between the fundamental frequency of a vibrating string and its various harmonics?

2. Make a diagram of a string vibrating simultaneously in its fundamental and fourth harmonic mode. Take the amplitude of the fundamental to be four times the amplitude of the fourth harmonic.

3. What three factors determine the frequency or pitch of a vibrating string? Briefly explain the effects on the string's frequency when each of these factors is altered one at a time.

PROBLEMS

1. An electric tuning fork with a frequency of 80.0 Hz sets a string vibrating in nodes and antinodes. See Figure 3 – 7. If the tension in the string is due to a 640-g mass hanging from a pulley at the far end, and five loops have an overall length of 114.5 cm, find (a) the wavelength λ in meters, (b) the tension in the cord in newtons, (c) the speed of the waves on the string, and (d) the mass per unit length of the string in kilograms/meter.

2. Make a table with headings given for Table 11 – 3, and using the data given in Table 11 – 2, fill in the columns for all five trials.

3. Use the data calculated in Problem 2. (a) Draw a graph for the data, plotting V, the velocity, vertically against the square root of the tension $\sqrt{F}$ horizontally. (b) What can you conclude from your graph? (c) Write down an equation for the result. See Figure 11 – 8.

PROJECT

Bring a stringed instrument, such as a violin, viola, or cello, to class. When properly tuned, the three instruments mentioned have strings with fundamental frequencies given by the following notes:

cello	C_2 66 Hz	G_2 99 Hz	D_3 148 Hz	A_3 220 Hz
viola	C_3 132 Hz	G_3 198 Hz	D_4 296 Hz	A_4 440 Hz
violin	G_3 198 Hz	D_4 297 Hz	A_4 440 Hz	E_5 660 Hz

Measure the length of one of the strings, and from the known frequency of the fundamental, calculate (a) the wavelength of the fundamental on the string, (b) the speed of these waves on the string, and (c) the tension in the string. Using Equation (11g) and your determinations of $\lambda, f,$ and F, calculate the total mass of the string in kilograms.

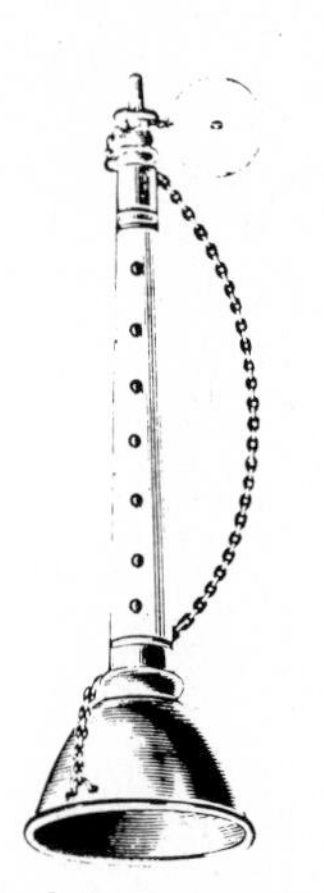

Chapter Twelve

VIBRATING AIR COLUMNS

Wind instruments constitute a large family of musical instruments employing vibrating air columns as their source of musical sounds. Sound waves emerging from all wind instruments are due to the vibrations of columns of air within the walls of tubes. There are various ways in which these air columns are set vibrating. These subjects will be introduced here and treated in more detail in Chapters 18 and 19.

12.1 Air column resonance

The fundamental principles involved in the vibration of an air column can be demonstrated by means of an experiment such as shown in Figure 12–1. A source of sound waves in the form of a vibrating tuning fork is held over the open end of a long hollow tube containing water. These waves, traveling down the tube with the speed of sound in air, are reflected from the water surface and travel back to the top. By raising and lowering the water level, points will be found where the air column will resonate to the frequency of the tuning fork, the sound will be greatly intensified, and the pipe will sing out with the pitch of the fork.

The first resonance will occur at N_1, where the water level is only a few centimeters from the top. A second resonance will occur at N_2, approximately three times the distance of N_1 from the top; a third resonance will occur at N_3, five times the distance of N_1 from the top; and so forth. These odd-numbered multiples of the distance of N_1 from the top are due to the experimental fact that the relatively rigid water surface forces a displacement node at the interface. An antinode is formed near the open end at the top.

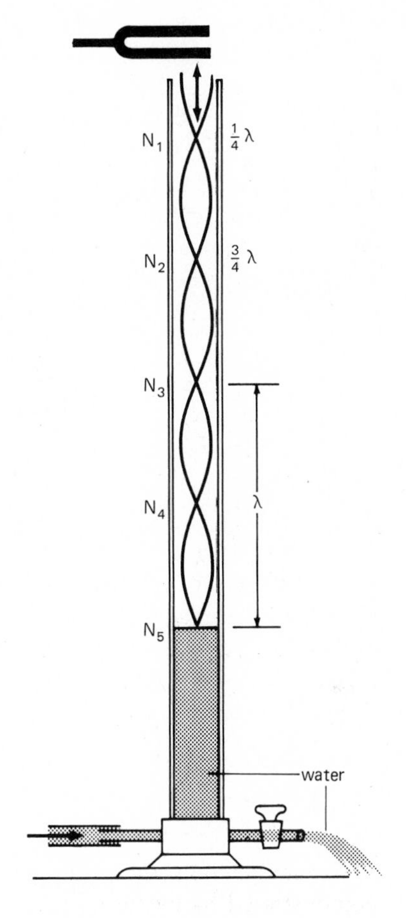

FIGURE 12–1
Sound waves from a tuning fork set up standing waves in an air column that has been adjusted by the water column to the proper length.

Since each half wavelength loop has a node at each end, we can see that the conditions above are satisfied only if the column corresponds to an integral number (including zero) of half wavelengths and one-quarter wavelength. Because the longitudinal standing waves in the air column are difficult to illustrate, transverse half wavelength loops and nodes are drawn to represent them. (See Figure 4–6.)

If a tuning fork of known frequency is used in the experiment above, the speed of sound in air can be calculated. Observe that the distance between each pair of nodes is $\lambda/2$, so that λ is the length of two loops, $N_3 - N_1$ or $N_4 - N_2$, as shown. In one laboratory experiment that was performed, the temperature was 27°C, the tuning fork had a frequency of 528 Hz, and the nodes were 32.9 cm apart. These give the following results:

$$\lambda = 65.8 \text{ cm} = 0.658 \text{ m}$$

Substitution of known quantities into Equation (3a) gives

$$V = f\lambda = 528 \text{ cyc/s} \times 0.658 \text{ m/cyc} = 347.4 \text{ m/s}$$

12.2 Vibration modes in pipes

Diagrams of the various modes in which air columns may vibrate in an open or closed **pipe** (cylindrical tube) are shown in Figure 12–2. Beginning at the left, a pipe open at both ends (open-open pipe) is shown vibrating with (1) a single node at the center and an antinode at both ends, (2) two nodes and three antinodes, and (3) three nodes and four antinodes, and so on. If the pipe

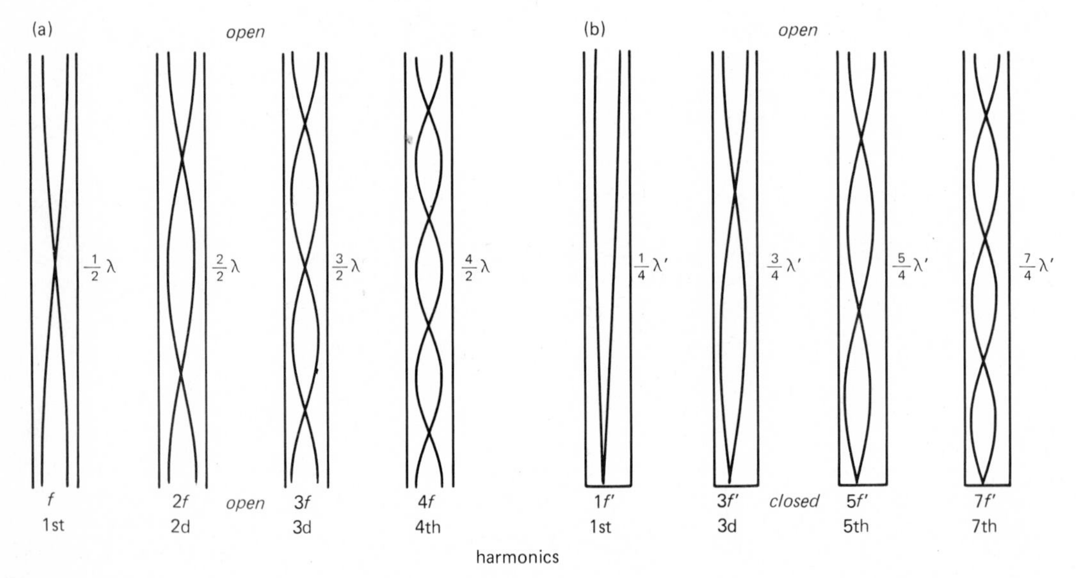

FIGURE 12–2
Air columns in pipes have natural modes of vibration that are harmonics of the fundamental. (a) Both ends open and (b) one end open and the other closed.

is open at one end and closed at the other, it may vibrate with (1) one node at the closed end and an antinode at the open end, (2) two nodes and two antinodes, (3) three nodes and three antinodes, and so on. In all vibrating air columns, a node always forms at a closed end, and an antinode always forms near an open end. If both ends of a pipe are closed, and a means for setting the enclosed air into vibration is provided, nodes will form at both ends and at one or more antinodes in between.

The different frequencies to which a given cylindrical air column may resonate and sing out are specific and fixed in value. They depend on the speed of air at the time, on the length of the pipe, and on whether the ends are open or closed. Suppose, for example, the pipes in Figure 12 – 2 are all 50 cm long and the speed of sound is 350 m/s. Using Equation (3a),

$$V = f\lambda \qquad [12a]$$

we find that they should vibrate with the following frequencies:

	f	$2f$	$3f$	$4f$	$5f$	
Both ends open	350 Hz	700 Hz	1050 Hz	1400 Hz	1750 Hz	. . .
	f	$3f$	$5f$	$7f$	$9f$	
One end open	175 Hz	525 Hz	875 Hz	1225 Hz	1575 Hz	. . .

If a pipe is open at both ends, the lowest possible resonant frequency f is produced by the fundamental mode. The higher frequencies with whole-numbered multiples of the fundamental are harmonics $2f$, $3f$, $4f$, $5f$, ... With a pipe closed at one end only, the lowest frequency, now an octave lower, is again called the fundamental f. The higher frequencies with odd-numbered multiples, $3f$, $5f$, $7f$, $9f$,. . ., are the odd harmonics. These vibration modes are all referred to as **natural modes of vibration.** It is important to note that with one end of the pipe closed, the even-numbered harmonics are absent.

12.3 Pressure nodes and antinodes

For standing longitudinal waves in air, vibration nodes are positions of no motion, while antinodes are positions of maximum vibration amplitude. Air pressure at these points along a resonating pipe is just the reverse of this. At each vibration antinode where the motion is a maximum, the pressure variation is a minimum. At vibration nodes where the motion is zero, the pressure variation is a maximum. See Section 4.3 and Figure 12 – 3.

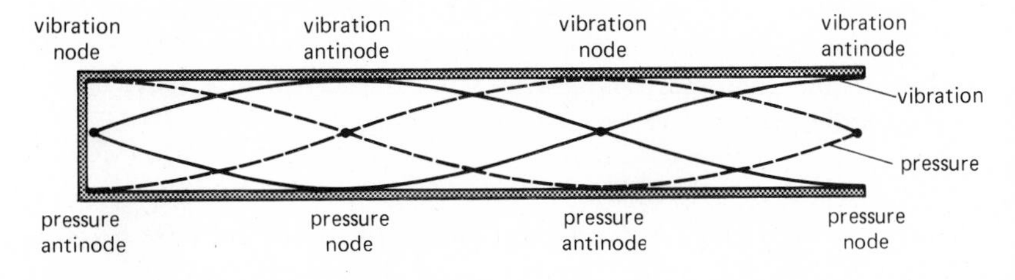

FIGURE 12 – 3 **Resonating air column showing vibration nodes and antinodes in relation to pressure nodes and antinodes.**

The closed end of a pipe must be a vibration node since the end prevents the air from moving. The open end is in contact with the outside air, which is maintained at constant pressure. The open end must therefore be a point of minimum pressure variation, a pressure node. It is then also a point of maximum motion, a vibration antinode.

An excellent experiment demonstrating pressure nodes and antinodes in a vibrating air column is diagramed in Figure 12–4. This device, called a **Rubens tube,** consists of a long straight pipe filled with illuminating gas. The gas enters the pipe through a plunger at the left and escapes through small holes drilled through the top at regular intervals. Sound waves from a tuning fork, or small organ pipe, enter the gas column by setting a thin paper sheet, stretched over the other end, into vibration. By sliding the plunger to the correct position, the air column will resonate, and the small gas flames will appear as shown in the diagram. The height contour formed by the yellow flame tips clearly shows the nodes and antinodes. By Bernoulli's principle, one would expect the highest flames to occur at the pressure antinodes.

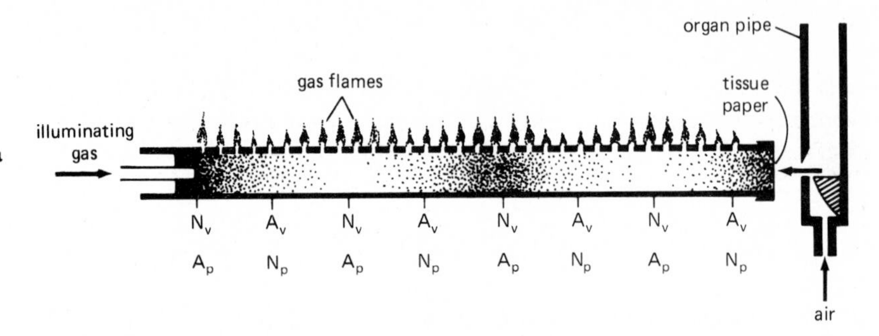

FIGURE 12–4 **Rubens tube experiment demonstrating standing waves in a long tube containing illuminating gas. Displacement and pressure nodes and antinodes are indicated. (With permission of D. Van Nostrand Company)**

Experiments show that the pitch of an open pipe is slightly lower than that indicated by the considerations above. The vibration antinode at any open end occurs slightly beyond the end. The pipe behaves as if its length is given approximately by the relation

$$l = l_0 + 0.30D \qquad [12b]$$

where l_0 is the actual pipe length and D is its diameter. This end correction assumes that the wavelength is large with respect to the diameter of the pipe. For high-frequency notes, the wavelength may be comparable to the pipe diameter, and some partials may be slightly sharp with respect to the fundamental and hence give rise to some dissonance.

12.4 Frequency of a vibrating gas column

The different notes sounded by many wind instruments are produced by varying the length of the air column. This is illustrated by the pipe organ, which uses ranks, or sets or pipes, with different lengths. See Figure 12–5. The shorter the pipe, the higher are the fundamental frequency and pitch; the longer the pipe, the lower are the fundamental frequency and pitch. The pipes

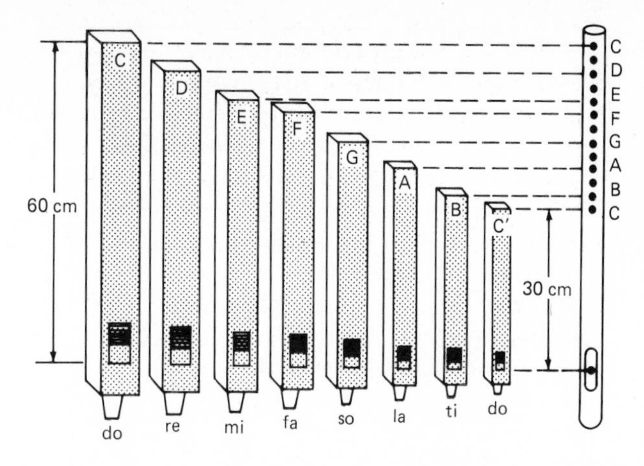

FIGURE 12–5 Organ pipes arranged in a musical scale. The longer the pipe, the lower are its fundamental frequency and pitch. The vibrating air column of the flute is terminated at various points by openings along the tube. (With permission of D. Van Nostrand Company)

of a big concert pipe organ vary in length from approximately 10 m (32 ft) for the lowest note to approximately 15 cm (6 in.) for the highest note. For the middle octave of the musical scale, extending in frequency from $C_4 = 264$ Hz to $C_5 = 528$ Hz, the open-ended pipes vary between 66 cm for C_4 to 33 cm for C_5, an octave higher. For the woodwinds like the flutes, the length of the air column is varied by openings along the side of the instrument. In many of the brasses, the vibrating air column is changed in length by (1) sliding a U-shaped tube along its body, as with the trombone, or (2) pressing valves in the straight section, as with the trumpet. (See Sections 19.6 and 19.7.)

Since a vibrating air column denotes the state of standing longitudinal waves, the frequency of vibration is found to depend upon two factors: the length of the air column and the speed of sound through it. The speed of longitudinal waves in a gas is given by Newton's formula, as modified by Laplace:

$$V = \sqrt{K \frac{p}{d}} \qquad [12c]$$

where K is a number representing the **compressibility** of a gas, p is the gas pressure, in newtons per square meter, and d is its density, in kilograms per cubic meter. The pressure p is close to standard atmospheric pressure, which is

$$p = 101{,}300 \text{ N/m}^2 \qquad [12d]$$

Compressibility is a measure of the ability of a substance to be compressed. Theoretical values of the compressibility of a few common gases are given in Table 12–1.

For all standing waves, the distance L between two consecutive nodes, which is the length of one loop, is equal to one-half a wavelength λ. See Equation (3c). Substituting the value of V from Equation (12c) into the general

TABLE 12-1: The Densities and Compressibilities of a Few Common Gases

GASES	CHEMICAL SYMBOL	DENSITY AT 0°C (kg/m³)	COMPRESSIBILITY NUMBER K
Carbon dioxide	CO_2	1.980	1.333
Sulfur dioxide	SO_2	2.927	1.333
Air	N_2, O_2	1.293	1.400
Oxygen	O_2	1.429	1.400
Nitrogen	N_2	1.250	1.400
Hydrogen	H_2	0.090	1.400
Carbon Monoxide	CO	1.250	1.400
Helium	He	0.180	1.667
Neon	Ne	0.900	1.667
Argon	A	1.784	1.667
Krypton	Kr	3.708	1.667
Xenon	Xe	5.851	1.667

wave equation $V = f\lambda$, and solving for the frequency of vibration, we obtain the general formula

$$f = \frac{1}{\lambda}\sqrt{K\frac{p}{d}} \qquad [12e]$$

Example

A straight section of pipe 2.50 m long is open at both ends. The pipe is filled with helium gas at standard temperature and pressure. Find (a) the speed of sound in helium, and (b) the frequency of the fundamental mode when it is set vibrating with standing longitudinal waves. Neglect end corrections.

Solution

The given quantities, as shown in Table 12–1, are, as follows: gas density $d = 0.180$ kg/m³ and the compressibility of a monatomic gas $K = 1.667$. The pressure, by Equation (12d), is $p = 101{,}300$ N/m². The unknown quantities are the speed of sound in helium and the frequency of the fundamental. (a) Using Equation (12c) gives, for the speed of sound,

$$V = \sqrt{\frac{1.667 \times 101{,}300 \text{ N/m}^2}{0.180 \text{ kg/m}^3}}$$

$$V = \sqrt{938{,}150 \text{ m}^2/\text{s}^2}$$

$$V = 969 \text{ m/s}$$

(b) To find the frequency, use Equation (3a). Since an antinode forms at both ends and a node at the center, the length of the pipe is equal to $\lambda/2$, or $\lambda = 2L$. By Equation (3a), we obtain

$$f = \frac{V}{\lambda} = \frac{969 \text{ m/s}}{2 \times 2.50 \text{ m}}$$

$$f = 194 \text{ c/s} = 194 \text{ Hz}$$

Examination of Equation (12e) shows that if either the air pressure or the density changes, the velocity of sound will change. As a result, the frequency of the vibrating air column will change. For example, if the instrument is taken into a warm room, the pressure may remain constant but the air density will drop. The velocity of sound will, therefore, increase, causing the instrument to go sharp. Since string and percussion instruments do not change with temperature, the winds must retune whenever the temperature changes significantly.

12.5 Fluid friction

When a gas or liquid is made to flow through a pipe or around an object, or an object is made to move through a previously stationary fluid, resistance to the motion manifests itself through the flow pattern in the fluid. The propulsion of automobiles, trains, and planes through the air and ships through the water impart energy to the surrounding fluid to set it in motion. According to the theory of relative motion, it makes no difference whether the object is considered moving and the previously undisturbed fluid is standing still, or whether the object is standing still and the fluid is made to flow around it. It is the rela-

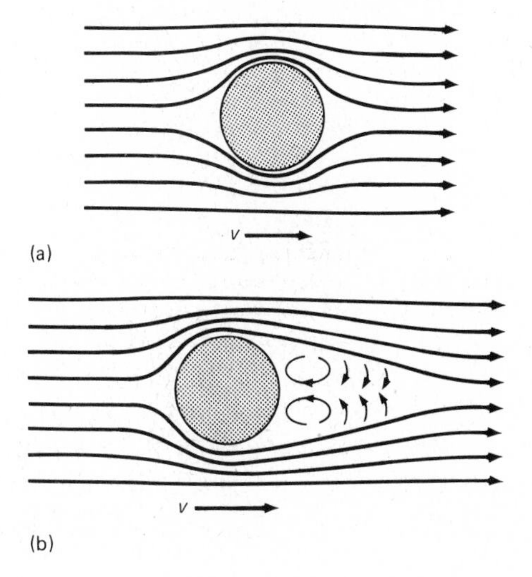

FIGURE 12-6
A liquid flows around a smooth round object: (a) low-velocity fluid showing laminar flow, (b) high-velocity fluid showing turbulent flow.

tive motion of the two that brings about flow patterns and therefore **fluid friction.**[1]

Laboratory experiments show that at relatively slow speeds, the flow of a fluid around an object is smooth and regular, as shown in Figure 12–6(a). This type of fluid pattern is called **laminar** or **streamline flow.** At higher speeds, the fluid flow is not smooth but becomes turbulent, and it is characterized by small **eddy currents** that form behind the object, as shown in diagram (b). The fluid has to move out and around the object quickly, and a continuous flow of energy is consumed by the formation of a trail of eddies left behind. This kind of fluid pattern is called **turbulent flow** and constitutes a greater loss of energy and therefore greater friction.

If the relative speed is increased still further, the eddies form alternately on one side, then on the other, creating a trail of whirlpools or **vortices.** See Figure 12–7. This trail of vortices is commonly called a **Karman trail.** The creation of Karman trails is demonstrated by the waving of a flag at the top of a flagpole, as well as by the flapping of the rope against the pole. These motions are graphic evidence of the vortices that follow each other alternately on one side, then on the other. When the wind howls through the trees, movable twigs and leaves generate sounds. For any given wind speed, smaller objects

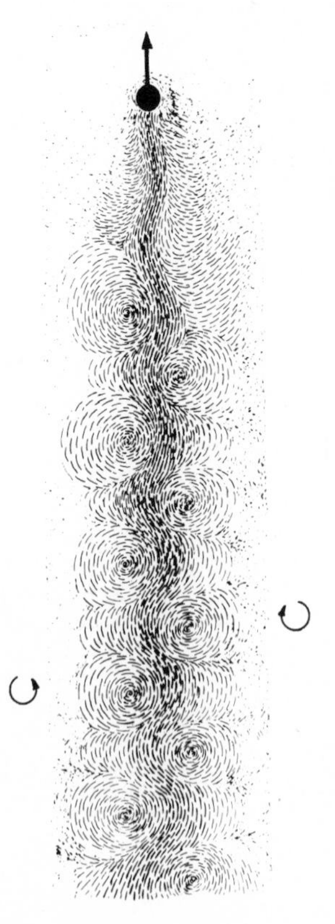

FIGURE 12–7
Eddies set up by pulling an obstacle through the air or through water form a *Karman trail* of vortices. (With permission of D. Van Nostrand Company)

give rise to higher-pitched sounds than large ones. As the wind speed varies, these pitches rise and fall. Similarly, a stretched rubber band or fine wire, when located in an open window where it catches the breeze, will give rise to a musical note. The impulse imparted by each vortex acts on the rubber band or wire, pushing it first to one side, then to the other. This is the principle of the **aeolian harp.** (See Project 3 at the end of the chapter.)

12.6 Edge tones

The vibrations of an air column in any kind of wind instrument are due to some kind of disturbance produced at or near one end of the enclosed air column. Such disturbances are often brought about by a steady flow of air over the sharp edge of a hole. This produces periodic vortices at the mouth of the instrument and constitutes the initial periodic disturbance required.

Consider a steady stream of air blown against a wedge, as shown in Figure 12–8(a). The wedge has the effect of splitting the Karman trail, as well as strongly modifying the nature of the vortices. The rate of vortex generation is related to the orifice-to-wedge distance as well as to the air speed. Thus the steady stream of air through the opening sweeps back and forth across the edge, producing periodic air pulses on each side. Produced with a frequency f, these pulses give rise to a musical note called an **edge tone.** The frequency can be increased by increasing the flow rate as well as by decreasing the orifice-to-wedge distance.

When the edge forms a portion of the open end of a resonant pipe, such

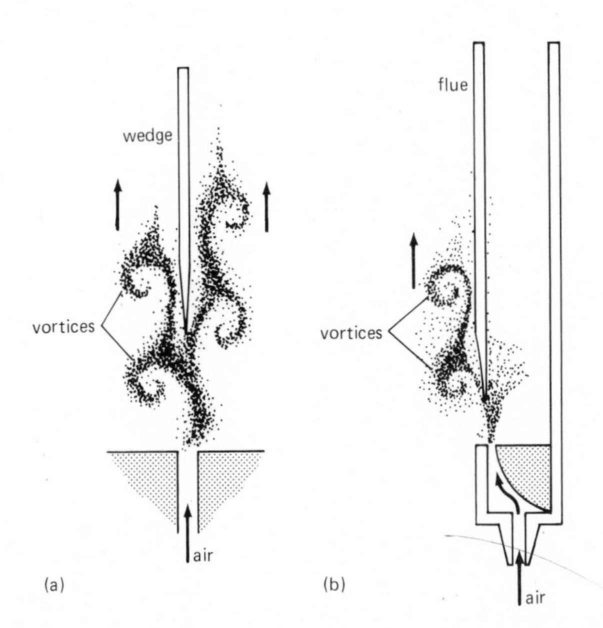

FIGURE 12–8
(a) Edge tones are produced when a gas jet interacts with a sharp wedge. (b) The edge tone mechanism is controlled by the standing wave in a resonant organ pipe.

as in Figure 12–8(b), the first puffs of air into the pipe will initiate the build-up of resonant modes of comparable frequency. One mode, say the fundamental, will build up most rapidly and cause air to flow in and out of the mouth at that resonant frequency. This in turn will carry the sheet of air from the windway in and out of synchronism. The air blown across the windway is now locked to the pipe resonance and continues to energize that mode. The edge tone is now controlled by the resonant mode and cannot take on any frequency. However, by overblowing and increasing wind speed substantially, the pipe can be forced to suddenly "jump" into one of its higher modes, in which the faster alternation of pulses creating the edge tone are synchronized with the higher pipe resonances. The pipe can, in principle, be controlled to vibrate in any of its available modes by blowing at the appropriate speed.

12.7 The pipe organ

A large pipe organ may consist of many sizes and types of pipes. See Figure 12–9. The diapason is a standard flue-type pipe open at both ends, as shown in diagram (b). A pipe organ may include three to four octaves of these principal pipes to form a complete organ. However, lower notes may require fairly long lengths. For example, C_1 requires a 5-m-long ("16-foot") pipe.* It is therefore common to **stop** (close) the end of pipes, thereby allowing them to play an octave lower than the equivalent open pipe. See Figure 12–10. The trade-off

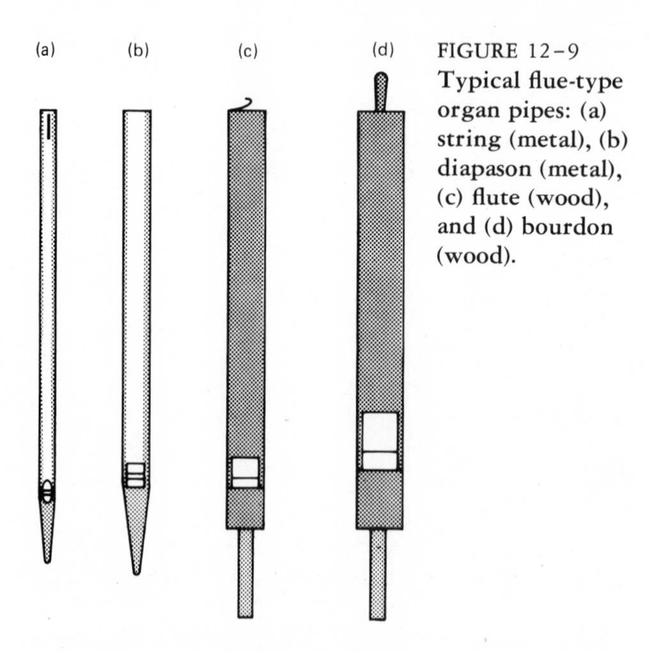

FIGURE 12–9
Typical flue-type organ pipes: (a) string (metal), (b) diapason (metal), (c) flute (wood), and (d) bourdon (wood).

*See Table 14–5 for the meaning and frequency of C_1.

for this situation is that the even partials will be missing, resulting in a different timbre.

Organ pipes can be round, square, conical, trumpet-shaped, and so on.[2] Some pipes have auxiliary components attached or have their openings shaped or notched. Each type of pipe will have its own characteristic timbre. It is common to build a composite organ employing several complete sets (ranks) of different types of pipe, allowing the organist a wide range of sounds. See Figure 12–11. Tuning to the proper pitch is done by adjusting some feature at the top of the pipe, which varies its effective length, as well as adjusting the edge-jet relationship. The desired timbre is accomplished by proper adjustment of all these variables, a process called **voicing.** (See Section 18.7.)

In addition to edge tone excitation, pipes can also be excited by vibrating reeds, as we will see in Chapter 18.

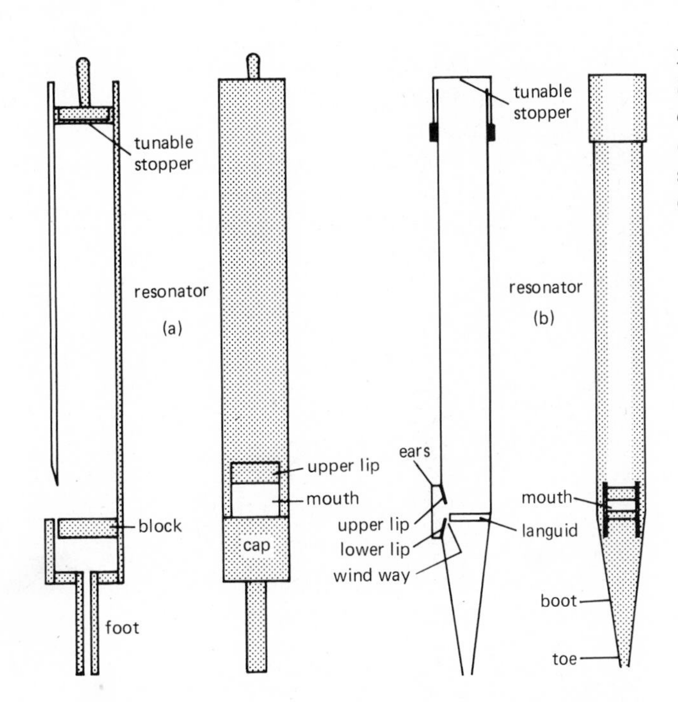

FIGURE 12–10 **Stopped flue-type organ pipe construction: (a) square wood and (b) round metal.**

12.8 Pipe dimensions

The diameter of a pipe does not affect any of the resonant frequencies of the pipe, except through the effect on the end correction. See Equation (12b). However, it does affect the ease with which the various resonant frequencies are excited. When the diameter is comparable with the length, as in the drum shell, the pipe does not resonate appreciably. As the **pipe scale** (ratio of diam-

FIGURE 12–11
A Ruffatti 89 rank electro-pneumatic pipe organ. The console with four keyboards is shown below. (Courtesy of Dave Christensen)

eter to length) decreases, good resonance occurs for the fundamental pitch. Further reduction of the pipe scale increases the air friction against the walls, making it more difficult to excite the fundamental mode and leaving it easier to excite the higher harmonics. Therefore, the range of harmonics that the pipe will favor can be controlled by appropriately adjusting the pipe scale. For some examples of this, see Table 12–2.

The favored harmonic region determines not only the resonances the player will most easily excite but also the amount of upper partials that accompany the played note. For example, a trumpet will normally be played in the 3–6 harmonics, but the trumpet will also resonate well into the higher partials, giving the trumpet its bright timbre.

The timbre of organ pipes is also determined by their pipe scale. "String" pipes, simulating the overtone-rich timbre of stringed instruments, have small pipe scales. Alternatively, "flute" pipes have a larger pipe scale, resulting in a purer fundamental tone. See Figure 12–9.

TABLE 12-2: **Relations Between the Pipe Scale Ratio and the Harmonics Favored**

INSTRUMENT	HARMONICS FAVORED	PIPE SCALE (DIAMETER/LENGTH)
Drum	–	1–2
Organ pipe (diapason)	1	0.07
Clarinet	1, 3, (5)	0.03
Trumpet	3, 4, 5, 6	0.01

12.9 The hummer

If one blows into a piece of corrugated tubing, such as an electric conduit or a section of flexible automobile hydraulic fluid tubing, a musical tone will be heard. See Figure 12–12. Its pitch will correspond to one of its resonant modes as an open-open pipe. Any of the harmonics from 1 to at least 10 can be played by simply increasing the air speed appropriately.[3]

What happens in the hummer is that small vortices are formed at the lip, which move down the pipe, moving against each corrugation in turn. When the collision rate matches a resonant frequency of the pipe, the pipe will sing out with the musical note.

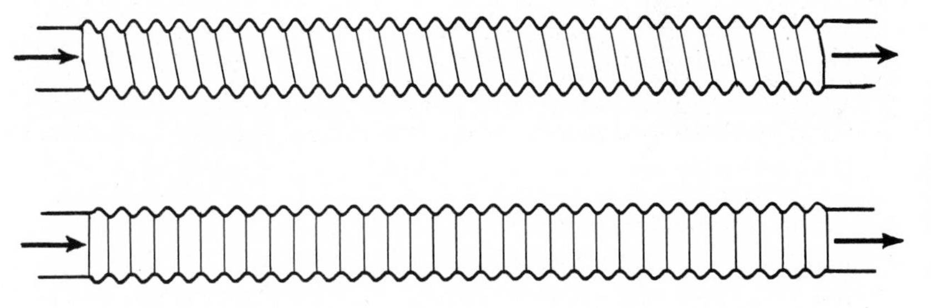

FIGURE 12–12 **A section of corrugated pipe will resonate at different frequencies, depending upon the speed of the wind through it.**

QUESTIONS

1. Briefly explain in your own words why the wind produces a howling sound as it passes through the trees.

2. Briefly explain in your own words why wind instruments go out of tune when the temperature changes.

3. Upon what four factors does the fundamental frequency of a vibrating air column depend?

4. Briefly explain in your own words how the fundamental frequency of a vibrating air column varies as each of the four factors in question is increased.

5. Blowing across the mouth of a bottle causes a tone to be produced, whose frequency depends upon the geometry of the bottle. Assuming the excitation mechanism to be similar to that of the organ pipe, explain in your own words how it works.

PROBLEMS

1. Find the speed of sound waves in helium gas (He) if the pressure is equivalent to standard atmospheric pressure.

2. Calculate the speed of sound in a tube containing pure hydrogen (H_2) if the pressure is equivalent to standard atmospheric pressure.

3. A long glass tube is filled with xenon at standard pressure, is 2.650 m long, and is open at both ends. What is (a) the speed of sound in the gas, and the frequency of (b) the fundamental and (c) the fourth harmonic?

4. If a bugle were straightened into a long straight tube, its length would be 2.65 m long. (a) Calculate the frequencies of the first six harmonics, assuming both ends are open. Assume the speed of sound to be 350 m/s. (b) To what notes on the musical scale do these belong? (See Table 14–5.)

5. An organ pipe filled with air at 0°C sounds middle C (C_4), having a frequency of 264 Hz. Assume the air pressure to be standard atmospheric pressure. Find (a) the speed of sound in air and (b) the length of the pipe if both ends are open.

6. Wind instruments usually warm up from the player's breath. Calculate the maximum change expected if the instrument is tuned to concert A (440 Hz) at the room temperature of 23°C and then warmed up to body temperature of 37.0°C.

PROJECTS

1. Bring a woodwind or brass instrument to class for you and your classmates to study. Determine the total length of the longest possible air column that can be set vibrating. How does this compare with the lowest possible obtainable note with this instrument? Assume both ends are open, then one end open and the other end closed.

2. Perform an experiment similar to the one described in Section 12.1. Use a tuning fork that has been marked with its rated frequency. If the apparatus shown in Figure 12–1 is not available, obtain a glass, metal, rubber, plastic, or cardboard tube of some reasonable length. Fill a deep basin or bucket with water, and immerse one end of the tube in the water. Hold a vibrating tuning fork over the open end at the top, and raise or lower the tube in an effort to find a resonant point. The tube will sing out loud and clear when resonance occurs. Measure the room temperature and calculate the speed of sound in air, using Equation (5c). Using Equation (3a), calculate the wave-

length of the sound waves, and compare with your measured air column. How do these two values compare? Briefly explain.

3. Construct an aeolian harp.[4]

NOTES

[1]See H. E. White, D. H. White, and M. Gould, *Physics, An Experimental Science* (New York: American Book Co., 1968), Chap. 40.

[2]G. A. Audsley, *The Organ of the Twentieth Century* (New York: Dover, 1970).

[3]F. S. Crawford, "Singing Corrugated Pipes," *American Journal of Physics* 42 (1974):278.

[4]D. D. Dorogi, "Build an Ancient Window Harp," *Popular Mechanics Do-It-Yourself Encyclopedia,* vol. 2 (New York: Hearst Books, 1968), p. 22.

Chapter Thirteen

VIBRATING BARS, PLATES, AND MEMBRANES

When a bundle of sticks of wood is dropped on the floor or a table, the sound that is heard is not described as a muscial sound but as a *noise.* If a single stick is dropped, the sound would be described in the same way. If, however, a set of sticks are arranged in the order of length or thickness and dropped one after the other, one observes each stick giving rise to a different but characteristic note. Furthermore, if the sticks are cut appropriately, the sounds they will produce will form a musical scale. (This experiment should be performed to appreciate the effect.) In this chapter we will explore the musical sounds produced by vibrating solid materials.

13.1 Vibrating bars

The musical sticks described above are usually referred to as **vibrating bars.** Many musical instruments use **bars** (rectangular cross sections), **rods** (round cross sections), **tubes** (hollow rods), or **reeds** (end clamped bars) as sources of musical sounds. These include the xylophone, marimba, triangle, saxophone, chimes, and many others.

Standing waves in a uniform bar or rod may consist of vibrations classified as one of three kinds:

transverse longitudinal torsional

All three of these are natural modes of vibration and depend for their frequency upon the size and shape of the bar or rod, the material of which it is made, how it is supported, and with the way in which it is set vibrating. See Figure 13–1.

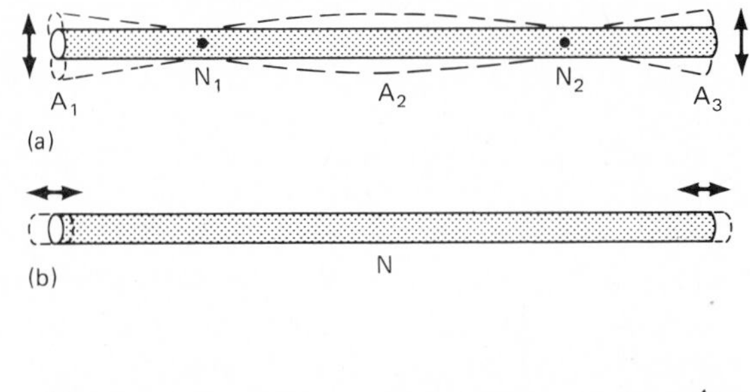

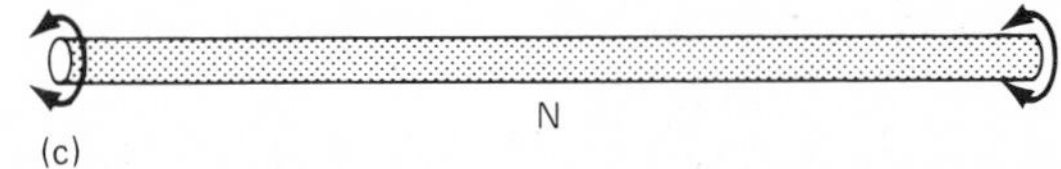

FIGURE 13–1
A rod shown vibrating in three different modes: (a) transverse, (b) longitudinal, and (c) torsional.

If one end of a uniform bar is clamped in a vise and the upper end is free, it may be set vibrating transversely in a number of standing wave patterns. See Figure 13–2. The simplest mode at the left (a) vibrates with the lowest frequency, called the fundamental; diagrams (b) and (c) show the bar vibrating in higher modes. Lord Rayleigh calculated and demonstrated that such a vibrating bar will have frequency relations given approximately by $1f$, $6.25f$, $17.5f$, $34.4f$, $56.5f$, ...[1] Furthermore, the nodes are not equally spaced since the upper node of the second mode is slightly more than one-fifth of the length from the free end, while the two upper nodes of the third mode are approxi-

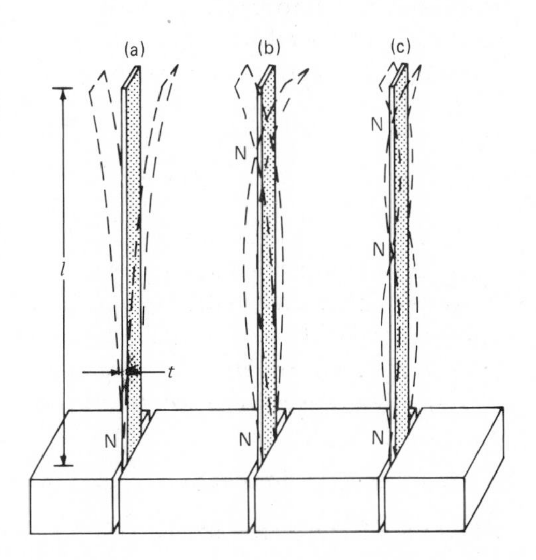

FIGURE 13–2
Rods rigidly clamped at one end and free at the other can be set vibrating with transverse standing wave patterns: (a) the fundamental mode, (b) the second mode, and (c) the third mode.

mately one-eighth and one-half the length from the free end, respectively. Lord Rayleigh also showed that the fundamental frequency of an unclamped uniform bar is given by the relation

$$f = k\frac{t}{l^2} \qquad [13a]$$

where t is the thickness parallel to the direction of bending, l is the length, and k is a constant involving the elasticity of the bar material.

The reeds of a number of musical instruments vibrate in much the same way as the bar shown in Figure 13–2(a) and are to be found in instruments such as the clarinet, saxophone, harmonica, accordion, and reed organ. Reeds are usually not uniform in thickness but are thinned down at one end.

Transverse standing waves may also be set up in a bar or rod supported at two points, about one-quarter of the length from the ends, by striking it a blow at or near the center. As illustrated by the bar in Figure 13–3, the center and two ends move up and down, forming nodes N_1 and N_2 at the supports and antinodes A_1, A_2, and A_3 at the center and ends. As in the case of vibrating strings, the longer the bar, the lower is the pitch; and the shorter the bar, the higher is the pitch. This mode of vibration is called the fundamental and its frequency is given by Equation (13a).

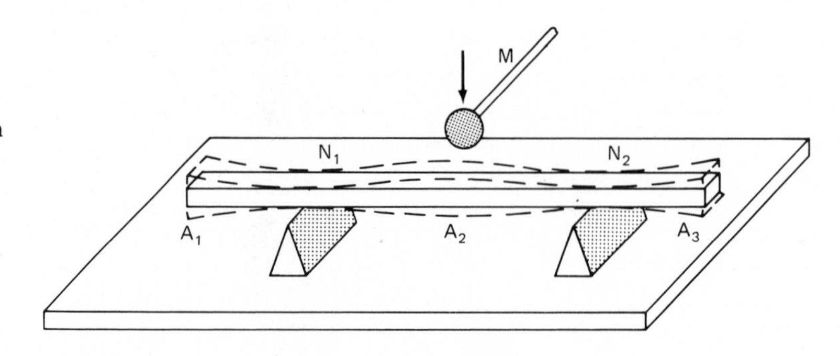

FIGURE 13–3
A rod resting on two supports and struck a blow with a mallet M can be set vibrating at its fundamental frequency with nodes N_1 and N_2 and antinodes at A_1, A_2, and A_3.

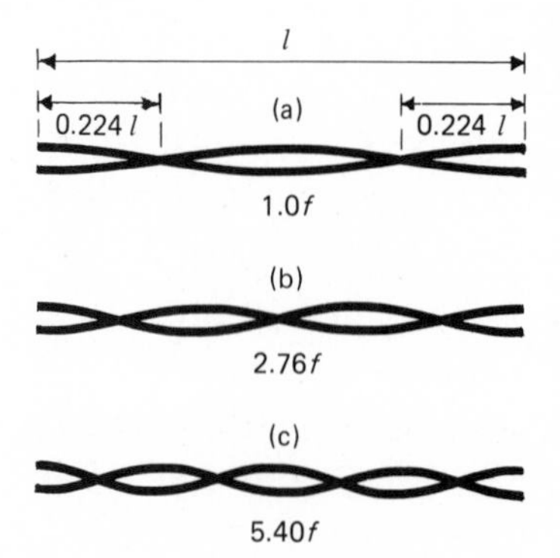

FIGURE 13–4
Uniform rods vibrating with standing transverse waves have nodes and antinodes that form symmetrical patterns: (a) fundamental, or first mode, (b) second mode, and (c) third mode.

If a bar is struck at a point some distance from the center, it will be set vibrating in modes other than those shown in Figure 13–4.[1] The fundamental can be suppressed by touching the bar at or near the center, allowing the higher modes to dominate. The first few higher modes have frequencies $2.76f$, $5.40f$, $8.93f$, $13.34f$, . . . , according to basic theory. Since these modes have frequencies which are not integral multiples of the fundamental, they are said to be **inharmonic**. Since the supports are not located at the nodes of the upper modes, those modes are somewhat suppressed. The emitted sound is therefore dominated by the fundamental.

13.2 Tuned bars

Orchestral bells are actually composed of metal bars of varying lengths and are mounted on a suitable frame. See Figure 13–5. They are uniform in cross section and supported at points $0.224\ l$ from each end. Struck with a hard mallet, the vibration lasts for some time but continuously diminishes in amplitude as the energy is dissipated through the emission of sound waves and internal friction. The shorter bars produce the higher notes, and the longer bars produce the lower notes.

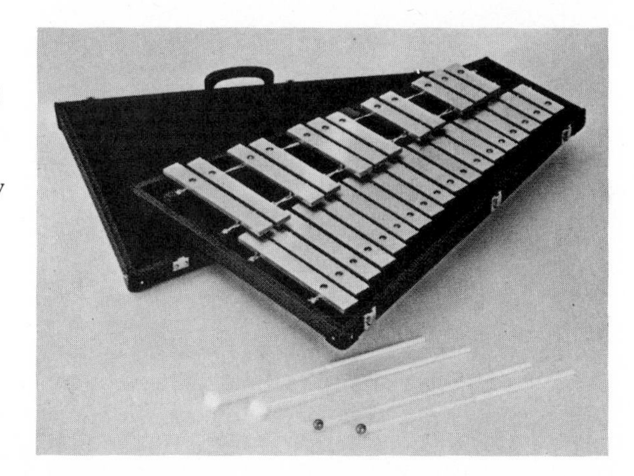

FIGURE 13–5 Orchestral bells, used in bands and orchestras, sound like the bells of a carillon. (Courtesy of Ludwig Drum Company)

The **bell lyra** is a portable version of the orchestral bells, with lighter bars mounted on a lyre-shaped frame. See Figure 13–6. Other vibrating bar instruments include the xylophone, marimba, and vibraphone. (See Section 20.5).

The beater (mallet) is important in determining the timbre of a vibrating bar. If a soft beater strikes the bar, it spreads out, extending the width and time of contact. See Figure 13–7. The higher-frequency overtones are therefore smothered before they can be sounded, resulting in a mellow, deep timbre. The bright overtone-rich sound of the orchestra bells is due in part to the use of hard, narrow beaters on the metal bars.

If a straight bar is slowly bent into the shape of the capital letter U, the nodes of the vibrating bar move down from the ends toward the shoulders, as shown in Figure 13–8. This process forms a tuning fork, a musical instrument that produces the purest of tones. Struck a hard blow, however, overtones can be produced, which are far from harmonics of the fundamental. See diagram (e). The center prong at the bottom of the tuning fork is called the **tine**, which serves to conduct the fork's vibrations to a soundboard. Touching the tine to a tabletop will set a large area vibrating with the same frequency and thereby intensify the emitted sound. Held in contact with such a soundboard, the law of conservation of energy requires a more rapid damping of the fork vibrations. (See A1.14.)

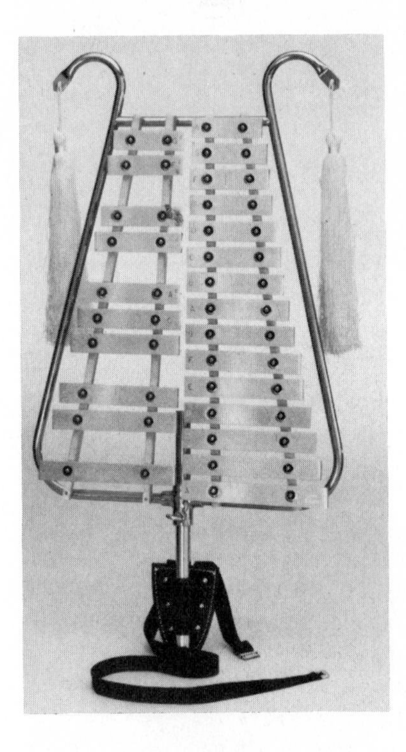

FIGURE 13–6
The bell lyra, or glockenspiel, played in a marching band, sounds like the bells in a carillon. (Courtesy of Ludwig Drum Company)

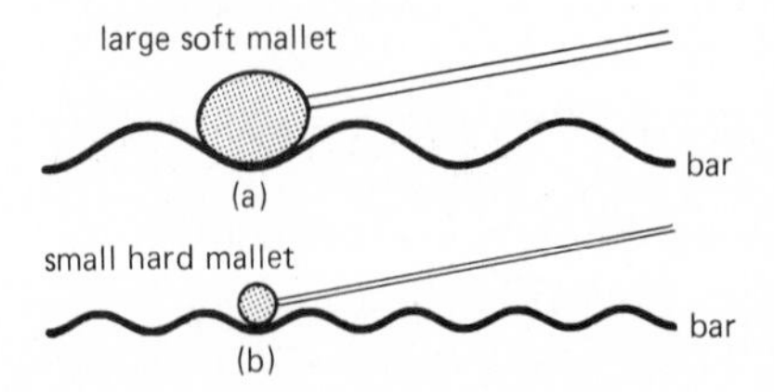

FIGURE 13–7
Mallets. The higher partials created will be suppressed more by (a) a soft mallet than (b) a hard mallet.

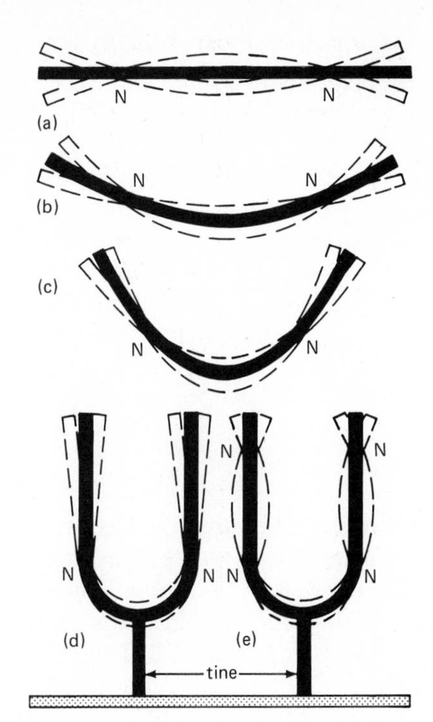

FIGURE 13-8
A straight rod bent into the shape of a capital letter U becomes a tuning fork. (a, b, c) A rod vibrating with its fundamental. (d) A tuning fork vibrating with its fundamental. (e) A tuning fork vibrating with its first partial, or second mode.

13.3 Longitudinal and torsional vibrations

Longitudinal standing waves and torsional standing waves can easily be produced in a rod. The experiments shown in Figure 13-9 serve as good demonstrations. A uniform rod with a round cross section and clamped at the center can be used to demonstrate both kinds of standing waves. A cloth soaked in alcohol, or rosin solution, is used to grasp the rod tightly near the clamp. A smooth uniform stroke toward the end will then make the rod sing out with

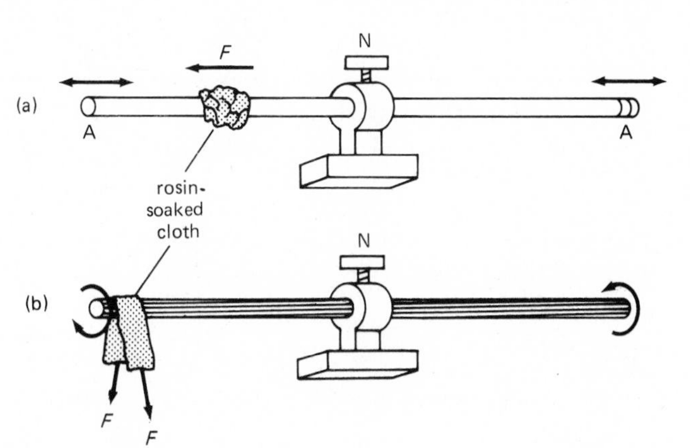

FIGURE 13-9
Longitudinal (a) and torsional (b) standing waves in a round rod clamped at the center with both ends free.

standing waves. The fundamental mode finds one node at the clamped center, with antinodes at both ends. Higher partials are produced but the fundamental is by far the most intense. In practice, an aluminum rod supported between your fingers at any nodal point will work.

If the soaked cloth is draped over one end of the rod and both cloth ends pulled down, one end harder than the other, causing it to slide, standing torsional waves can be generated. See Figure 13–9(b). Oscillations such as these correspond to some of the higher overtones in vibrating bar instruments[1] but are generally of minor importance. The longitudinal mode is used in an experiment performed to measure the wavelength of sound waves; see Figure 4–9.

13.4 Vibrating plates

Plates can be thought of as two-dimensional bars. The frequencies of vibration are determined by the rigidity of the plate, its density, its dimensions, and the striking technique. The sound emitted by a vibrating cymbal plate, or vibrating steel drumhead, is, in general, due to its emission of a number of inharmonic frequencies. These frequencies are due to the complicated modes of vibration over the surface of the metal disk.

Typical patterns of center-clamped, vibrating, square and round plates, each having a characteristic frequency, are shown in Figure 13–10. These are but a few of the many possible modes of vibration. The method of producing these nodal patterns was demonstrated in the sixteenth century by Chladni, a German physicist. See Figure 13–11. A square or circular metal disk is clamped at the center, and sand or salt is sprinkled over the surface. While touching the rim at one or two points like N_1 and N_2, a cello bow is drawn down over the edge at another point. Nodes will be formed at the stationary points N_1, N_2, N_3, . . ., and antinodes in the regions L_1, L_2, L_3, ... The grains of sand or salt bounce away from the vibrating regions and into the nodal lines where there is little or no motion. At one instant the regions

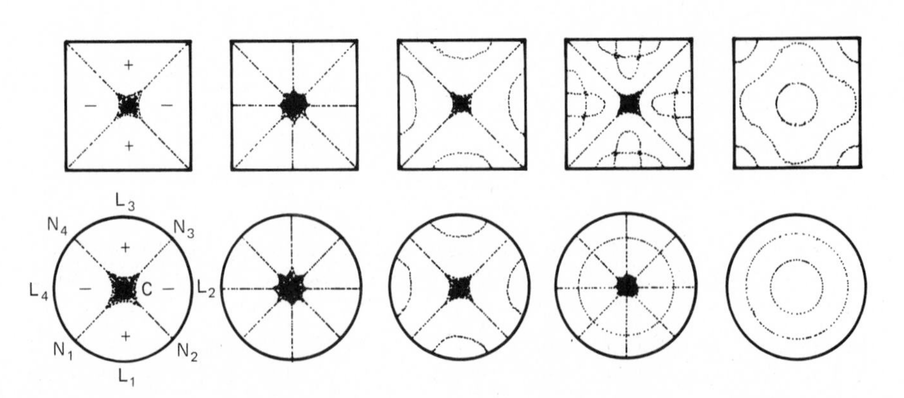

FIGURE 13–10
Chladni figures showing the nodes and loops of vibrating plates. These are but a few of the many two-dimensional standing wave patterns that can be obtained. (With permission of D. Van Nostrand Company)

marked + are up and the regions marked − are down. Half a vibration later these positions are reversed. The patterns formed in this way are called **Chladni figures.**

Chladni figures can be seen by a large audience by using a 30-cm-square, or 30-cm-diameter, plate, made of bronze, aluminum, or glass, about 3.0 mm thick and painted on the top with a dull black paint. Salt or sand can best be spread by using a kitchen salt shaker. The use of white sand mixed with fluorescent pigment will enhance the visual effect.[2] A large mirror placed at about 45° to the vertical enables the audience to effectively "look down" on the patterns as they form.

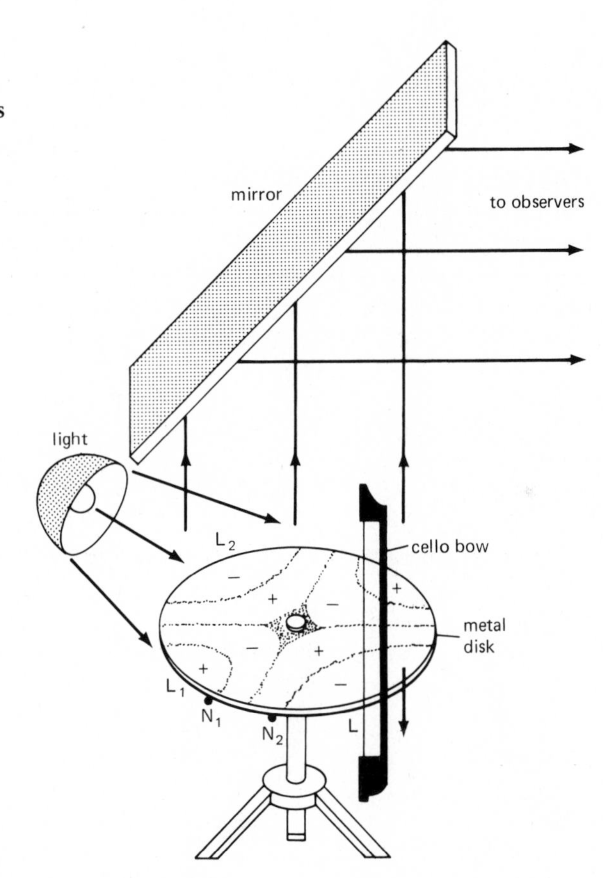

FIGURE 13–11
Apparatus for observing Chladni's sand figures on a vibrating plate.

Vibrating plates constitute two-dimensional standing wave patterns. The more complex the pattern, the higher is the frequency. Since the center is clamped, a node always forms there. Being free at the edges, antinodes form there, as well as in other places. The clash of cymbal plates will excite several of these modes at the same time, resulting in the production of many inharmonic frequencies. Chladni has also shown that, for two plates of similar material, the frequencies will vary inversely with the overall dimensions. For example, if both plates have a similar shape, and one is twice as thick as and twice the diameter of the other, the larger will produce the same pattern an octave lower than the other.

Plates clamped or hinged at the perimeter are often called **diaphragms,** and they are usually much thinner than center-clamped plates. Examples of diaphragm motion are found in steel drums (see Section 20.5.) and in wood vibrations of stringed instruments (see Chapter 16).

13.5 Bells

In some respects, a bell is similar to the cymbal plate. When struck a blow by the clapper, a bell is set vibrating, with nodes and antinodes distributed in a symmetrical pattern around the rim and over the surface. See Figure 13–12. The vibration of the rim is illustrated in Figure 13–13(a); the experiment is shown in diagram (b). Small cork balls are suspended by threads, just touching the outside edge of a punch bowl. A cello bow drawn across the edge of the rim of the bowl will set the rim vibrating in nodes and loops. Nodes are always even in number, just as they are with vibrating plates; alternate loops move in opposite directions. (For an alternative experiment, use a wineglass. The rim of a thin wineglass is readily set vibrating by wetting the forefinger and rubbing the top of the rim with smooth strokes parallel to the lip.)

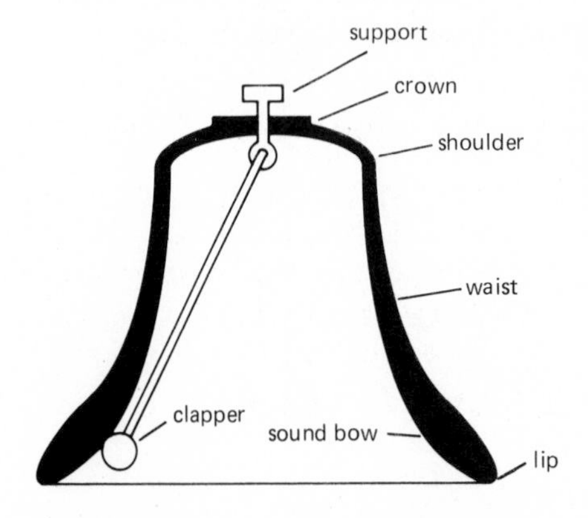

FIGURE 13–12
Cross-sectional diagram of a bell like the ones used in most of the world's carillons.

During the Middle Ages, most towns had a tall tower near their center containing one or more bells. These bells were used to control the lives of the people: to call them to a fire, to work, to church, to war. As time passed, interest grew in the development of musical bells and systems (carillons) of tuned bells. Craftsmen discovered that by proper casting and shaping, the lower partials, which determine the pitch and timbre of a bell, could be brought into some sort of harmonic relationship, resulting in a more musical quality. The development of such musical bells is attributed largely to the work of Frans Hemony and others in Belgium; hence Flemish bells are usually considered as a standard of tonal quality.

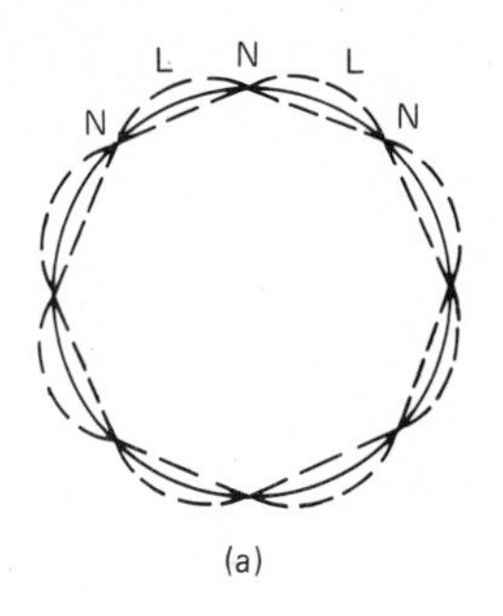

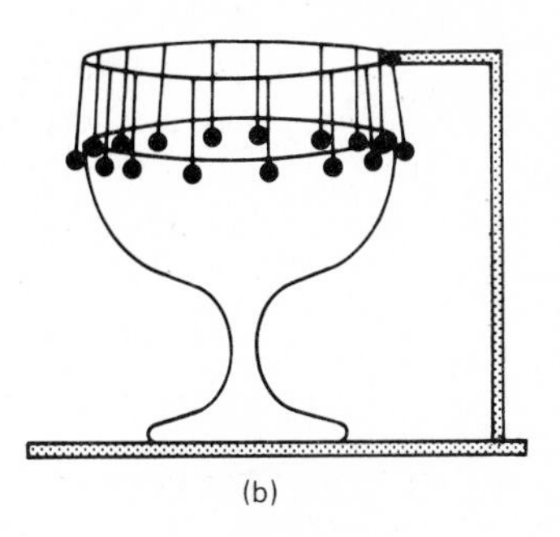

FIGURE 13–13 The rim of a bell or a thin glass bowl vibrates with nodes and loops (a), as illustrated by a simple experiment (b). (With permission of D. Van Nostrand Company)

Although bells differ in a number of subtle and debatable ways, there seem to be some common features. The sound of a bell is characterized by a large number of partials. Some of the upper partials harmonize (approximately) with the fundamental, but most do not. Assuming the lowest mode (**hum tune**) to be C_4, the traditional bell is designed to have higher partials at C_5, $E\flat_5$, G_5, C_6, E_6, and C_7.* The minor third, $C_5 - E\flat_5$, is desirable to give the bell its characteristic plaintive tone. The pitch is determined by C_5, C_6, G_6, and C_7, which form the harmonic series based on C_5.[1] In practice, the various partials do not match the harmonic series perfectly. The pitch (**strike tone**) will not necessarily coincide with the partial around C_5 but is apparently determined by some sort of an average fit to the nearly harmonic series.[3, 4] The clarity of the pitch seems to depend upon how closely the partials match the harmonic series.

The medieval city of Bruges, Belgium, has a central tower whose belfry contains one of the finest existing carillons. The 47 bells, most of which were cast in 1748, span four octaves, without the C♯ and D♯ in the lowest octave. This carillon is played by an all-mechanical automated system in which a rotating drum causes hammers to strike the bells in the proper sequence. See Figure 13–14. Alternatively, the carillon can be played manually from a keyboard consisting of levers connected by wires to the clappers.

13.6 Vibrating membranes

Drums consist of a flexible uniform membrane, called the head, stretched tightly over a circular frame by a number of screws or clamps distributed uniformly around the periphery. Such a **membrane** acts as a two-dimensional

*See Table 14–5 for the frequencies of the notes shown.

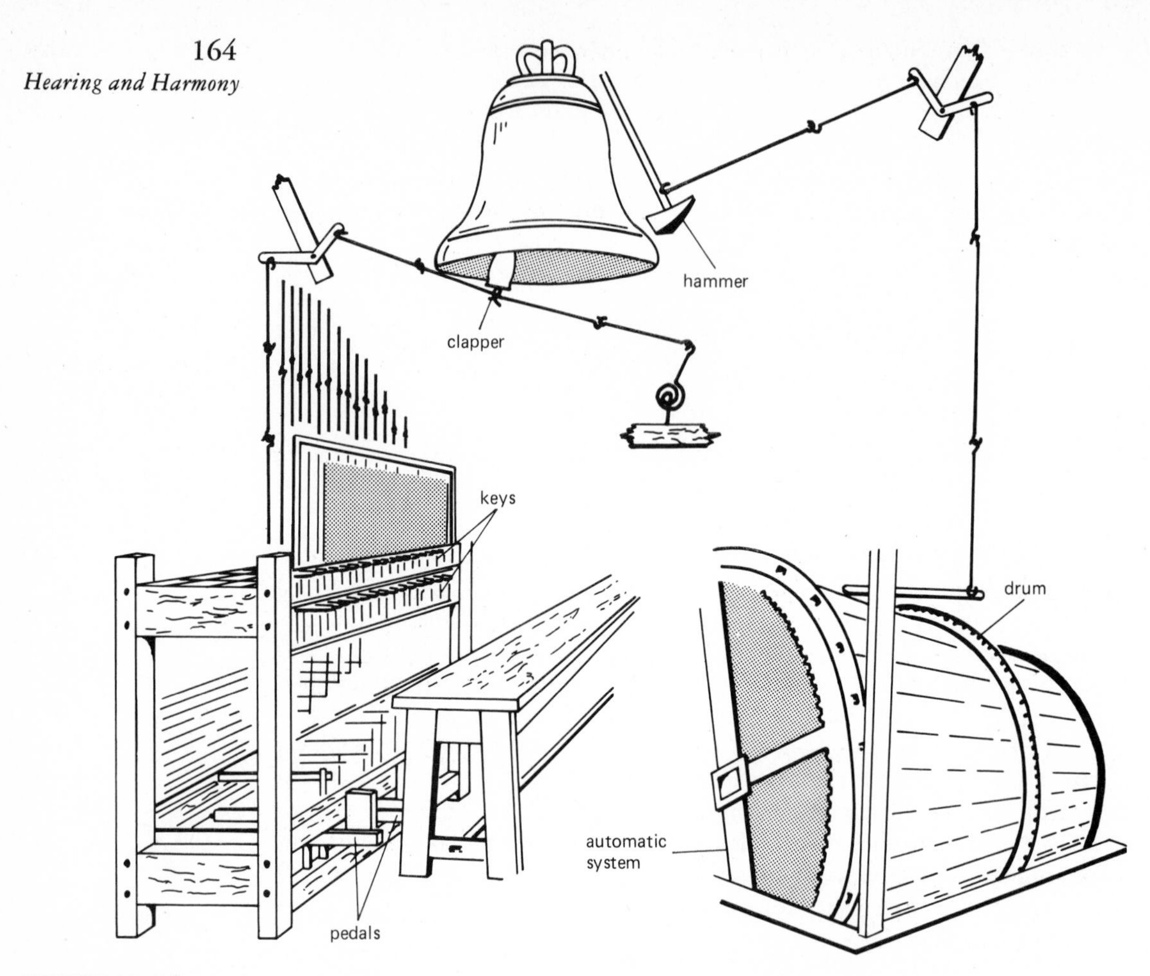

FIGURE 13–14
Mechanical system that operates the carillon in the bell tower of Bruges, Belgium.

vibrating string, since the frequencies are determined primarily by applied tension rather than stiffness. However, the partials for vibrating membranes are not harmonically related, as they are for vibrating strings.

The head of a drum is set into vibration by striking it a blow with a stick that consists of a wooden shaft with a felt ball at the far end. Sounds are produced by striking the head at various distances from the center. Several of the simpler modes of vibration are shown in Figure 13–15. For the lowest frequency, called the fundamental, all parts of the head are vibrating in phase, with a circular node at the edge. The second mode, with a frequency of $1.59f$, has one nodal line across a diagonal. This node divides the head into two identical sections, with a node around the periphery. The third mode, with a frequency of $2.13f$, has two nodal lines at right angles and one nodal ring; the fourth mode, of frequency $2.29f$, has two nodal lines and two nodal rings; and so on.

The frequency f of the fundamental is given by the relation

$$f = \frac{0.766}{D}\sqrt{\frac{T}{d}} \qquad [13b]$$

where D is the diameter of the head, in meters, T is the tension in the membrane, in newtons/meter, and d is the surface density of the membrane, in kilograms/meter^2.

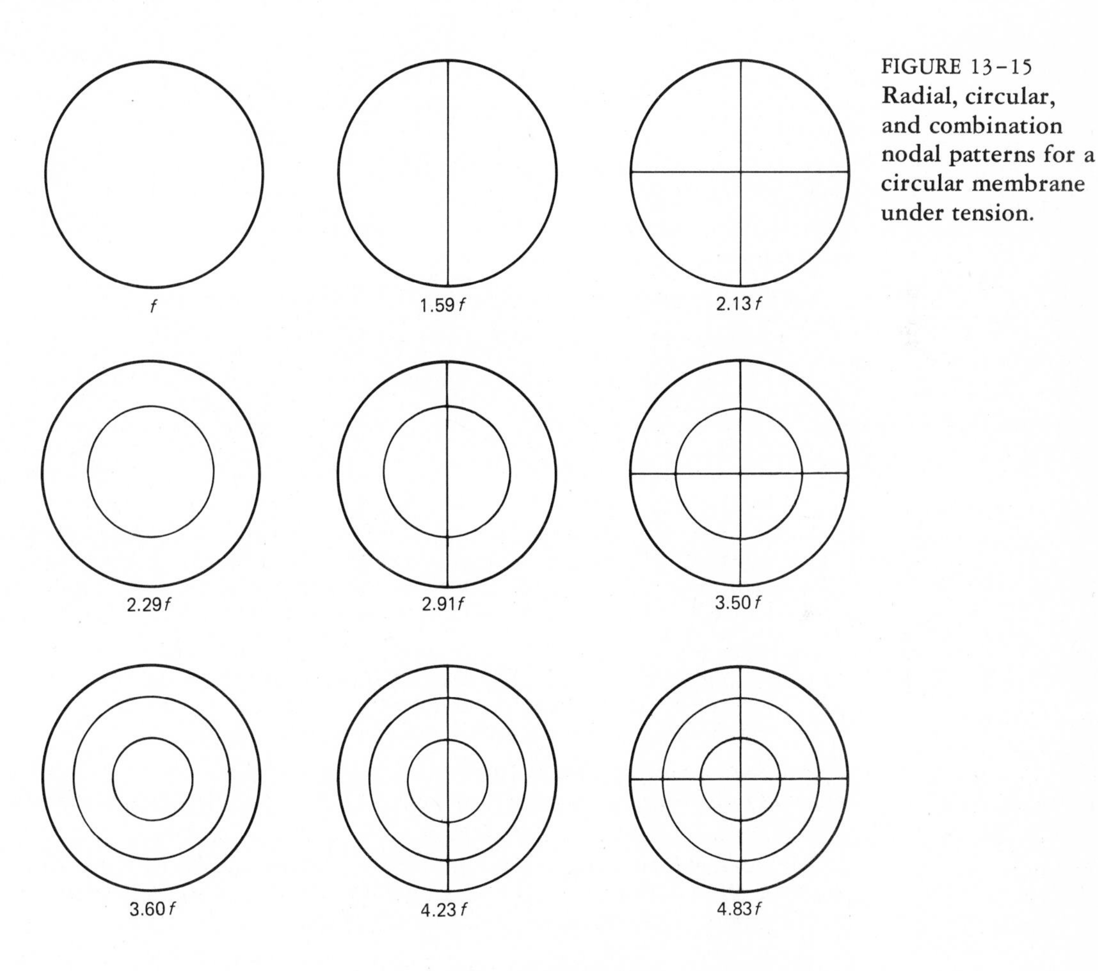

FIGURE 13-15 Radial, circular, and combination nodal patterns for a circular membrane under tension.

13.7 Laboratory experiment

Obtain a uniform strip of wood, free of knots, with approximately the following dimensions: thickness, 0.6 cm; width, 3.0 cm; length, 200 cm. Put the wood strip in a vertical position and clamp one end in a vice. Set the strip vibrating as shown in Figure 13-2 and with a stopwatch determine the time required for ten complete vibrations. Now determine the time required for ten vibrations when using several shorter lengths (move the strip down in the vice jaws about 10 cm each time). Continue these measurements until the strip vibrations are too fast to be counted. Using graph paper, draw a graph, plotting the frequency horizontally and the reciprocal of the length, $1/l$, vertically. Is the graph a straight line? Now plot the frequency horizontally and the reciprocal of l^2 vertically. Is the graph nearly straight? Explain your results.

QUESTIONS

1. Name six musical instruments employing vibrating rods or reeds as their sources of sound.

2. How does the frequency of a vibrating plate depend upon the number of nodal lines in its pattern?

3. Describe the difference between longitudinal standing waves and torsional standing waves in a rod of circular cross section.

4. Make a diagram of the vibrating rim of a bell or a glass bowl.

5. Describe an experiment demonstrating Chladni figures.

6. What is the difference between a membrane and a plate clamped or hinged at the rim? What determines the vibrating frequencies in each case?

PROBLEMS

1. The longest bar in a set of orchestra bells is 30.0 cm, and its fundamental frequency is 198 Hz. Find the lengths of other bars in the set if they have frequencies of (a) 220 Hz, (b) 264 Hz, (c) 330 Hz, (d) 396 Hz, and (e) 440 Hz. Assume all bars have the same thickness and cross section.

2. One end of a uniform rod is clamped so that part of it extends above the vise jaws. See Figure 13–2. The free end above is set vibrating. If the fundamental vibration mode has a frequency of 132 Hz, find the frequency of (a) the second mode, (b) the third mode, and (c) the fourth mode.

3. One bar on a xylophone is uniform in cross section and has a fundamental frequency of 264 Hz. Find the frequency of (a) the second mode and (b) the third mode.

4. The fundamental pitch of a circular plate, clamped at the center with the periphery free, has a fundamental frequency of 66.0 Hz. Calculate the frequencies of the next two lowest frequency modes.

PROJECTS

1. Obtain a circular disk of sheet brass, aluminum, or steel, about 3 mm thick and about 30 cm in diameter. Clamp it at the center with a bolt, washers, and nuts, and then in a vice, as shown in Figure 13–11. Sprinkle salt over the disk, and with a cello bow set the disk vibrating as described in Section 13.4. How many different patterns can you find?

2. Select a uniform strip of wood about 3 mm thick, 3 cm wide, and 2m long. Cut eight pieces from this stick [with lengths as given by Equation (13a)] that will produce frequencies proportional to the numbers 24, 27, 30, 32, 36, 40, 45, and 48. Make the longest stick about 30 cm long. Lay these sticks on a table and arrange them in the order of length. Now drop them one after the other in order. How would you describe the effect?

3. Cut sticks of wood to the same length (20 cm, for example), about 3 cm wide, and plane them to various thicknesses so that they are tuned to the musical scale. See Table 14–5. Practice playing a simple tune by dropping them in the appropriate order.

NOTES

[1]T. D. Rossing, "Acoustics of Percussion Instruments, Part I," *Physics Teacher* 14 (December 1976):546; and "Part II," *Physics Teacher* 15 (May 1977):278.

[2]R. C. Nicklin, "Colorful Chladni," *Physics Teacher* 11 (May 1973):312.

[3]F. H. Slaymaker and W. F. Meeker, "Measurements of Tonal Characteristics of Carillon Bells," *Journal of the Acoustical Society of America* 26 (1954):512.

[4]A. H. Benade, *Fundamentals of Musical Acoustics* (New York: Oxford University Press, 1976), sec. 5.6.

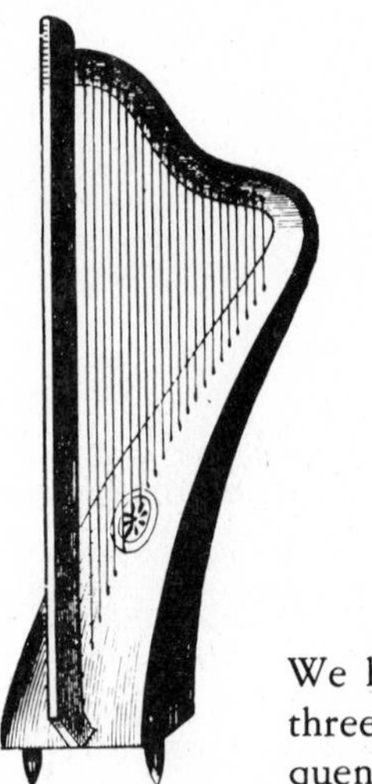

Chapter Fourteen

MUSICAL INTERVALS AND SCALES

We have seen in previous chapters that one of the three principal aspects of musical sounds is their frequency or pitch. We have also seen that frequency is a physical description of any periodic vibration and is given by the number of complete cycles per second, whereas pitch is a subjective description of sound with respect to its position on a musical scale. During the last several centuries, many musical scales have been developed, but few remain in use today. When students of music, science, or mathematics begin studying the frequencies of notes on two or three of the commonly recognized scales, and the consonance or harmony of tone combinations and intervals, they soon become entranced by small whole numbers. Before long they are into the fascinating subject of "the magic of numbers," which is called **numerology**. It is hoped that before you have finished this chapter, you too will have had some pleasure in dealing with small integers.[1]

14.1 Frequency and pitch

That frequency and pitch are closely related one to the other can be demonstrated in many different ways. Every boy or girl who has owned a bicycle has at one time or another fastened a piece of cardboard to the forks of the front wheel so that one end projects between the spokes. When the wheel turns, the spokes strike the card and make it vibrate. The faster one rides, the faster the card vibrates, and the higher is the pitch of the sound.

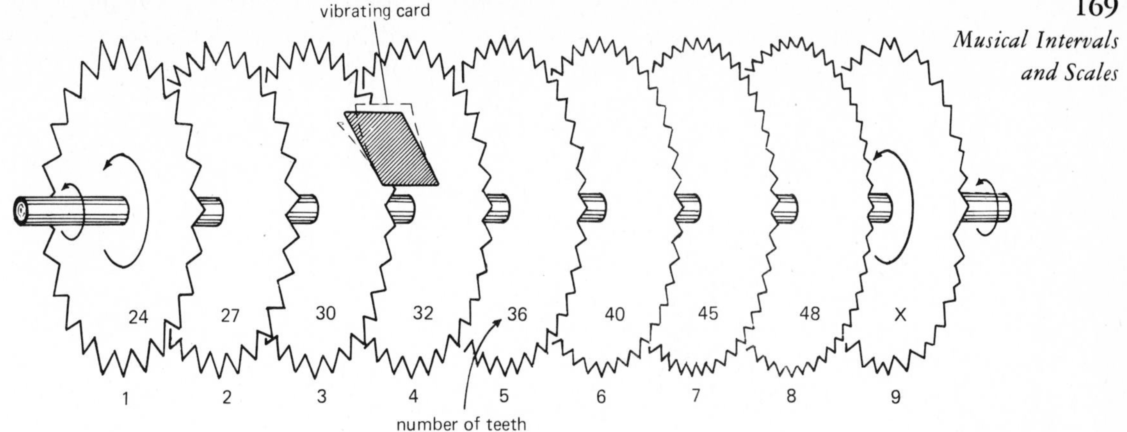

FIGURE 14–1
A classroom demonstration using rotating toothed disks and a piece of cardboard to discover the relationship between frequency and pitch. The end disk produces a noise.

A similar experiment, converting this principle into a musical scale, is shown in Figure 14–1. Nine disks mounted on a single shaft are rotated at constant speed by an induction or synchronous motor. As the disks turn, a piece of cardboard, or thin plastic sheet, is held in the hand so that one end touches the teeth, disk after disk. The card will vibrate periodically, and the first eight toothed disks will produce a musical scale. When the card is held to the teeth of the ninth disk, a strange sound will be heard. Such a sound without periodicity is called a **noise**, since it in no way resembles a musical note, and it must be heard to appreciate the effect. Noise can also be considered to be any undesirable sound.[2]

Each of the first eight disks in Figure 14–1 must have equally spaced teeth; the number in each disk is given by the following sequence:

Disk number	1	2	3	4	5	6	7	8	
Number of teeth	24	27	30	32	36	40	45	48	[14a]

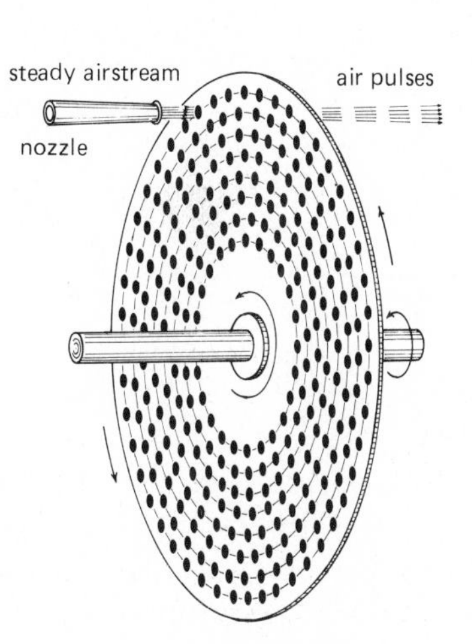

FIGURE 14–2
A rotating disk with rings of regularly spaced holes produces a musical scale when an airstream passes through them in consecutive order.

A second demonstration relating pitch and frequency is shown in Figure 14–2. A thin metal disk with eight concentric rings of holes is mounted on a shaft that turns at constant speed by an induction or synchronous motor. An airstream from a compressed air tank is blown from a nozzle through ring after ring of the equally spaced holes. If the number of holes in consecutive rings are those given by Equation (14a), the sounds will be recognized as a musical scale.

We may summarize the results of these two experiments by the following: As the frequency rises, the pitch rises; and as the frequency falls, the pitch falls.[3]

14.2 The Pythagorean scale

Although the ancient Greek philosopher Pythagoras is best known for his theorem in geometry, he also explored the principles of music. His greatest contribution to music was his discovery that **harmony** and **consonance** are associated with the smallest whole numbers, 1, 2, 3, and 4. For his experiments he is said to have used a monochord—a hollow box with a single string stretched between two supports near the ends, as shown in Figure 14–3 and Figure 1–1. A third support called a bridge could be moved to any point in between, dividing the two string segments into any desired ratio. Pythagoras found that when the two segments were set into vibration at the same time, the resultant sound was most pleasant to hear if their lengths were in the ratios 1:1, 1:2, 2:3, and 3:4. Two such sounds are said to be **consonant**, or **harmonious**. Pythagoras did not know, however, that it was the frequencies of the strings that were consonant. Today we know that the frequency of a stretched string under constant tension is inversely proportional to its length; see Equation (11g).

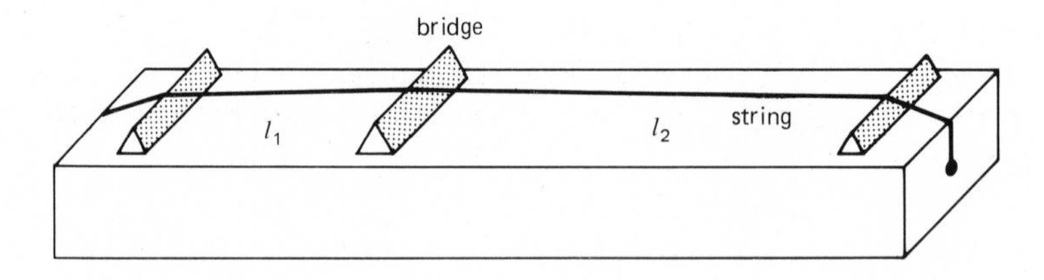

FIGURE 14–3
A Pythagorean monochord for demonstrating harmonic relations between string segments.

$$f \propto \frac{1}{L}$$

The smallest whole numbers for the frequency ratios form the principal intervals for the **Pythagorean musical scale**. They are as follows:

Pythagorean intervals

Frequency ratio	1:1	1:2	2:3	3:4	
Interval name	unison	octave	fifth	fourth	[14b]

Time graphs for two notes having the frequency ratio 1 : 1 are shown in Figures 8 – 2 and 8 – 3. Other time graphs with the frequency ratio 1 : 2 are given in Figures 8 – 5, 8 – 6, and 11 – 5(a).

In music, an **interval** between any two tones with frequencies f_1 and f_2 is given by the ratio of the two frequencies, f_2/f_1, and not by the difference, $f_2 - f_1$. To add two intervals is to multiply their frequency ratios, and to subtract intervals from each other is to divide one frequency ratio by the other. For example, to go up a fifth from any frequency f_1 to f_2, we obtain the frequency ratio $f_2/f_1 = \frac{3}{2}$. To go up a fourth higher in pitch, from f_1 to f_3, we multiply $\frac{4}{3}$ by $\frac{3}{2}$ and obtain

$$\frac{4}{3} \times \frac{3}{2} = 2f \qquad [14c]$$

which is exactly one octave.

The Pythagorean scale is based upon these consonant interval ratios: the **fourth**, the **fifth**, and the **octave**. If we start with the note C, which we will assume has the frequency f, and go up a fourth, we obtain a frequency $(\frac{1}{1} \times \frac{4}{3})f = \frac{4}{3}f$, a note that we call F. Starting again at note C, if we go up a fifth or an octave, we obtain $(\frac{1}{1} \times \frac{3}{2})f = \frac{3}{2}f$ or $(\frac{1}{1} \times \frac{2}{1})f = 2f$, the notes we call G and C′. We now have a scale of four notes.

C	F	G	C′
f	$\frac{4}{3}f$	$\frac{3}{2}f$	$2f$

[14d]

If we now go down a fourth from G, we obtain $(\frac{3}{2} \div \frac{4}{3})f = \frac{9}{8}f$, a note we call D. Going up a fifth from D, we obtain $(\frac{9}{8} \times \frac{3}{2})f = \frac{27}{16}f$, a note we call A. We now have six notes.

C	D	F	G	A	C′
f	$\frac{9}{8}f$	$\frac{4}{3}f$	$\frac{3}{2}f$	$\frac{27}{16}f$	$2f$

[14e]

If we now repeat the interval and add it to D, we obtain $(\frac{9}{8} \times \frac{9}{8})f = \frac{81}{64}f$, a note we call E. If we add this same interval to A, we obtain $(\frac{27}{16} \times \frac{9}{8})f = \frac{243}{128}f$, a note we call B, thereby completing a scale of eight notes called the **Pythagorean scale**. See Table 14 – 1.

TABLE 14 – 1: The Pythagorean Scale

SCALE NOTES	C		D		E		F		G		A		B		C′
FREQUENCIES	f		$\frac{9}{8}f$		$\frac{81}{64}f$		$\frac{4}{3}f$		$\frac{3}{2}f$		$\frac{27}{16}f$		$\frac{243}{128}f$		$2f$
INTERVAL RATIOS		$\frac{9}{8}$		$\frac{9}{8}$		$\frac{256}{243}$		$\frac{9}{8}$		$\frac{9}{8}$		$\frac{9}{8}$		$\frac{256}{243}$	

In the table observe that all but two intervals are the same, $\frac{9}{8}$. The remaining two intervals E to F and B to C are equal, but not as big as the others. These last two are called **Pythagorean diatonic semitones**. Their intervals are best seen on the Pythagorean scale chart shown in Figure 14–4. Starting at C, there are two whole tones, followed by one semitone, three whole tones, and one semitone. The actual frequencies of the notes, based on the standard frequency of A = 440 Hz, are given at the bottom of the chart. (See Section 14.5.) This is the middle octave of the musical scale, which, by comparison with the piano keyboard of 88 keys, contains approximately seven octaves.

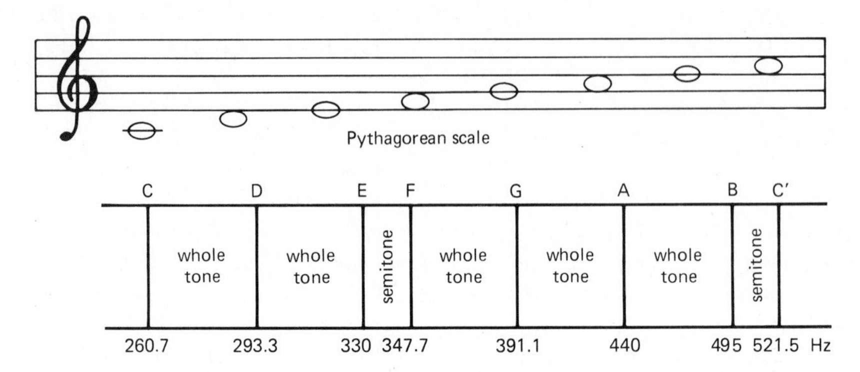

FIGURE 14–4
The Pythagorean diatonic scale with frequencies of the middle octave, based upon the standard frequency A = 440 Hz.

14.3 Pitch nomenclature

Many systems have been used for designating octaves, five of which are given in Table 14–2. The arrangement shown in the first row has been recommended by the United States Standards Association, and it is the one used in this book. The second row has some merit since it is centered around the middle octave of the piano keyboard, and it uses integer subscripts for lower octaves and superscripts for higher octaves. The third row used by Helmholtz, and still used by some authors, is such a mixture that it is difficult to remember. The fourth and fifth rows have been used for years by piano and organ manufacturers, respectively.

TABLE 14–2: Some Octave Nomenclatures

USA STANDARD	C_1	C_2	C_3	C_4	C_5	C_6	C_7	C_8
ALTERNATIVE	C_3	C_2	C_1	C	C^1	C^2	C^3	C^4
HELMHOLTZ	CC	C	c	c′	c″	c‴	c⁗	c^{v}
PIANO	C_4	C_{16}	C_{28}	C_{40}	C_{52}	C_{64}	C_{76}	C_{88}
ORGAN	CCC	CC	C	c^1	c^2	c^3	c^4	c^5

The frequencies for seven octaves of the Pythagorean scale are given in Table 14–3. Starting with any note in the scale and dividing by 2 gives the frequency of the corresponding note one octave below, and multiplying by 2 gives the frequency of the corresponding note one octave above. *This easy to remember rule holds for all scales.*

TABLE 14-3: Frequencies (in Hz) for Seven Octaves of the Pythagorean Diatonic Scale

FIRST	SECOND	THIRD	FOURTH	FIFTH	SIXTH	SEVENTH
C_1 32.6	C_2 65.2	C_3 130.4	C_4 260.7	C_5 521.5	C_6 1043	C_7 2086
D_1 36.7	D_2 73.3	D_3 146.7	D_4 293.3	D_5 586.7	D_6 1173	D_7 2347
E_1 41.3	E_2 82.5	E_3 165	E_4 330	E_5 660	E_6 1320	E_7 2640
F_1 43.4	F_2 86.9	F_3 173.8	F_4 347.7	F_5 695.3	F_6 1391	F_7 2781
G_1 48.9	G_2 97.8	G_3 195.6	G_4 391.1	G_5 782.2	G_6 1564	G_7 3129
A_1 55	A_2 110	A_3 220	A_4 440	A_5 880	A_6 1760	A_7 3520
B_1 61.9	B_2 123.8	B_3 247.5	B_4 495	B_5 990	B_6 1980	B_7 3960
C_2 65.2	C_3 130.4	C_4 260.7	C_5 521.5	C_6 1043	C_7 2086	C_8 4172

14.4 The just diatonic scale

We have seen in the preceding section how the Pythagorean scale is built up by using consonant intervals of octaves, fifths, and fourths, with frequency ratios 2:1, 3:2, and 4:3, respectively. In this respect, the just diatonic scale and the Pythagorean scale are identical. From here on, however, we make some slight changes for the just scale. We now add intervals with frequency ratios given by the next smaller whole numbers, 5:4 and 6:5. These consonant intervals are called the **major third**, $\frac{5}{4}$, and the **minor third**, $\frac{6}{5}$. Their addition to those of Equation (14d) has been the foundation of Western music for several hundred years. Such a scale, first introduced by Zarlino in 1558, is called the **just diatonic scale** and is given in Table 14-4.

TABLE 14-4: The Just Diatonic Scale

SCALE NOTES	C		D		E		F		G		A		B		C
FREQUENCIES	$1f$		$\frac{9}{8}f$		$\frac{5}{4}f$		$\frac{4}{3}f$		$\frac{3}{2}f$		$\frac{5}{3}f$		$\frac{15}{8}f$		$2f$
INTERVAL RATIOS		$\frac{9}{8}$		$\frac{10}{9}$		$\frac{16}{15}$		$\frac{9}{8}$		$\frac{10}{9}$		$\frac{9}{8}$		$\frac{16}{15}$	
SCALE NUMBERS	24		27		30		32		36		40		45		48

Observe in the table that throughout the octave there are three different interval ratios between consecutive notes: **just major tones** with the frequency ratio $\frac{9}{8}$, **just minor tones** with the ratio $\frac{10}{9}$, and **just semitones** with the ratio $\frac{16}{15}$. The just semitone is slightly larger than the Pythagorean semitone.

The middle octave of the just diatonic scale is shown by the chart in Figure 14-5. The scale numbers in Table 14-4 are those given in Equation (14a), and they can be used to calculate the frequencies of all notes in all octaves of the just diatonic scale. If these numbers are multiplied by 11, they give the frequencies of the middle octave. See Table 14-5. If they are multiplied by 22, they give the next higher octave; if they are multiplied by 5.5, they give the next lower octave; and so forth.

The chord CEG, a **major triad,** is seen to correspond to a frequency ratio of 4:5:6. The same is true for the sequence FAC and GBD. Chords played in the key of C major, F major, and G major are accurately consonant. Moreover, if C, F, or G is taken as a fundamental, the harmonic series based on that frequency will correspond to notes of the just diatonic scale through the sixth harmonic. Unfortunately, chords played in a number of other keys are not consonant, resulting in the limited usefulness of this scale. Further search leads to the conclusion that the scale that is perfect in all keys is impossible to construct.[4] For a given piece of music, however, a "best" scale can usually be found.

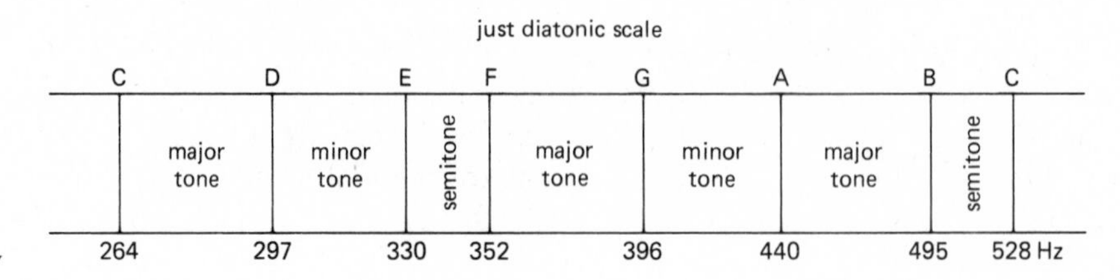

FIGURE 14-5 The just diatonic scale with frequencies of the middle octave, based upon the standard frequency A = 440 Hz.

TABLE 14-5: Frequencies (in Hz) for Seven Octaves of the Just Diatonic Scale

FIRST	SECOND	THIRD	FOURTH	FIFTH	SIXTH	SEVENTH
C_1 33.0	C_2 66.0	C_3 132	C_4 264	C_5 528	C_6 1056	C_7 2112
D_1 37.1	D_2 74.3	D_3 148.5	D_4 297	D_5 594	D_6 1188	D_7 2376
E_1 41.3	E_2 82.5	E_3 165	E_4 330	E_5 660	E_6 1320	E_7 2640
F_1 44.0	F_2 88.0	F_3 176	F_4 352	F_5 704	F_6 1408	F_7 2816
G_1 49.5	G_2 99.0	G_3 198	G_4 396	G_5 792	G_6 1584	G_7 3168
A_1 55.0	A_2 110.0	A_3 220	A_4 440	A_5 880	A_6 1760	A_7 3520
B_1 61.9	B_2 123.8	B_3 247.5	B_4 495	B_5 990	B_6 1980	B_7 3960
C_2 66.0	C_3 132.0	C_4 264	C_5 528	C_6 1056	C_7 2112	C_8 4224

14.5 A standard of pitch

For several hundred years, the pitch of the notes of the musical scales has been determined by specifying A_4 as the standard of pitch. The frequency of this standard note has varied widely and changed so frequently that no set value could really be called standard.

In Handel's time (1685–1759), A_4 was determined by his personal tuning fork, which had a frequency of 422.5 Hz. Since the brilliance of string instruments, like the violin family, appears to increase with higher frequencies, the standard A gradually went up in value until at the end of the nineteenth century it had reached 461 Hz in the United States and 455 Hz in England. Since a change in standard pitch imposes major problems on musicians and instrument manufacturers, a fixed value became more and more essential. Finally in 1953, the International Standards Organization recommended that $A_4 = 440$ Hz be adopted as the standard frequency for music throughout the world. Unfortunately, not all musicians followed the recommendation, and there still exist some orchestral groups that tune their instruments to A = 442 Hz or A = 444 Hz.

Opera singers today are singing Beethoven and Mozart arias about a semitone higher than the pitch for which they were written.[5] This necessitates tuning the accompanying string instruments by increasing their string tensions by nearly 12 percent. To do this, some of the violins made by the old Italian masters have had to be strengthened, which means that their tone quality is not the same.[6]

Some years ago a scientific scale was developed that was based upon C being given by integral powers of the number 2. On this scale, middle C has a frequency of 256 Hz. There are many of such tuning forks still around and commercially available, and many high school and college science teachers still take middle C as 256 Hz.

The standard of pitch for most symphony orchestras today is taken to be A = 440 Hz. The oboist usually carries a tuning fork or pitch pipe to sound the correct pitch. The other musicians then tune to the oboe.

14.6 The chromatic scale

The chromatic scale, based upon the just diatonic scale, includes the sharps and flats for each note. The difference between the frequency of any note of the just scale and its sharp or flat is given by the ratios 25/24 or 24/25, respectively. This relatively small interval, which is commonly used in music, is given by the difference between the just diatonic minor tone and a just diatonic semitone, that is $\frac{10}{9} \div \frac{16}{15} = \frac{25}{24}$.

A graph or chart for one octave of the chromatic scale is given in Figure 14–6. The horizontal scale is a plot of the frequency ratios for each octave; the numbers shown underneath are frequencies for the middle octave only.

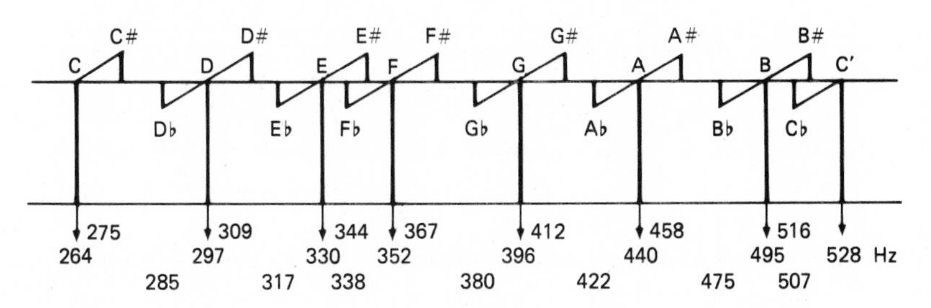

FIGURE 14–6 The middle octave of the just chromatic scale, based upon A = 440 Hz.

Observe, for example, that the frequency of C♯ is 275 Hz, and that for D♭ is 285 Hz. The difference is 10 Hz, an interval easily recognized at this pitch by nearly everyone.

14.7 The equal-tempered scale

The piano is constructed and tuned to a scale that can be looked upon as a happy compromise of the chromatic scale, and it is called the **equal-tempered scale**. See Figure 14–7. This term, usually shortened to **tempered scale**, comes from the way in which each octave is divided into twelve equal-inter-

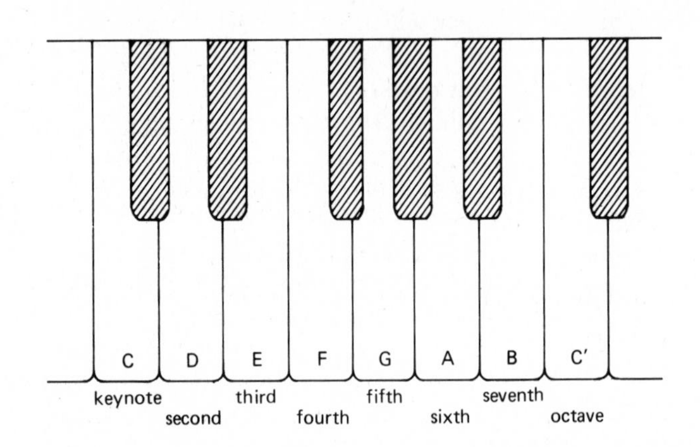

FIGURE 14–7
One octave of the piano keyboard showing the keynote C and the succeeding notes.

val ratios, as illustrated in Figure 14–8. The actual frequencies are standardized on concert A ($A_4 = 440$ Hz). Observe that all spacing between tones (using black as well as white piano keys) are equal. The white keys have frequency ratios that divide the octave into five whole tones and two halftones, or a total of twelve halftones. Observe that black keys are between two white keys, and that each has a dual role of one sharp and one flat.

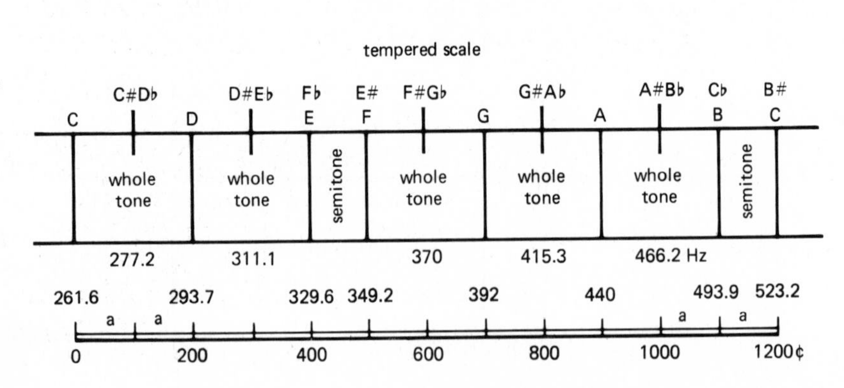

FIGURE 14–8
The tempered scale with the frequencies of the black and white keys, based upon A = 440 Hz. The scale of frequency ratios, in cents, is shown below.

By plotting these twelve steps as equal, as is done in Figure 14–8, the interval C–C♯ is the same as the interval C♯–D, and so on. Let us call this interval *a*. If we start with C as unity, the relative frequency of C♯ is *a*; D is *a* times this, or a^2, and D♯ is *a* times this, or a^3, and so forth. When C′ is reached, we have for the entire octave

$$a^{12} = 2 \qquad [14f]$$

where *a* is the twelfth root of 2, or

$$a = 2^{1/12} = 1.059463^*$$

This is the interval of the **tempered semitone**; the frequency of any tone, multiplied by this number, gives the frequency of a note one semitone higher. In the same way, any frequency divided by this number gives the frequency of a note one semitone lower. For example, $A_4 = 440$ Hz, and multiplied by 1.059463, this gives 466.2 Hz as the frequency of A_4♯ or B_4♭.

All frequencies for the white keys of the piano, in all seven octaves of the tempered scale, are given in Table 14–6.

TABLE 14–6: Frequencies (in Hz) for Seven Octaves of the Tempered Scale

FIRST	SECOND	THIRD	FOURTH	FIFTH	SIXTH	SEVENTH
C_1 32.7	C_2 65.4	C_3 130.8	C_4 261.6	C_5 523.3	C_6 1046.5	C_7 2093.0
D_1 36.7	D_2 73.4	D_3 146.8	D_4 293.7	D_5 587.3	D_6 1174.7	D_7 2349.3
E_1 41.2	E_2 82.4	E_3 164.8	E_4 329.6	E_5 659.3	E_6 1318.5	E_7 2637.0
F_1 43.7	F_2 87.3	F_3 174.6	F_4 349.2	F_5 698.5	F_6 1396.9	F_7 2793.8
G_1 49.0	G_2 98.0	G_3 196.0	G_4 392.0	G_5 784.0	G_6 1568.0	G_7 3136.0
A_1 55.0	A_2 110	A_3 220.0	A_4 440	A_5 880.0	A_6 1760.0	A_7 3520.0
B_1 61.7	B_2 123.5	B_3 246.9	B_4 493.9	B_5 987.8	B_6 1975.5	B_7 3951.1
C_2 65.4	C_3 130.8	C_4 261.6	C_5 523.3	C_6 1046.5	C_7 2093.0	C_8 4186.0

There are two major reasons for constructing and tuning the piano to the tempered scale. First, sharps and flats can be combined in a single action. Second, the pianist can play equally well in any key. In so doing, any given composition can be played within the range of a person's singing voice. In other words, any note on the keyboard can be taken as the keynote of the musical scale.

Although the notes of the tempered scale are not as harmonious as they are on the just or Pythagorean scales, they are very close, particularly to the Pythagorean scale. As a result of these benefits, orchestral instruments generally have adapted to the tempered scale.

One notable limitation of the tempered scale is the large departure of the

*Using a pocket calculator, enter 2, press the [x^y] key, and read 2. Now enter 0.08333333 (which is equal to 1/12), press the [=] key, and read 1.059463. (This sequence may not work with every calculator.)

major and minor thirds and sixths from the just scale. Consequently, the tempered thirds and sixths appear to be "off," a limitation that the practicing musician is very much aware of.

14.8 Frequency ratio

To distinguish between two tones with large or small frequency ratios, a unit called the **cent** has been introduced. The cent, as one might suspect, divides each of the twelve intervals *a* of the tempered scale into 100 equal parts. The cent is abbreviated by the symbol ¢.

$$1¢ = (2)^{1/1200} \qquad [14g]$$

which, upon evaluation, becomes 1.00057779. This small unit permits one to *add* and *subtract* frequency ratios, rather than multiply and divide fractions. A scale of cents is shown at the bottom of Figure 14–8.

Suppose we wish to find the number of cents n in any interval with a frequency ratio R. Equation (14g) then becomes

$$¢^n = 2^{n/1200} = R \qquad [14h]$$

We now take the logarithm of both sides of the equation and obtain

$$\log 2^{n/1200} = \log R$$

from which we can write

$$\frac{n}{1200} \log 2 = \log R$$

or

$$n = \frac{1200}{\log 2} \log R$$

giving

$$n = 3986 \log R \qquad [14i]$$

This result says that the logarithm of any interval ratio R, multiplied by the constant 3986, gives the interval ratio n in cents. For example, to find the number of cents in the perfect fourth, with the frequency ratio $R = \frac{4}{3}$, we obtain,

$$n_1 = 3986 \log \frac{4}{3} = 3986 \times 0.12494 = 498¢$$

To find the number of cents in the perfect fifth, with its ratio of $R = \frac{3}{2}$, we find

$$n_2 = 3986 \log \frac{3}{2} = 3986 \times 0.17609 = 702¢$$

Adding these two intervals, we obtain

$$n_1 + n_2 = 498 + 702 = 1200¢$$

which is exactly one octave, as it must be.

14.9 Melody and memory

A series of notes played or sung in succession produces a **melody**. If the melody is sung in a reverberant environment, such as a concert hall, each note will overlap with the reverberation of the last, allowing the singer to make concurrent comparisons of successive notes. The timing of the overlap is important. If the succession of notes is too rapid, several reverberating notes will be audible, causing confusion. But without the overlap, such as provided by a reverberant environment, the ability to correctly set intervals between successive notes deteriorates with time beyond about half a second. The ability to set intervals is further inhibited by background noise or the production of extraneous notes.[8] This is perhaps one reason why musicians prefer a reverberant environment, particularly one matched to the pace of the music (see also Chapters 27 and 28).

Some people have the ability to name the pitch of any musical note, long after other identified notes have been heard. This ability is called **absolute pitch** or **perfect pitch**. There are two opposing theories that try to explain this mysterious ability. One group claims that almost everyone has some ability to discern pitch and that improvement is a matter of training. The other group claims that absolute pitch has little to do with training and is due to a gift of nature. The origin of this ability is still unknown.

14.10 Perception of pitch

As we have seen, pitch is a subjective or psychophysical sensation. For a pure tone, the pitch is determined almost entirely by the frequency of the sound wave. However, the pitch also depends somewhat on loudness. If a tuning fork of frequency A_2 (110 Hz) is moved close to the ear, the pitch is usually perceived to drop by as much as a quarter tone. Alternatively, the pitch of a tone with a frequency above 1000 Hz is often found to rise slightly as loudness is increased. Moreover, for low frequencies at low intensities, a tone can be varied in frequency by about 5 percent without the listener being aware of it.

For an impure tone, the pitch sensation will depend upon the timbre in a complex way.[9] Suppose, for example, that C_3, C_4, and C_5 are simultaneously sounded. Studies have shown that the pitch is usually discerned as that of the lowest tone, C_3 in this case. If the C_3 were entirely absent, the perceived pitch would be C_4. For the case of a predominant C_4, with a slight presence of C_3, the pitch is indefinite. Such a tone would be recognized as a C but not clearly identified as middle or low C (See Project 1).

Another common situation is a tone involving a fundamental and several of its higher harmonics. For example, take a tone composed of harmonics of 100, 200, 300, 400, and 500 Hz. Such a tone would have a pitch G_2, characteristic of its fundamental frequency of 100 Hz. Curiously, even if only the 200-,

300-, and 500-Hz partials were present, for example, the timbre would change, but the pitch would still be discerned as 100 Hz. We will see why in the next chapter.

14.11 Intonation

Intonation may be defined as the production of tones that agree in pitch or frequency with those specified for a particular musical scale. Manufacturers have the problem of making instruments that conform to the musical scale in which they will be played. This is not too difficult for string instruments, for they are readily tuned by tightening or loosening the strings. With wind instruments, however, the increased speed of sound waves through the air at higher temperatures produces a considerable increase in pitch.[10] This amounts to about 3¢ for each degree Celsius rise in temperature. However, the players of many brass or wind instruments do have some control over the pitch of the notes they produce. An experienced player can "lip" the frequencies of notes up or down over a small range and adjust to the desired pitch. Vocalists too have control over the pitch of the notes they sing, and this is done by varying the tensions in the vocal cords.

Helmholtz claimed that players of string instruments, unaccompanied by a keyboard instrument like a piano, actually played according to the musical scale of just intonation. His claim was based upon observations of the frequencies played by a famous violinist of that time. And not long ago there

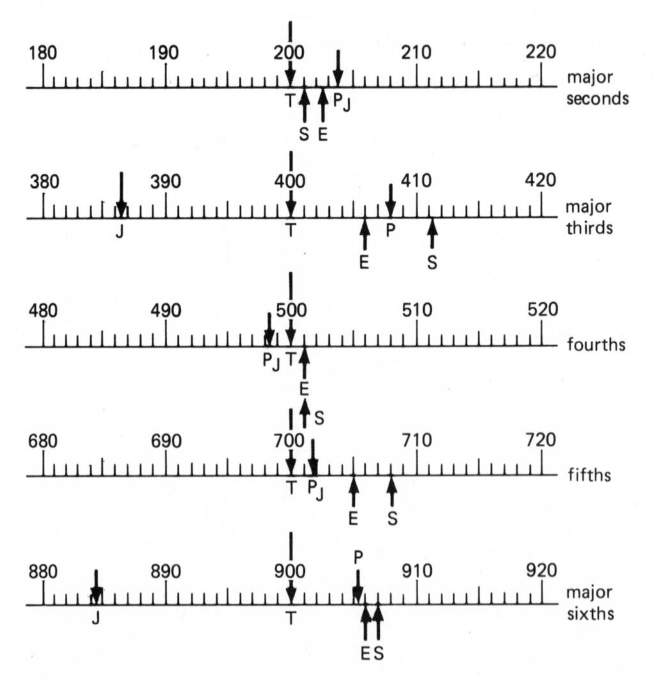

FIGURE 14–9 **Interval separations (in cents) found in intonation experiments using solo (S) and ensemble (E) singers. Also shown are Pythagorean (P), just (J), and tempered (T) intervals.**

arose some speculation concerning the intonation of modern musicians in performance upon a relatively pitch-free instrument such as a violin or cello. Some believe that a century and a half of tempered tuning, as is found on the modern piano keyboard, has schooled musicians to perform in equal temperament. Others feel that harmonic tradition and natural instincts force us to hear intervals in tune with a specific harmonic scale.

Recent experiments to settle this long-standing controversy have been performed by Nickerson.[11] Six string quartets, attached to six different colleges or universities, were used to perform a melodic passage in solo (S) and in harmonic ensemble (E). The music played was from Haydn's *Emperor Quartet,* which was composed in four variations. Their renditions were recorded and analyzed by modern electronic equipment, with the results as shown in Figure 14–9. Let us summarize the results for the Pythagorean (P), just (J), and tempered (T) intervals. When fifths are compared, P and J intervals are slightly preferred over the T interval. The reverse is found to be true for the fourth. For the major third and major sixth, the P intervals are preferred, while for the major seconds there is little difference between all three. Thus there seems to be no simple pattern evident. The perception of consonance is apparently not dictated precisely by number relations. Undoubtedly consonance is determined somewhat by training and experience and may depend upon the piece played. More study is needed on this interesting puzzle.

14.12 Notes

It has been seen that a continuous **tone** can be characterized objectively by giving its frequency, intensity, and waveform. Individual **notes** are characterized also by their beginnings, duration, and ends. These **transient characteristics** are usually referred to as **attack, sustain,** and **decay.** See Figure 14–10. These intensity variations, particularly the attack, are as important as wave-

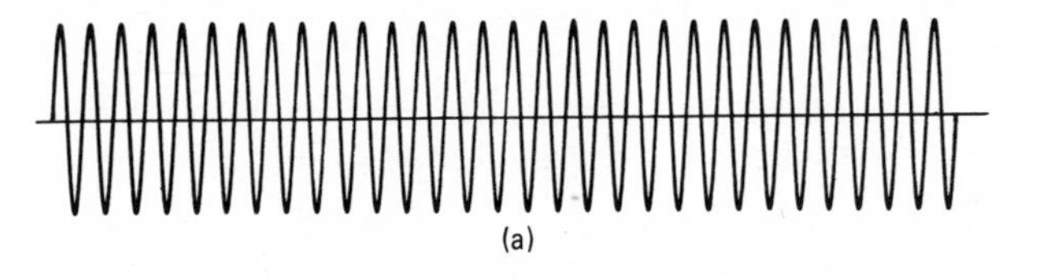

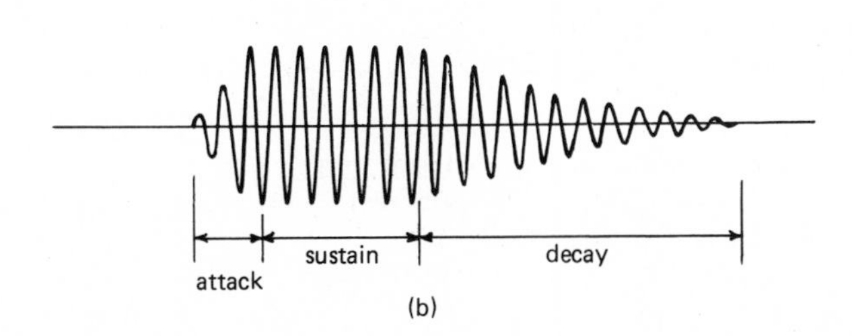

FIGURE 14–10
(a) A pure tone characterized by its frequency and constant amplitude. (b) A note is characterized by its transient characteristics as well.

form in determining the characteristic sound of each musical instrument. A percussive note, such as a plucked string, has a sudden attack, followed by a slower decay. (See Section 16.7.) A horn usually produces a relatively slow attack, followed by a modest sustain. Magnetic tape recordings of various instruments, with attack and decay clipped off, show surprising similarities between quite different instruments. (See Project 2.) Conversely, electronic synthesizers can be used to change a plucked sound to a horn sound simply by lengthening the attack and shortening the decay. (See Section 24.6.)

QUESTIONS

1. Name the four harmonic dyads whose frequency ratio is given by the smallest whole numbers. What are their ratio numbers?

2. Make a chart for one octave of the Pythagorean diatonic scale, showing (a) the scale notes, (b) the interval ratios, and (c) the frequencies in terms of f for the keynote C.

3. Make a chart for one octave of the just diatonic scale, giving (a) the scale notes, (b) the frequency numbers, (c) the interval ratios, and (d) the frequencies in terms of f for the keynote C.

4. What is (a) a chromatic scale, (b) a perfect major chord, (c) a harmonic triad, and (d) a unit called the cent?

5. What is (a) intonation, (b) absolute pitch, (c) a harmonic tetrad, and (d) standard pitch?

PROBLEMS

1. Find the frequency ratio in cents for the following dyads on the just chromatic scale: (a) C♯ – C, (b) D♯ – D, (c) D – C, (d) E – C, (e) G – E, and (f) C′ – B.

2. (a) Multiply each of the numbers in the second row of Equation (14a) by 11, and write down the frequencies along with their scale notes. (b) Multiply each of the numbers in the second row of Equation (14a) by 22, and write them down directly under the corresponding frequencies in part (a). (c) Calculate the difference between corresponding pairs of notes on these two scales, in cents.

3. (a) Find the frequency ratio between the two frequencies $f_1 = 256$ Hz and $f_2 = 320$ Hz. (b) Add the interval of a fifth to f_2 to obtain f_3, and find the frequency ratio f_3/f_1. (c) Find the frequency of f_3.

4. (a) Find the frequency ratio between the two frequencies $f_1 = 320$ Hz and $f_2 = 576$ Hz. (b) If we go down from f_2 by an interval of a fourth, find the frequency ratio f_3/f_1. (c) Find the frequency of f_3.

PROJECTS

1. By combining tones from two or three adjacent octaves, a note can be produced which is definite in overall pitch but is not clearly related to a particular octave range. A sequence of such notes can appear to rise on the musical scale and yet reach only the starting note after one octave. Continuing upward, the sequence will never actually reach the next octave but will contin-

ually repeat the same scale. This paradox, attributed to Roger N. Shepard,[12] must be heard to be appreciated.

The notes in Table 14–7 can be created with the tone generators of an electronic synthesizer, combined and recorded onto magnetic tape. An oscilloscope will help adjust the relative amplitudes indicated in the table.

TABLE 14–7: **Frequency Mixtures for an Indefinite Note Sequence**

C		C_4 (50%)	C_5 (50%)
D	D_3 (3%)	D_4 (60%)	D_5 (37%)
E	E_3 (6%)	E_4 (65%)	E_5 (29%)
F	F_3 (10%)	F_4 (70%)	F_5 (20%)
G	G_3 (15%)	G_4 (70%)	G_5 (15%)
A	A_3 (25%)	A_4 (65%)	A_5 (10%)
B	B_3 (35%)	B_4 (60%)	B_5 (5%)
C	C_4 (50%)	C_5 (50%)	

2. Record single notes from a variety of instruments, using a tape recorder. Clip out the attack and decay of each note, and determine whether or not others can identify the instruments by listening to only the sustained portion.

3. Can you sing a note as high in frequency as your lowest whistle? Try this aurally first. Check with an oscilloscope. The results will be surprising.[13]

NOTES

[1]J. Jeans, *Science and Music* (London: Cambridge University Press, 1949).

[2]H. A. Frederick, "American Tentative Standard Acoustical Terminology," *Journal of the Acoustical Society of America* 9 (1937):60 (for many definitions of the terms used in sound and music).

[3]H. Helmholtz, *Sensations of Tone* (reprint ed., New York: Dover, 1954).

[4]D. E. Hall, "Quantitative Evaluation of Musical Scale Tunings," *American Journal of Physics* 42 (July 1974):543.

[5]G. Hendricks, "The Case of the Disappearing High C's," *Etude* 67 (1949):47.

[6]F. A. Saunders, "A Scientific Search for the Secret of Stradivarius," *Journal of Franklin Institute* 29 (1940):1.

[7]A. H. Benade, *Fundamentals of Musical Acoustics* (New York: Oxford University Press, 1976).

[8]D. Deutsch, "Music and Memory," *Psychology Today* 6 (1972):87.

[9]F. Wightman and D. Green, "The Perception of Pitch," *American Scientist* 64 (1974):208.

[10]R. W. Young, "Dependence of Tuning of Wind Instruments on Temperature," *Journal of the Acoustical Society of America* 17 (1946):187.

[11]J. F. Nickerson, "Intonation of Solo and Ensemble Performance of the Same Melody," *Journal of the Acoustical Society of America* 21 (1949):593.

[12]R. Shepard, "Circularity in Judgments of Relative Pitch," *Journal of the Acoustical Society of America* 36 (1964):2346.

[13]F. S. Crawford, "Which Octave Do You Whistle Dixie In?" *American Journal of Physics* 41 (1973):1010.

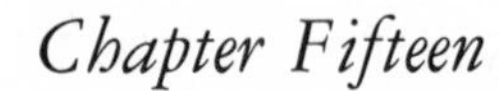

Chapter Fifteen

COMBINATION TONES AND AURAL HARMONICS

In Section 10.3, we studied the effects of simultaneously sounding two pure tones of slightly different frequency, f_1 and f_2. From the resultant sound pulsations, we identified the *beat frequency*

$$f_{BF} = f_2 - f_1 \qquad [15a]$$

and the width of the *critical band* Δf_{CB} of the basilar membrane in the inner ear. In this chapter, we will consider the effects of extending these same experiments to greater frequency differences.

15.1 Combination tones

Suppose we set one sound generator to produce a pure tone of fixed frequency f_1 and a second generator to simultaneously sound a second pure tone f_2, which we vary over the wide range in frequency from f_1 up to the octave $2f_1$.[1] In addition to the primary tones f_1 and f_2, other sound effects reveal themselves as additional pitch sensations. Each of these additional tones have frequencies different from f_1 and f_2 and are easily determined by the pitch-matching methods of Goldstein.[2]

Imagine now that we perform the experiment just described—of slowly raising and then lowering the frequency of f_2. Between unison at f_1 and the octave at $2f_1$, we listen carefully for beats and other frequencies different from f_1 and f_2. As f_2 is sounded and starts upward from f_1, we first hear the beats described in Section 10.3. The beat frequency appears as a fluctuation in the intensity of the average frequency $(f_2 + f_1)/2$. Soon this beating gives way to a roughness of the sound, and before long, when the beat frequency reaches 20 to 30 Hz, we hear a separate tone of rising pitch. This particular combination tone is called the **difference tone**, and its frequency is

$$f_{DT} = f_2 - f_1 \qquad [15b]$$

As f_2 rises, the pitch of this difference tone starts from a very low pitch and *steadily rises* as f_2 approaches the octave $2f_1$.

As f_2 starts at the octave $2f_1$ and *slowly decreases* in pitch, we hear another tone, starting from a very low pitch and *slowly rising*. By listening carefully, just after f_2 has passed the halfway mark of $\frac{3}{2}f_1$, still another tone can be heard. This additional tone starts very low in pitch and rises. These two additional frequencies are also **combination tones** and are given by

$$f_{C1} = 2f_1 - f_2 \qquad f_{C2} = 3f_1 - 2f_2 \qquad [15c]$$

Both of these tones are best heard when the intensities of f_1 and f_2 are relatively low in value and f_2 lies between about $1.1f_1$ and $1.3f_1$. At high intensities of f_1 and f_2, these same tones can be heard as low-pitched tones near the octave $2f_1$ and the fifth $\frac{3}{2}f_1$.

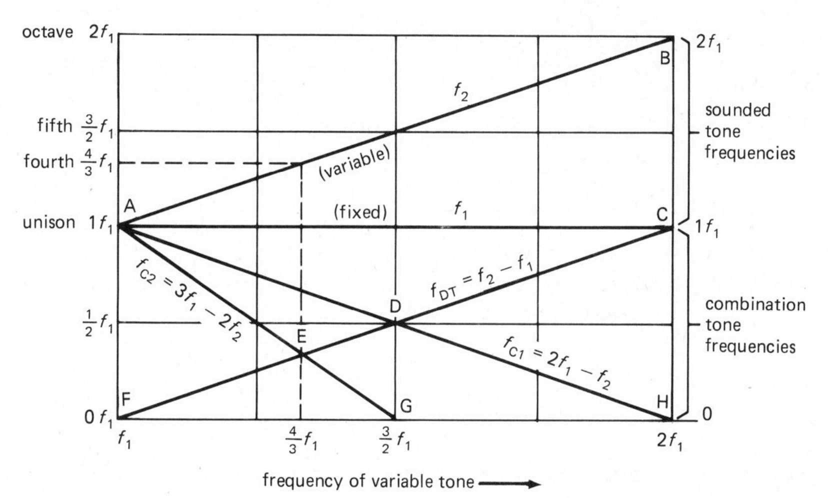

FIGURE 15-1 Graph of three combination tones that can be heard by the normal human ear as the result of sounding two pure tones with frequencies f_1 (fixed) and f_2 (variable from f_1 to $2f_1$).

Note from the graph in Figure 15-1 that as f_2 rises from f_1 at A to $2f_1$ at B, the difference tone f_{DT} rises from zero at F to f_1 at C, and the combination tones f_{C1} and f_{C2} fall in frequency from f_1 at A to zero at H, and f_1 at A to zero at G, respectively.

Example

Let two pure tones F_4 and C_5 from the just diatonic scale be sounded simultaneously. Find the difference tone f_{DT} and the combination tones f_{C1} and f_{C2}. To what notes on the just scale do these tones belong?

Solution

From Table 14–5, we find $F_4 = 352$ Hz and $C_5 = 528$ Hz. Calling these f_1 and f_2 and substituting directly into Equations (15b) and (15c), we obtain

$$f_{DT} = f_2 - f_1 = 528 - 352 = 176 \text{ Hz}$$
$$f_{C1} = 2f_1 - f_2 = 704 - 528 = 176 \text{ Hz}$$
$$f_{C2} = 3f_1 - 2f_2 = 1056 - 1056 = 0 \text{ Hz}$$

In this example, both the difference tone f_{DT} and the combination tone f_{C1} produce the same frequency of 176 Hz, which means that they reinforce each other. This tone corresponds to the *missing fundamental* at F_3 on the just scale.

15.2 Nonlinear systems

The combination tones described in the preceding section are not present in the sounds that enter the ear. It is believed that they are created as the result of some part (or parts) of the neural system that responds nonlinearly to the periodic sound pressure stimulus. Such a system, wherever it is seated, is called a **nonlinear system.**

Nonlinear systems are found in our everyday life in certain mechanical devices and electronic circuits. For example, consider the mechanical system in Figure 15–2. In the diagram, A is a uniform strip of spring steel, clamped at the lower end B on one side of a block of wood, as shown. If we now apply a force F to the right and measure the bending displacement, we find it obeys **Hooke's law,** which states that the displacement is proportional to, and in the same direction as, the force F. (See Section 2.4.) This means that if we double the applied force, we obtain double the displacement, and if we triple the

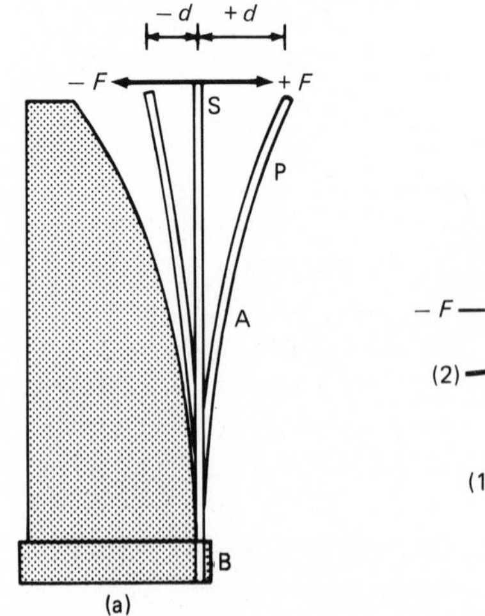

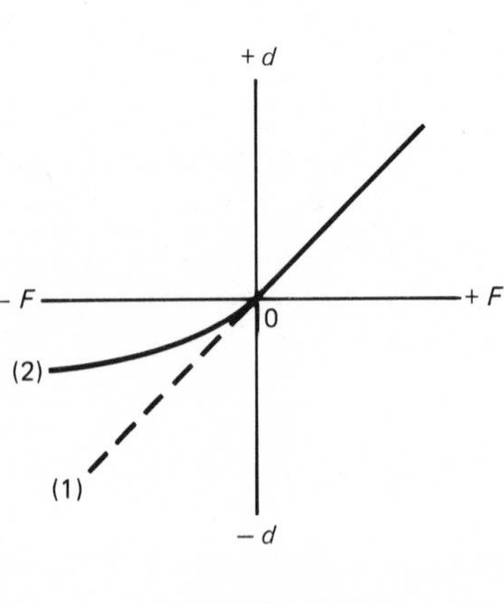

FIGURE 15–2 **Diagram of a metal strip vibrator whose displacement *d* is nonlinear with respect to the applied force *F*.** (a) **Vibrating strip with one-sided backing and** (b) **graph of the nonlinear response.**

force we triple the displacement, and so forth. Graphically, this produces a straight line, as shown at the upper right in diagram (b), and we say the displacement obeys Hooke's law.

If we now apply a force F to the left, we find the displacement is not proportional to the applied force F but decreases more slowly in the opposite direction. As a graph, we obtain a curve, as shown at the lower left of diagram (b). Had the curved block of wood not been there, we would have obtained a continuation of the straight line, shown as dashed. The straight line (1) represents a linear response to a variable force F, and the curved line (2) represents a nonlinear response to the same force F.

Suppose we now impose the pressure of a sinusoidal sound wave on the end of a leaf spring. The force exerted will push the spring back and forth. If the curved wood block were not there, the spring would vibrate linearly, as far to one side as the other, and the vibration would be SHM, as shown in Figure 15 – 3(a). If the curved wood block is in place, the spring will vibrate nonlinearly, and a graph of the motion would be asymmetric, as shown in diagram (b). As a result, a sinusoidal input gives a nonsinusoidal output.

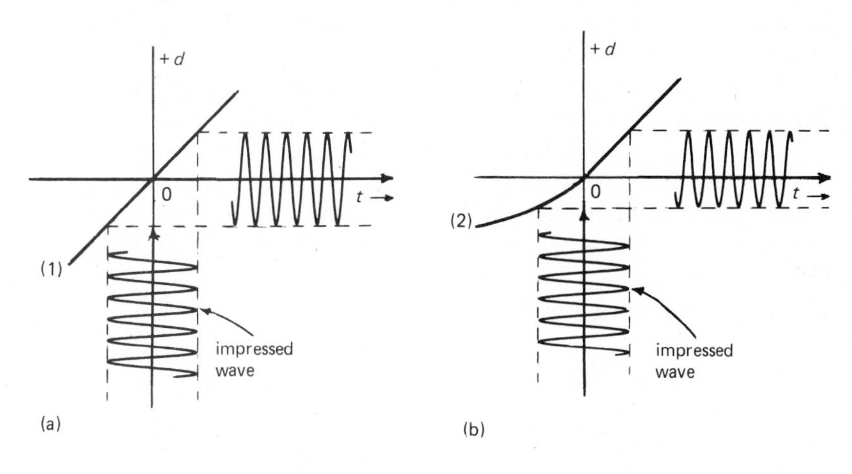

FIGURE 15 – 3
Graphs showing the difference between linear and nonlinear responses to a impressed wave from a pure tone: (a) normal linear vibration response, (b) distorted, nonlinear vibration response.

If one applies Fourier's theorem to this output, or to any periodic but nonsinusoidal function, the results can be expressed as a series of harmonics f, $2f$, $3f$, . . ., with the curve repetition rate as the fundamental. (See Section 8.8.) Experimentally, the ear responds linearly to the sounding of all pure tones at low intensities (up to 30 dB) and one should not hear harmonics above the fundamental. Above 50 dB, however, the ear responds nonlinearly, and one actually hears the fundamental frequency $1f$, as well as one or more of the lowest harmonics $2f$, $3f$, $4f$, and so on. From this brief discussion, we assume that somewhere within the perception chain, a nonlinear system comes into play at high intensities, which generates harmonics from any pure tone impressed upon it. This is the subject of the next section.

15.3 Aural harmonics

For many years it was believed that the nonlinear response of the ear was mechanical in origin and due largely to the middle ear. Recent investigations have shown, however, that the mechanical parts of the ear, up to and beyond the oval window, are linear in their response to all intensities.[3] It would appear, therefore, that any nonlinearity lies within the neural system of the cochlea itself, and recent research[4] points to electrochemical processes that are related in a complicated way to the frequency distribution of vibrational energy in or around the hairlike cilia or neurons.

Whatever the origin of nonlinearity may be, the ear remains linear at low intensity levels and begins to be overloaded—that is, nonlinear—at the following intensity levels and frequencies:

30 dB at 350 Hz
50 dB at 1000 Hz
55 dB at 5000 Hz

Aural harmonics begin to be heard at these intensities and increase in number and intensity as the intensity of the primary tone increases. A graph of the aural harmonics of a loud fundamental of 95 dB at 250 Hz is shown in Figure 15–4.

Experimental justification for aural harmonics is given in the diagrams of Figures 10–5, and 10–6, and 10–7. There the masking effect of one tone by another is presented, and we observe that sharp dips in the masking curves occur at frequencies corresponding to the aural harmonics of the masking tone. If aural harmonics did not exist, these curves would be smooth, like

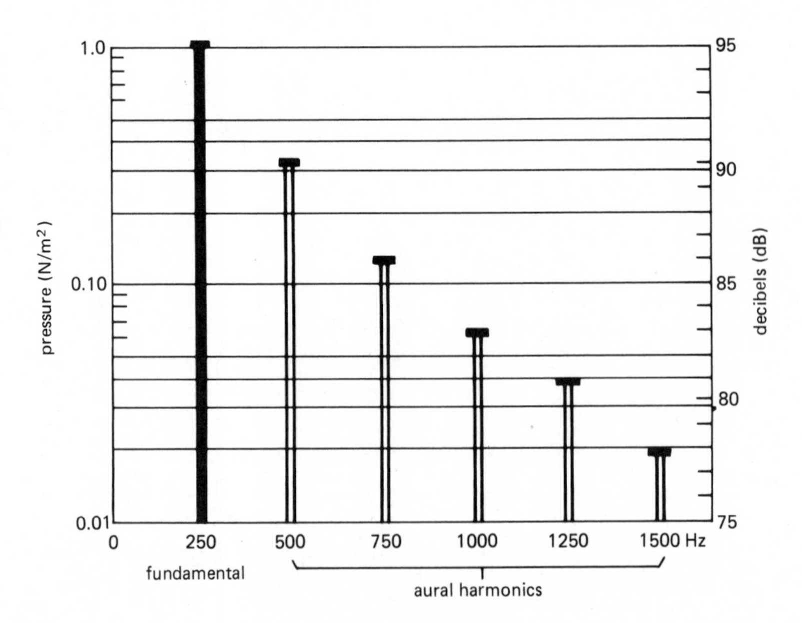

FIGURE 15–4 Aural harmonics heard by the human ear as the result of a pure tone of frequency 250 Hz, sounded at the high level of 95 dB.

Figure 10–5. There are also small dips in these curves, which would be shown in a more detailed diagram. As the masked tone f_2 increases in frequency and approaches the masking tone f_1, or any of its harmonics $2f_1, 3f_1, 4f_1, \ldots$, the curves tend to rise and then drop suddenly and reach a minimum at the exact frequencies.[5] Beats between f_2 and the aural harmonics of f_1 make the masked tone detectable at lower intensity levels. These results offer most conclusive evidence of aural harmonics.

15.4 Phase changes can be heard

Since the time of Helmholtz it has been believed that changes in phase between the component frequencies of any complex musical sound have no detectable effect upon its timbre. It has been assumed that changes in the intensity of any one or more partials change the timbre, but that changes in their relative phases without changes in amplitudes could not be detected. Recently, however, a number of investigators have found detectable differences in tone quality. In 1934, for example, Chapin and Firestone[6] found that phase changes may give rise to perceptible changes in both timbre and loudness.

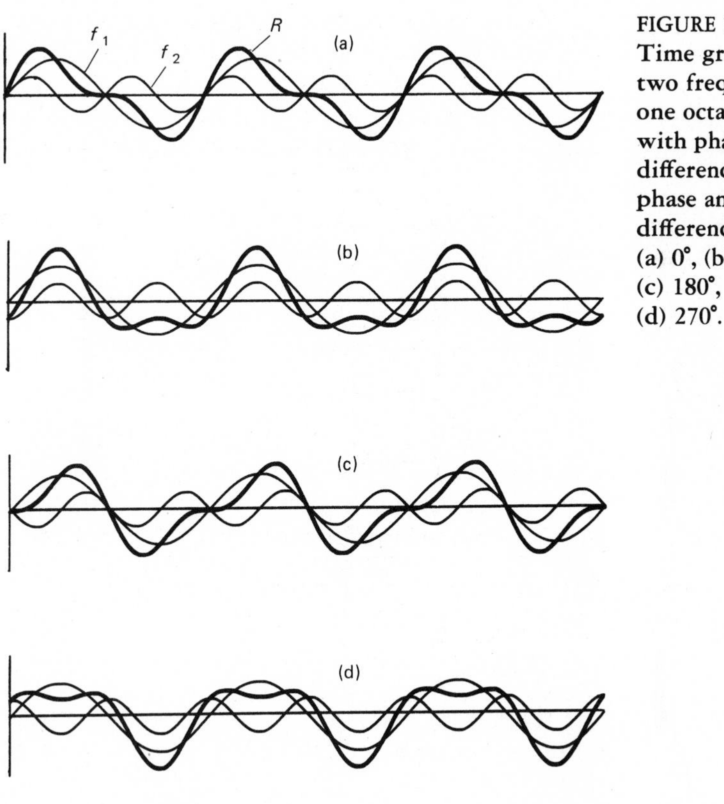

FIGURE 15–5 **Time graphs for two frequencies one octave apart with phase angle differences. Initial phase angle difference $\Delta\phi_0$ (a) 0°, (b) 90°, (c) 180°, and (d) 270°.**

Suppose we again perform the experiment described in Section 15.1: sounding two pure tones f_1 and f_2 at constant but low intensities and then varying the frequency of f_2. As we pass through unison, $f_2 = f_1$, we hear strong **first-order beats**, the frequency of which is given by Equation (15a). As f_2 increases and we slowly pass through the octave $f_2 = 2f_1$, we again hear beats on both sides of $2f_1$. As an equation, we can write

$$f_{BF2} = f_2 - 2f_1 \qquad [15d]$$

where f_{BF2} represents the beat frequency, in hertz. Some people call these **second-order beats** and describe them as pulsations in tone quality, while others call them **subjective beats**.

Suppose we now observe what happens to a time graph of the two tones when their frequencies are exactly one octave apart, $f_2 = 2f_1$, and we change their phase angles. Four graphs are shown in Figure 15–5, where the differences in phase angle $\Delta\phi$ are 0°, 90°, 180°, and 270°, respectively. When we observe these on an oscilloscope screen, and the two frequencies are exactly in tune, whatever pattern is being observed will remain stationary. If we increase f_2 by 1 Hz, so that $f_{BF2} = \pm 1$, the sound trace seen on the screen will change slowly by going through the complete cycle of patterns once each second. These subjective beats clearly indicate that our ears are capable of detecting changes in vibration pattern forms. The effect is not very pronounced, however, and if a small interval of time elapses between waveforms like (a), (b), (c), and (d), few people would detect any difference.

15.5 Periodicity tones: the missing fundamental

For many years it was believed that consonance in pure tone combinations was to be attributed to the added frequencies of difference tones described in Section 15.1. Recent experiments have shown, however, that at normal intensity levels, difference tones and combination tones are very weak and do not contribute to sound quality.[7] To hear difference tones requires a nonlinear response of the ear, and we have seen in Section 15.3 that the threshold of nonlinearity is from 30 to 55 dB.

Consider two pure tones forming a perfect fifth with the frequency ratio $\frac{3}{2}$. A time graph of this combination, Figure 15–6(a), shows two vibrations f_1 and f_2 of equal amplitude and initial phase angles of 90° along with their resultant vibration curve R. Since time is plotted horizontally, the periods are shown by the horizontal intervals T_1 for f_1 and T_2 for f_2. The pattern **repetition rate** is given by T_0, and it corresponds to a frequency of f_0, where

$$f_0 = \frac{1}{2} f_1 \qquad [15e]$$

We have seen in Equation (2e) that the frequency is equal to the reciprocal of the period, $f = 1/T$. This repetition rate is called the fundamental frequency, or the **missing fundamental.** For some reason or another, our ears are sensitive to this tone. We know this is so because of the results of experimentation. If a group of people are asked to identify the *subjective pitch sensation* of a short melody containing a number of fifths as two-tone stimuli, most of them will select a frequency that lies one octave below f_1. When they are told to listen for the missing fundamental one octave below f_1, they all agree it is definitely there.

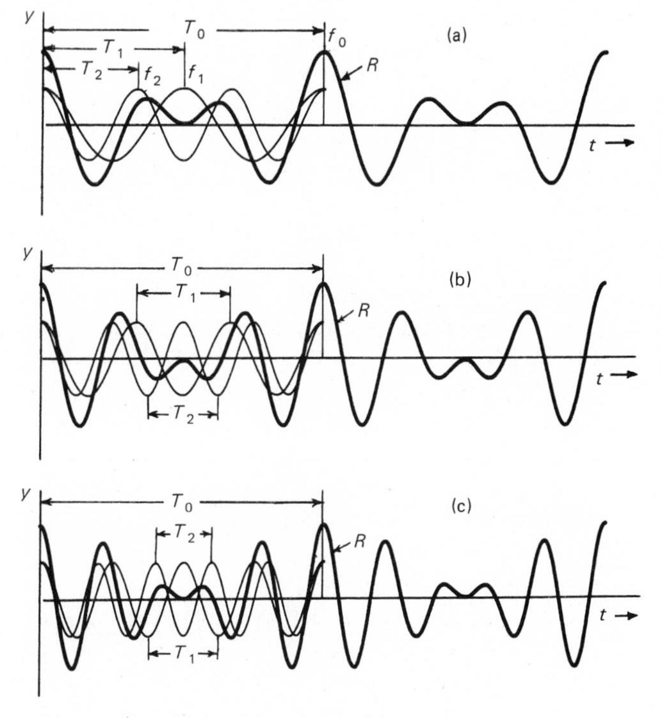

FIGURE 15-6
Time graphs for two-tone stimuli of (a) a perfect fifth, (b) a perfect fourth, and (c) a major third, showing the same repetition rate T_0 in terms of the period T_1 and T_2 for f_1 and f_2, respectively.

The pitch sensation of f_0 is called the **periodicity pitch** or **subjective pitch.** Experimentally, the frequency is not present in the endolymph of the cochlea at normal sound levels, while combination tones definitely are. It has been shown by Small that even though the region of the basilar membrane may be filled with frequencies in the region of f_0, and any number of imposed frequencies masked, the note f_0 will still be heard.[8] If a perfect fifth is sounded, as well as another pure tone f_3 of slightly different frequency than f_0, no beat sensation can be detected. It is even more remarkable that if f_1 only is fed into one ear, and f_2 only into the other, the missing fundamental is heard.[9] This would tend to show that subjective pitch is developed in our neural processing system, and that the brain recognizes the repetition rate in sound patterns.

If we include other consonant dyads with the perfect fifth, such as a perfect fourth, $f_2/f_1 = \frac{4}{3}$, a major third, $f_2/f_1 = \frac{5}{4}$, and a minor third, $f_2/f_1 = \frac{6}{5}$, we find

$$\begin{array}{cccc} 3:2 & 4:3 & 5:4 & 6:5 \\ f_0 = \frac{1}{2}f_1 & f_0 = \frac{1}{3}f_1 & f_0 = \frac{1}{4}f_1 & f_0 = \frac{1}{5}f_1 \\ \text{perfect fifth} & \text{perfect fourth} & \text{major third} & \text{minor third} \end{array} \qquad [15f]$$

Observe that these repetition rates correspond to frequencies and musical intervals identical to corresponding difference tones. Little wonder that for many years it was believed that subjective pitches were due to difference

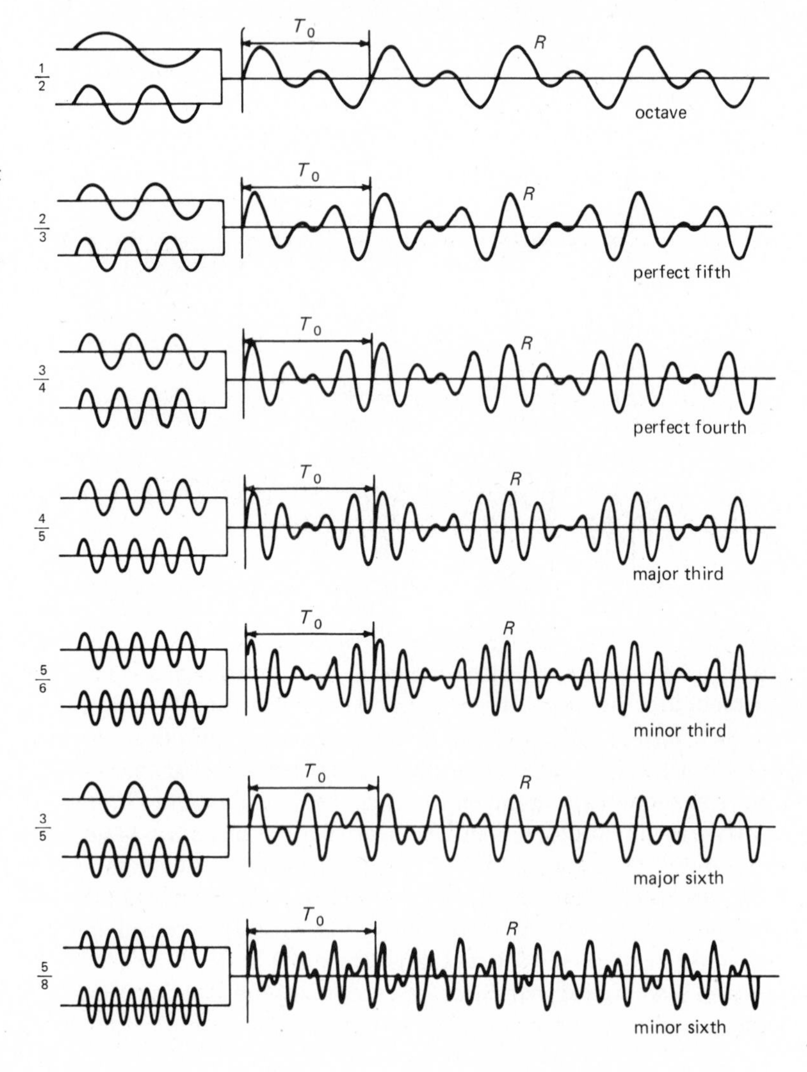

FIGURE 15-7 Time graphs for consonant dyads (at the left) and their combined vibration modes (at the right). Observe that the repetition rate is the same for all of these dyads.

tones. Observe also in Figure 15–7 that in all seven dyads the repetition rate appears with the same time T_0 and corresponds to the same missing fundamental f_0.

The experimental fact that our ears are capable of identifying subjective pitch frequencies and intensities in the musical range of 30 to 70 dB, and about 50 to 15,000 Hz, is quite amazing. A convincing example of subjective pitch sensations (the missing fundamental) is given by the fact that one can hear the correct pitch of base tones from an inexpensive transister radio in spite of the fact that all frequencies below 100 to 150 Hz are cut off by poor electronic circuits and small speakers.

15.6 Consonant and dissonant dyads

Musically oriented mathematicians have been intrigued particularly by the just diatonic scale. This interest arises from the relation between diatonic tone intervals and their frequency ratios. Table 15–1 lists some of particular interest.

TABLE 15–1: **Consonant and Dissonant Intervals within One Octave**

	Interval	Ratio			
PERFECT CONSONANCES	Octave	2:1	C_5C_4	D_7D_6	A_6A_5
	Fifth	3:2	G_4C_4	B_5E_5	D_4G_3
	Fourth	4:3	F_4C_4	A_3E_3	C_4G_3
IMPERFECT CONSONANCES	Major third	5:4	E_4C_4	A_5F_5	B_4G_4
	Minor third	6:5	G_4E_4	C_5A_4	G_6E_6
	Major sixth	5:3	B_5D_5	A_4C_4	E_4G_3
	Minor sixth	8:5	C_5E_4	F_4A_3	C_4E_3
DISSONANT INTERVALS	Second	9:8	D_4C_4	D_3C_3	D_5C_5
	Major seventh	15:8	B_4C_4	B_5C_5	B_3C_3
	Minor seventh	16:9	C_4D_3	C_3D_2	C_5D_4

Sounding the pairs shown in Table 15–1 with tone generators indicates that **consonance** is associated with frequency ratios given by small whole numbers. The smaller the integers expressing the ratio of any two notes, the more harmonious, or pleasing, is the sound to the ear. The larger the integers, the more **discordant**, or **dissonant**, is the combination.

Time graphs of the separate notes, and their combinations for the seven consonant intervals, are given in Figure 15–7. Observe the progressive complexity of the combined vibration forms as the ratio numbers become larger. All three of the perfect consonance ratios are common to the Pythagorean and just diatonic scales.

The subjective nature of consonance and dissonance is still a lively subject of debate. Musical training certainly seems to account for part of what listeners consider to be consonant or dissonant. Over the last three centuries, listeners have become more accustomed to imperfect consonance and dissonance in music.

However, studies with untrained listeners[10] have shown general agreement that dyads between unison and minor thirds are decidedly unpleasant. The cause is attributed to the roughness that occurs when two tones, or combinations of their harmonics, fall within the critical band on the basilar membrane (see Section 10.3).

Unpleasant or not, dissonance is still of musical value if its purpose is to produce an unsettling or jarring experience.

15.7 Chords

The simultaneous sounding of two or more notes, each of which forms a consonant interval with the others, constitutes a **chord**. An added restriction for a chord is that the highest and the lowest notes are not to be more than one octave apart. We have seen previously that two notes sounded together form a *dyad.* The perfect consonances, *octave, fifth,* and *fourth* are *harmonic dyads.*

Three notes sounded together constitute a *triad,* and four notes a *tetrad.* Musicians generally agree that there are *six harmonic triads,* and these are listed with examples in Table 15–2. The first chord tabulated is called a **perfect major chord,** the third one a **perfect minor chord.** Time graphs for each of the separate pure tones, in each of the first two triads, are given in Figure 15–8. Graphs for each of these particular combinations are given by curves (a) and (b) at the right. It is common practice to add the next higher octave to each triad, to form the **harmonic tetrads** CEGC′ and CFAC′, respectively. Graphs for these added pure tones, and their combinations with the triads, are shown by curves (a) and (b), respectively.

TABLE 15–2:
Harmonic Triads

INTERVALS	FREQUENCY RATIOS	EXAMPLE
Major third followed by a minor third	4:5:6	C E G
Fourth followed by a major third	3:4:5	C F A
Minor third followed by a major third	5:6, 4:5	E G B
Minor third followed by a fourth	5:6, 3:4	E G C′
Major third followed by a fourth	4:5, 3:4	C E A
Fourth followed by a minor third	3:4, 5:6	E A C′

15.8 Periodicity tones in dyads and chords

We have seen in Section 15.5 that two-tone combinations produce pitch sensations that harmonize with the notes of the primary sound waves. When a

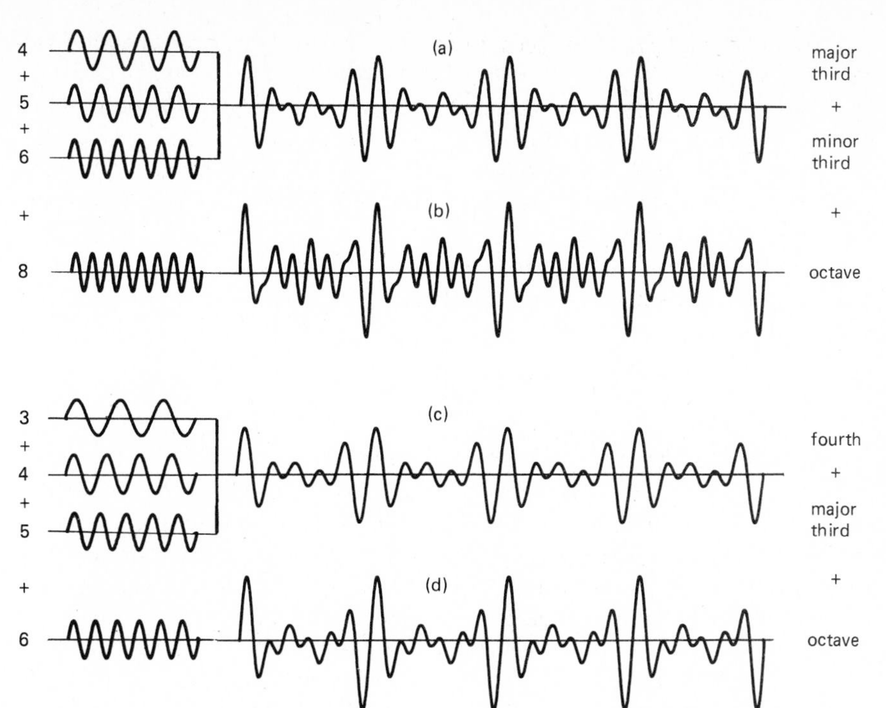

FIGURE 15-8
Time graphs for major triads (a) CEG, and (c) CFA, with octaves added to form major tetrads (b) CEGC′, and (d) CFAC′. The individual tone components for each chord are shown at the left.

perfect fifth is sounded, one hears three frequencies: the original pure tones f_1 and f_2 and the *missing fundamental,* which we call the *periodicity tone* f_0. These are the first three harmonics of the fundamental f_0.

Example

If the dyad C_4G_4 on the just diatonic scale is sounded, what is the frequency of the periodicity tone, and what are the frequency ratios of the three notes?

Solution

From Table 14-5 we find $C_4 = 264$ Hz and $G_4 = 396$ Hz. Using Equation (15e), we obtain

$$f_0 = \frac{1}{2} f_1 = \frac{264}{2} = 132 \text{ Hz}$$

which, on the just scale, is C_3, one octave below C_4. The three notes to be heard are

periodicity tone f_0	$C_3 = 132$ Hz	1st harmonic
primary tone f_1	$C_4 = 264$ Hz	2d harmonic
primary tone f_2	$G_4 = 396$ Hz	3d harmonic

Suppose we now sound a perfect major chord, using the three notes from the middle octave of the just diatonic scale. The frequencies are $C_4 = 264$ Hz, $E_4 = 330$ Hz, and $G_4 = 396$ Hz. The periodicity tones between each of the three pairs of pure tones sounded are

$$\text{for the major third } C_4E_4\text{: } f_0 = \frac{1}{4} f_1 = \frac{264}{4} = 66 \text{ Hz}$$

$$\text{for the minor third } E_4G_4\text{: } f_0 = \frac{1}{5} f_1 = \frac{330}{5} = 66 \text{ Hz}$$

$$\text{for the fifth } C_4G_4\text{: } f_0 = \frac{1}{2} f_1 = \frac{264}{2} = 132 \text{ Hz}$$

If we now add the octave $C_5 = 528$ Hz to these to form the tetrad

$C_4\ E_4\ G_4\ C_5$

we hear six frequencies in all. They are

periodicity tone	$C_2 = 66$ Hz	1st harmonic
periodicity tone	$C_3 = 132$ Hz	2d harmonic
primary tone	$C_4 = 264$ Hz	4th harmonic
primary tone	$E_4 = 330$ Hz	5th harmonic
primary tone	$G_4 = 396$ Hz	6th harmonic
primary tone	$C_5 = 528$ Hz	8th harmonic

Note that the third and seventh harmonics are missing. If difference tones could be heard, the third harmonic arising from the interval C_5E_4 would give 198 Hz. As a periodicity tone it is absent. The seventh harmonic, which would have a frequency of 446 Hz, is not heard. If it were present for any reason, it would not harmonize with the others.

Similar sets of harmonics arise with the perfect minor chord EGB and its tetrad EGBE′.

If you, the student, have gained the impression that there is still much to be learned about how we hear when pure tones and their combinations are sounded, you are quite right. There is still much to be learned, and perhaps you may help all of us by making new discoveries.

QUESTIONS

1. Define or briefly explain in your own words each of the following: (a) difference tone, (b) combination tone, (c) aural harmonics, (d) missing fundamental, and (e) periodicity tone.

2. Explain what must be done to hear combination tones. Draw a graph for all three and label them f_{C1}, f_{C2}, and f_{C3}.

3. How would you go about hearing a periodicity tone? Could you establish a periodicity tone's existence by simultaneously sounding another note of nearly the same frequency and listening for a beat frequency?

4. Can you devise a nonlinear mechanical system different from the one shown in Figure 15–2? Make a diagram of your best idea, and draw a response curve corresponding in principle to the one shown in Figure 15–2 (b).

1. Two pure tones C_5 and G_5, with frequencies from the Pythagorean diatonic scale, are sounded simultaneously. Find (a) the frequencies of the three combination tones and (b) the notes on the Pythagorean scale to which these tones belong.

2. Two pure tones C_5 and E_5 from the just diatonic scale are sounded simultaneously. Find (a) the frequencies of the three combination tones f_{C1}, f_{C2}, and f_{C3}, (b) the notes to which they correspond, and (c) the harmonics to which all the audible notes belong.

3. The following dyads, using frequencies from the just diatonic scale, are sounded: C_5E_5, G_5C_6, G_6E_6, and A_3E_3. Find (a) the frequency of the missing fundamental for each pair and (b) the notes on the just diatonic scale to which these notes belong.

4. The harmonic triad $C_4F_4A_4$ is sounded, using the frequencies of the just diatonic scale. Find (a) the frequencies of the periodicity tones produced by each pair of notes, (b) the notes on the just scale to which these frequencies belong, and (c) the harmonic numbers for all the notes heard by the ear.

5. Solve Problem 4 for the harmonic tetrad $C_4F_4A_4C_5$ if the frequencies are those of the just diatonic scale.

6. The harmonic tetrad CEGA′ is sounded, using frequencies of the middle octave of the just diatonic scale. Find (a) the frequencies of the periodicity tones produced by each pair of notes and (b) the notes on the just scale to which these frequencies belong.

PROJECT

Difference tones between musical instruments are most audible if instruments produce fairly pure tones, as does the flute. Try to compose and play some music for two flutes for which the difference tone plays a melody.[11] Playing in the upper registers often works best.

NOTES

[1]J. G. Roederer, *Introduction to the Physics and Psychophysics of Music,* 2d ed. (New York: Springer-Verlag, 1975).

[2]J. L. Goldstein, "Aural Combination Tones," in *Frequency Analysis and Periodicity Detection in Hearing,* ed. R. Plomp and G. F. Smoorenbur (Leiden: A. W. Suithoff, 1970).

[3]E. G. Weaver and M. Lawrence, *Physiological Acoustics* (Princeton: Princeton University Press, 1954).

[4]G. von Bekesy, *Experiments in Hearing* (New York: McGraw-Hill, 1960), p. 355.

[5]H. Fletcher, *Speech and Hearing in Communication* (Huntington, New York: Kreiger Publishing, 1972), p. 155.

[6]E. K. Chapin and F. A. Firestone, "Influence of Phase on Tone Quality and Loudness," *Journal of the Acoustical Society of America* 5 (1934): 173.

[7]R. Plomp, "Detectability Threshold for Combination Tones," *Journal of the Acoustical Society of America* 37 (1965): 110.

[8]A. M. Small, "Periodicity Pitch," in *Foundations of Modern Auditory Theory,* ed. J. V. Tobias (New York: Academic Press, 1970).

[9]A. J. M. Houtsma and J. L. Goldstein, "Perception of Musical Intervals:Evidence for Central Origin of the Pitch of Complex Tones," *Journal of the Acoustical Society of America* 51 (1972): 520.

[10]R. Plomp and W. J. M. Levelt, "Tonal Consonance and Critical Bandwidth," *Journal of the Acoustical Society of America* 38 (1965): 548.

[11]S. E. Stickney and T. J. Englart, "The Ghost Flute," *Physics Teacher* 5 (December 1967): 518.

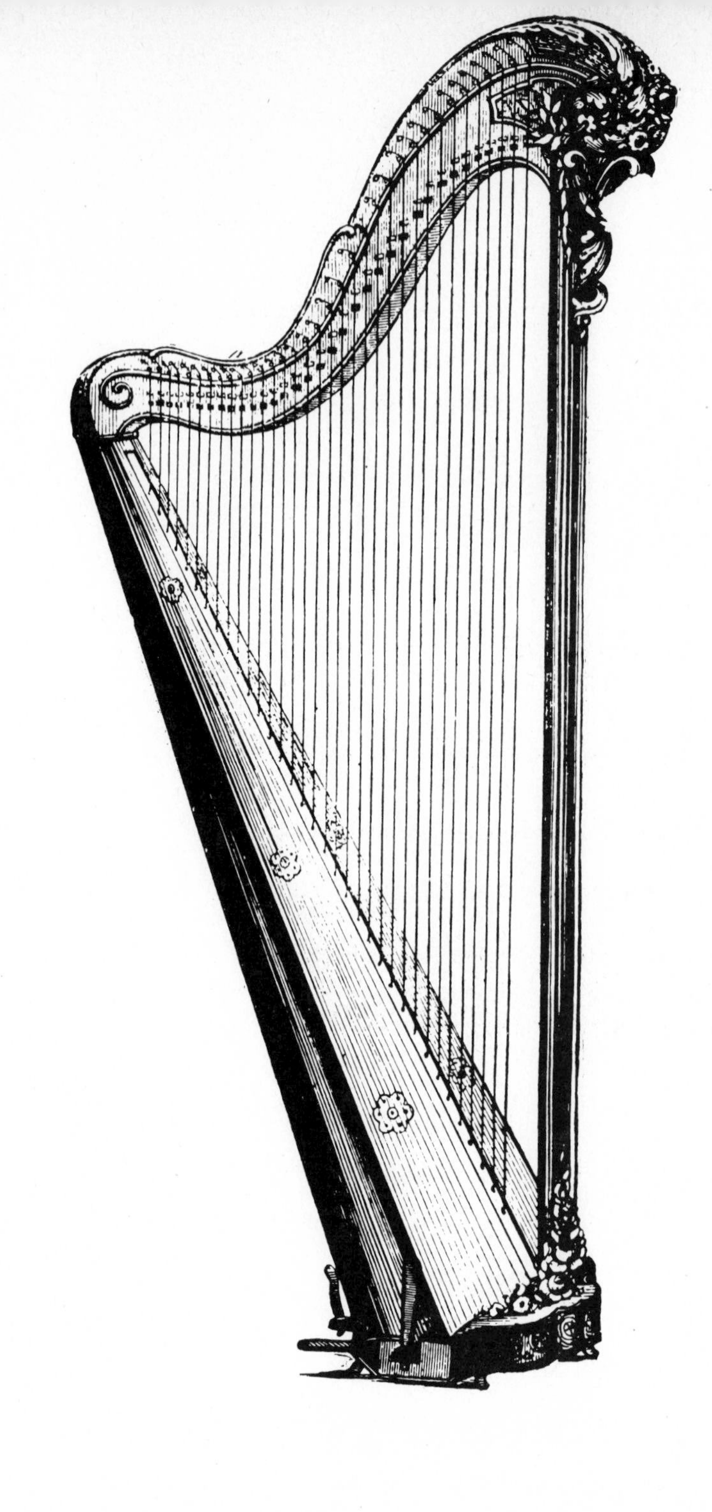

PART THREE

Musical Instruments

Chapter Sixteen

STRINGED INSTRUMENT DESIGN

The origin of stringed instruments is lost in antiquity. Most ancient cultures seem to have used simple stringed instruments made of a few strings stretched between two supporting mounts, with later modifications to include some sort of resonant cavity, such as in a gourd. Over the ages the forms of stringed instruments were diversified and acoustically improved. In this chapter we will examine the general principles of stringed instrument design.

16.1 Instrument design

The creation of the fine violins by master craftsmen during the eighteenth century was not due to advances in science but to a history of improvement by trial and error. Secrets of good design were passed on from father to son and craftsman to craftsman (see Figure 16–1). Only during the past few decades has our technical understanding caught up with advancement through trial and error.[1, 2]

A musical instrument has two major functions: to generate vibrations and to transform them to audible sound. Almost all the components of an instrument are highly functional, each doing its part to carry out the purpose of the instrument. Even the fancy scrolls of the violin bridge and the f-holes serve functional purposes in aiding the production of pleasing sound.

FIGURE 16-1
A front view of a violin showing the four strings, bridge, and two f-holes (after the Guarnerius pattern). (Courtesy of Scherl & Roth)

16.2 The vibrating string

It was shown by Mersenne (see Section 11.3) that the frequency of a vibrating string is

1. Inversely proportional to its length.
2. Proportional to the square root of the tension applied to the string.
3. Inversely proportional to its diameter.
4. Inversely proportional to the square root of its density.

In addition to Mersenne's four factors, the frequency of a stretched string can also be affected slightly by its rigidity, by the way in which it is held at the two ends, and by the flexing of the end supports. These latter effects tend to make the overtones slightly sharp. (See Table 11 – 1.) String nonuniformity, usually due to wear, will also cause variation from harmonicity. Good design, therefore, requires use of uniform flexible strings and rigid supports. Still, the strings will always have some rigidity, and one anchorage must move in order to transfer vibration to the instrument soundboard.

In designing an instrument with many strings of different frequency, one could vary any one of the parameters. For example, a guitar could be made with six identical strings, varying only in their tension. However, the highest string may require a tension that exceeds the breaking strength, whereas the lowest string may require so little tension that it will not adequately sustain a plucked note. As shown in Figure 16 – 2, the different frequencies of open-string vibration are actually achieved mainly by varying diameter and density. For example, the low-pitched E string of the guitar has the largest diameter, and it is wound with wire to increase its density as well as its diameter. An open-string instrument such as a harp will typically span several octaves. (See Figure 17 – 8.) To provide such a wide range of frequencies, all four basic parameters are varied.

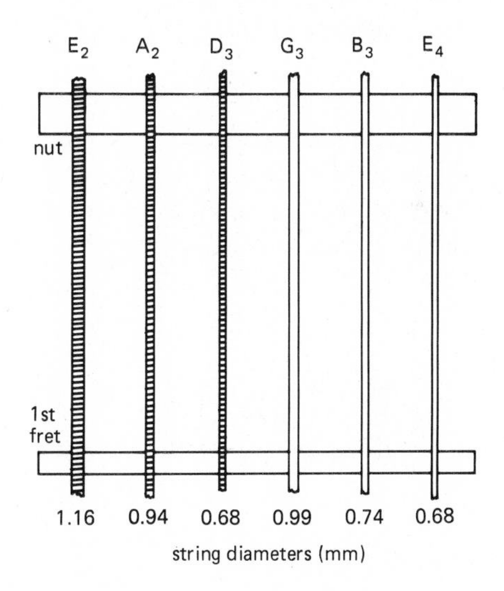

FIGURE 16 – 2
Detailed structures of guitar strings.

16.3 Few-string instruments

In designing a portable instrument, one cannot afford the luxury of having an open string for each note played, as is done with the piano. It is therefore necessary to use less strings, and so produce a range of pitches from each string. This can be done by varying one of Mersenne's four basic parameters during performance.

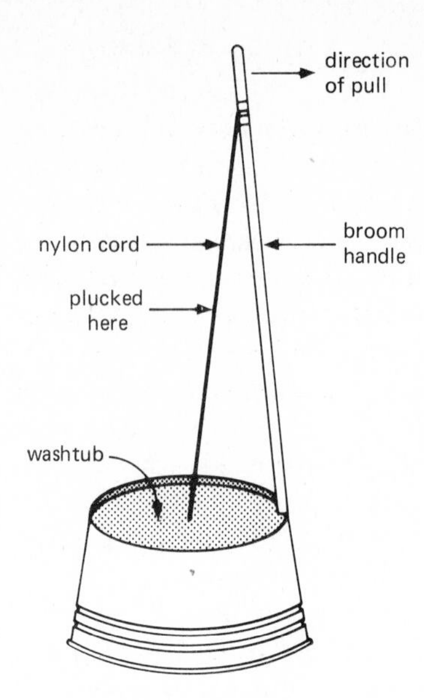

FIGURE 16–3
The wash tub bass produces a pitch that can be varied over two octaves, depending upon the tension that can be applied to the cord.

The string diameter and density cannot be conveniently varied during play. The tension, however, can be changed, within limits, and is often varied by tightening or loosening pegs (Scruggs banjo technique) or by stretching strings laterally with a finger, giving a continuous change in pitch (portamento). Instruments specifically designed to produce pitch variations by tension changes include the sitar and harp, as well as the washtub bass. See Figure 16–3. By far the most common method of varying pitch during performance is by changing string length. Shortening the vibrating length of a string is usually accomplished by pressing any portion of the string against a fingerboard, as with a violin.

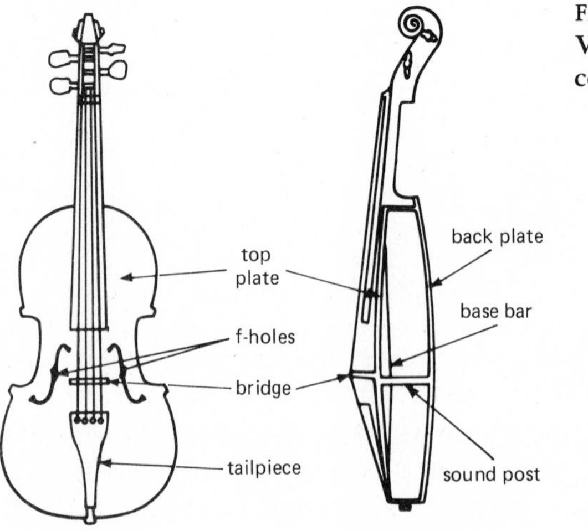

FIGURE 16–4
Violin construction.

16.4 The violin

Of the bowed stringed instruments, the violin has probably reached the highest stage of perfection. The violin uses four strings stretched across a high bridge, which in turn rests on the convex top plate of a sound box, as shown in Figure 16–4. The strings are normally made of catgut (sheep intestines) or steel, and they are usually wire-wound with aluminum or silver. Properly tuned, these four strings—G_3, D_4, A_4, and E_5—are a musical fifth apart and have frequencies of 198, 295, 440 and 660 Hz, respectively. The vibrating length of string between the bridge and the stopped point on the fingerboard can be varied continuously due to the absence of frets.

16.5 Timbre of a vibrating string

Let us assume that a performer wishes to play a high C (C_5) on a violin. By pressing the A_4 string against the fingerboard at the right place, the free length of string will sound C_5 as its fundamental pitch if it is plucked. It will also sound many of its overtones. In fact, it is virtually impossible to avoid sounding overtones on a stringed instrument. As seen in Figure 16–5, the shape the string takes as it is plucked is not a smooth sine shape but a ramp shape (see Figure 8–9). According to Fourier's theorem (Section 8.8), a wave that is not a sine wave is equivalent to a sum of harmonics, each with appropriate amplitude. Therefore, a plucked string will vibrate at its fundamental frequency as well as at several harmonic frequencies, all at the same time. (See Section 11.2.)

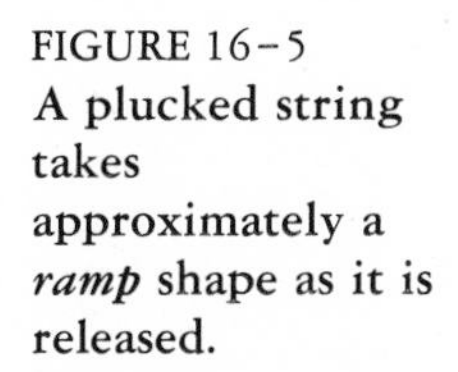

FIGURE 16–5
A plucked string takes approximately a *ramp* shape as it is released.

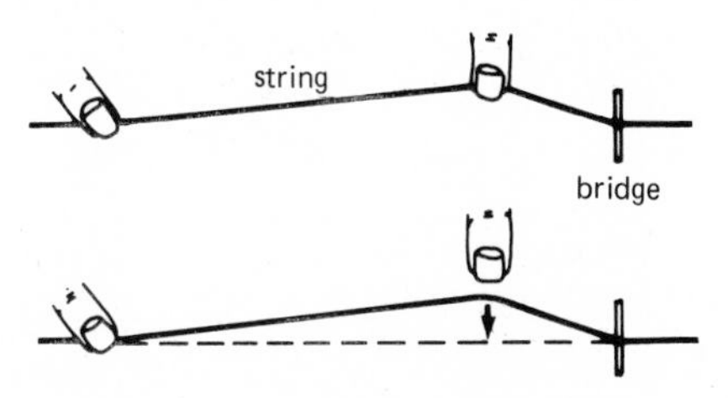

The timbre of a plucked string depends upon the relative amplitude of vibration of the harmonics. If the string is plucked in the middle, its triangular shape does not look too different from the fundamental sine wave. This means that the fundamental pitch will be strongly excited compared to the higher harmonics. Due to the symmetry of the string, even harmonics should be totally absent, as was shown in Figure 8–12. The resultant tone is most easily described by the frequency spectrum bar chart shown in Figure 16–6(a). If the amplitude of the fundamental (C_5) is called 1, the amplitude of the third, fifth, and seventh harmonics are $\frac{1}{9}$, $\frac{1}{25}$, and $\frac{1}{49}$, respectively. The amplitudes of the even harmonics are zero. Plucking a string near the middle will therefore give a fairly pure tone, with little harmonic content. Although this may be desirable from a harp, it will not be desirable for a guitar, which needs brightness. A guitar is therefore picked at about one-fourth of its length. The relative content of higher harmonics is increased as shown in Figure 16–6(b).

To get a "hard" sound, the string is activated even closer to the bridge, as shown in Figure 16–6(c). This results in even greater enhancement of higher harmonics. The bright or rich sound, which characterizes most bowed string instruments, is achieved by bowing 10–11% of the distance from the bridge. Bowed even closer (sul-pointicello), the timbre becomes "raspy."

The timbre, as well as the transient response, also depends on the method of excitation.[3] Striking produces more high-harmonic excitation than plucking. Bowing results in yet a different response.

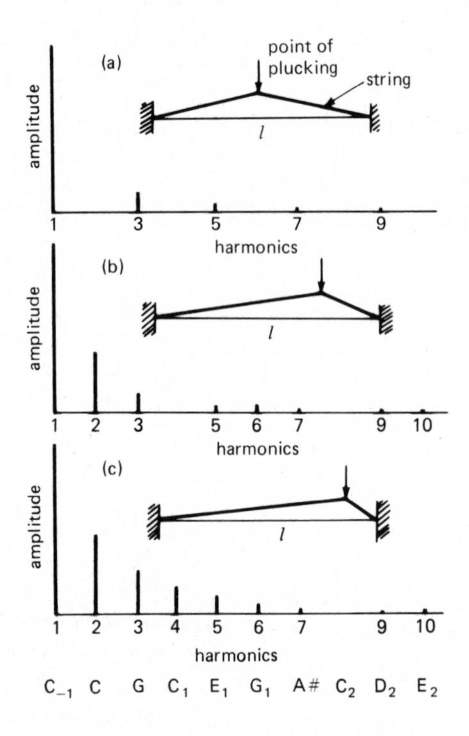

FIGURE 16–6
Harmonic content depends upon the initial shape of a plucked string.

16.6 Selecting harmonics

If a string is plucked at one-half of its length, the second harmonic is absent (as well as its multiples). If it is plucked at one-fourth of its length, the fourth harmonic is absent. One can see that, in general, a harmonic will be absent if it has a node at the point of plucking. In fact, a harmonic is more likely to be strongly sounded if it is plucked at its maximum or antinode. For example, the point of plucking at one-fourth of the length corresponds to the antinode of the second harmonic and the node of the fourth. (See Figure 11-2.) This results in strong sounding of the second harmonic and the absence of the fourth, as seen in Figure 16-6(b).

An opposite procedure ("playing in harmonies") is to lightly touch a string with one finger as it is plucked. The result, as shown in Figure 16-7, is that only those harmonics with a node at the point of touch can be excited. (See Section 11.2.) For example, if the C_3 string with a fundamental of 132 Hz is touched lightly at one-third of its open length, the available harmonics are 3, 6, 9, ... These have the frequencies 396, 792, and 1188 Hz, corresponding to the notes G_4, G_5, and D_6. These notes form a natural harmonic series based on 396 Hz or G_4. This technique produces high bell-like tones from otherwise long strings.

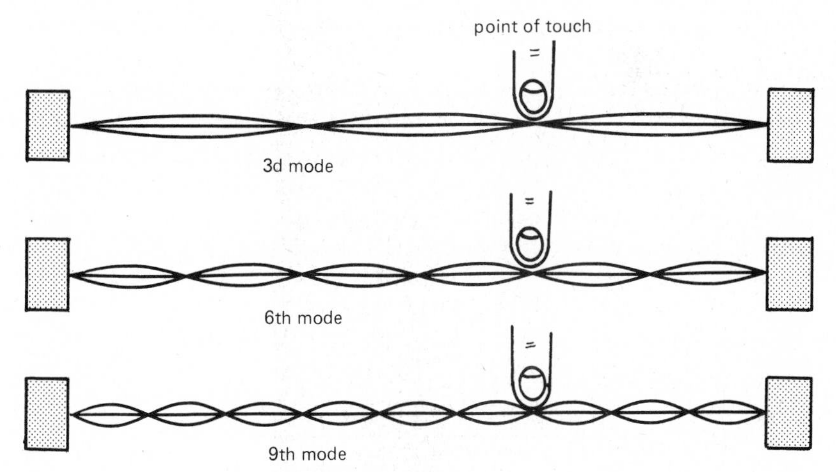

FIGURE 16-7
By lightly touching a vibrating string, only those modes with a node at the point of touch will be allowed.

16.7 Percussive string excitation

A string can be plucked as in a guitar, repetitively plucked as in a mandolin, or hammered as in a piano. The rise in sound output, or attack, is very sudden. However, its decay is slow, typically decreasing to half amplitude in less than a second. Although sound continues to be emitted long after that interval, its level is low enough so that notes may quickly follow each other to give a sense of continuity.

The characteristic timbre of an audible note is determined to a considerable extent by these *transient characteristics,* the *attack* and *decay* of its harmonics. See Figure 16-8. Moreover, the material from which the strings are made will determine the transient characteristics. For example, using catgut strings, the high harmonics will quickly decay, leaving the low harmonics sounding longer. The more brilliant sound of steel strings is due to the longer decay time of their high harmonics.

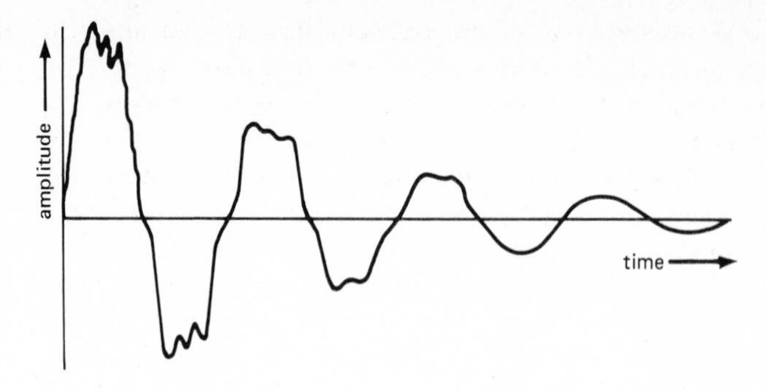

FIGURE 16–8
Transient response of a plucked string. Higher harmonics decay more rapidly than the fundamental. (Rate of decay of higher modes is exaggerated.)

16.8 Bowing

By drawing a horsehair bow across a string, the string will vibrate, producing a sustained or continuous tone. The horsehair is normally coated with rosin, a wood sap product, which increases the friction between the bow and the string. As the bow is drawn across the string, the string is first pulled laterally in the direction of the bow motion. Then the string slips back and stops, allowing the bow once again to engage and pull the string laterally. This alternate pull and slip is forced by the fundamental vibration of the string to synchronize with the period of vibration.[3, 4, 5] See Figure 16–9.

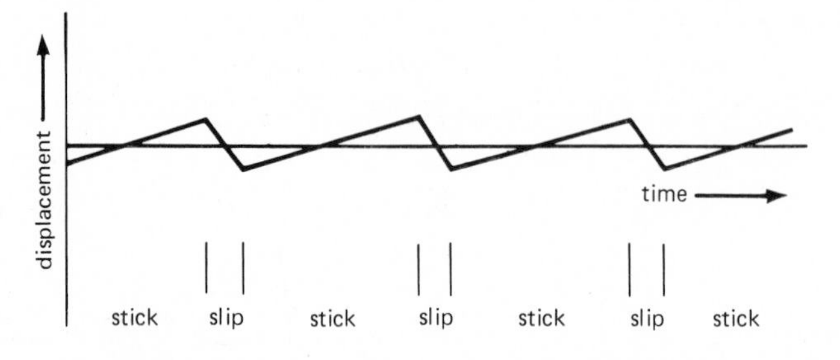

FIGURE 16–9
As the bow is moved upward, the string is pulled uniformly, and then it disengages and snaps back. (Actual observations show minor variations for these straight-line forms.)

There are two types of frictional force acting between two surfaces. When the surfaces are engaged, they are held together by **static friction.** The friction that acts during motion, called **dynamic friction,** is normally slightly less than static friction. A rosined surface has the unusual feature that its dynamic friction is considerably less than its static friction, especially at high slipping speeds. This allows the string to slip back freely after each disengagement with the bow. If the motion of the bow is not perpendicular to the strings, longitudinal vibrations can be set up. These are much higher in frequency than transverse vibrations and account for the squeaky sounds often made by beginning violin students.

Increasing the bow pressure—the force between bow and string—will usually cause a change in timbre but will not increase the loudness appreciably. Loudness can be increased by increasing the bowing speed, because the time period of each pull is set by the natural frequency of the string. The faster bow will therefore pull the string farther with each vibration, increasing the amplitude of vibration. Increasing bowing speed not only increases the loudness but changes the timbre as well. Most violinists have found, however, that by increasing the bow pressure at the same time that the speed is increased, they can increase the loudness without appreciably changing the timbre.

16.9 Body resonance

A vibrating string will put out little sound if it is firmly anchored at each end to a solid surface. If it is mounted to some sort of broad flexible surface, however, the sound output will increase dramatically. This principle is used in the design of string instruments.

One end of the vibrating portion of each violin string terminates at the **bridge,** which rests on two feet midway between the f-holes. As the strings of the violin vibrate, the bridge is caused to move back and forth, as well as up and down. See Figure 16–10. This motion is transferred to the top plate of the instrument. A short maple **sound post** is snugly wedged between the top and back plates of the sound box, close to the foot of the bridge on the high-string side. It serves in part to transmit the vibrations from the bridge to the back plate. Under the bass-string side of the bridge, a spruce **bass bar** runs lengthwise, giving support to the otherwise frail front plate. The sideways

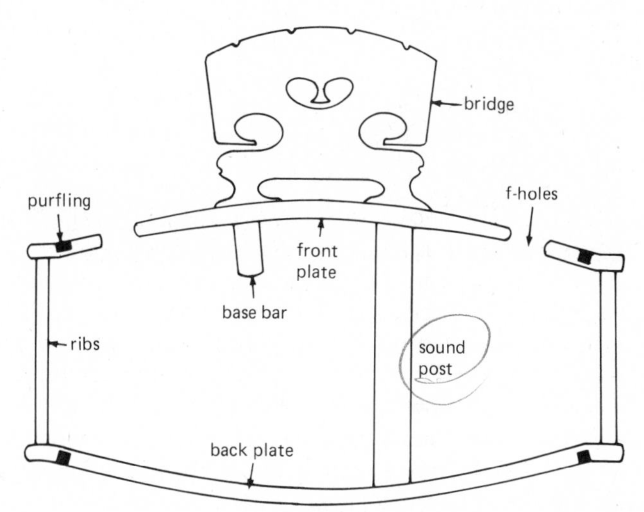

FIGURE 16–10
Section through a violin at the bridge.

vibrations of the lower strings cause a rocking vibration of the bridge, pivoted on the sound post, causing the bass bar to move up and down, carrying a sizable part of the top plate with it.

The top and back plates of the violin each have natural resonant modes of vibration. Some of these have been photographed using hologram interferometry.[6] See Figure 16–11. The lowest, or fundamental, frequency is that heard if the surface is tapped with the finger. This main wood resonance is caused by the surface moving up and down as a whole, as shown in Figure 16–11(a). Higher frequency notes resonate with more complex motions, as shown in Figures 16–11(b)–(d). If the string frequencies are at or near these natural vibrational frequencies of the body surfaces, the body resonates with the strings, resulting in a louder tone.

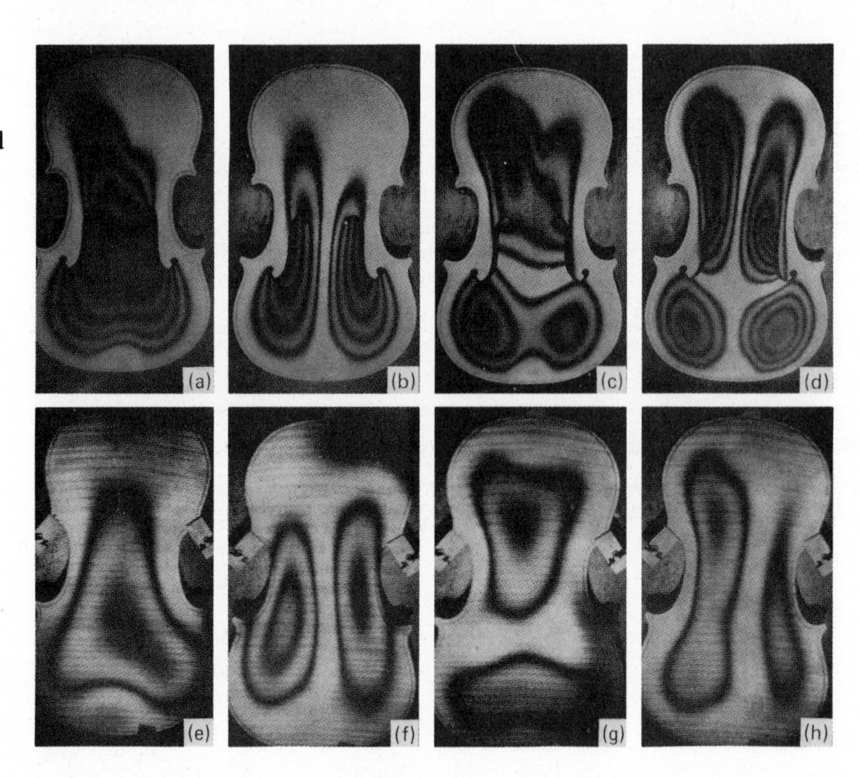

FIGURE 16–11
Wood resonance modes of violin plates, as illustrated using holographic interferometry. Top plate vibrations with f-holes and bass bar: (a) 465 Hz, (b) 600 Hz, (c) 820 Hz, (d) 910 Hz. Some back plate vibrations of a complete violin: (e) 580 Hz, (f) 770 Hz, (g) 870 Hz, (h) 970 Hz. (Courtesy of N.-E. Molin, H. Sundin, and E. Jansson)

The f-holes of the violin serve not only to connect the interior air resonance with the outside but also to make the spruce top plate more flexible. This flexibility results in a relatively low value for the main wood resonance, about 440 Hz or concert A, the note to which the second string is tuned. The maple back plate is stiffer and has a somewhat higher resonant frequency. The bridge itself is designed to resonate at even higher frequencies.

16.10 Cavity resonance

Various standing waves can be set up in the air within the body of an instrument. Again, certain resonant frequencies exist that give an enhanced response to corresponding string frequencies. A violin has a strong fundamental resonant frequency of the air in the sound box at 260 to 290 Hz, near middle C. The frequency of this cavity resonance is determined primarily by the interior volume and the size of the f-holes. This main air tone is typically about a fifth lower than the wood tone for a good violin. The air tone can be heard by blowing across the f-holes.

The body cavity has a weaker air resonance near 500 Hz and other higher and weaker resonances. The frequencies of these major resonances are crucial in determining the quality of a violin. It seems that master craftsmen of the Cremona school, Guaneri, Stradaveri, and others, created instruments for which all the various resonant frequencies were well separated, spanning the full playing range. Therefore, almost any note sounded would excite one of the resonances. In contrast, if resonances coincided, there would be notes with strong resonant response, separated by regions of low response. Such an instrument would have poorer sound quality.

16.11 The energy chain

The production of musical sound from a stringed instrument involves a sequence of several steps. Strings are caused to vibrate, transferring the vibration energy through the bridge to the body. This vibration is distributed to other portions of the instrument, causing the vibration of various surfaces as well as the generation of air modes within the instrument box. Acoustical waves are radiated through the holes in the box, as well as from the surfaces directly. Although the various modes of vibration can be identified separately, they all work together to produce the resultant resonant response curve of the violin.

The acoustical impedance match (see Section 5.5) at the bridge, between the strings and the body, is moderately poor. If it were good, no reflection would occur in the string and no standing waves could be established. If the match were very poor, no energy would be transmitted, and the audible sound would be low. The average stringed instrument reflects most of the energy in the string but allows enough to pass through the bridge to produce the desired sound.

For a percussively excited string, the rate of decay will be due mostly to the rate the energy feeds through the bridge. By improving the impedance match at the bridge, the produced sound will be louder, but the note will decay faster. Likewise, an increase in decay time will result in a fainter sound. The desired compromise is achieved by the angle the strings bend over the bridge (down-bearing force), the bridge design, and the body thickness, as well as structure in the vicinity of the bridge, taking into account the rest of the instrument characteristics.

Because of major bridge, body, and cavity resonances, and their interaction, there are certain broad frequency ranges called *formants* at which good response is produced. Each instrument has its characteristic formants. (See Section 8.10.)

Many members of the violin family, particularly the cello, have specific frequencies at which the motions of the entire instrument conspire to produce a loud wobbly sound called the **wolf note**. This is a narrow resonance at about F♯ on the cello G or C string, and it is carefully avoided by the performers. Its cause has been attributed to beating between pairs of natural vibrations.[7]

There are many features involved in the production of music from a stringed instrument. A fine musical instrument is one in which these features are well interrelated.

QUESTIONS

1. List five open-string instruments.

2. List two or three situations in which a musician changes a string's tension during performance to change pitch.

3. Why would a string with little tension sound "dead"?

4. Explain why faster bowing speed creates louder sound.

5. Do the resonant frequencies of a guitar surface form a harmonic series? Explain. [See W. B. Savage, *Problems for Musical Acoustics* (New York: Oxford University Press, 1977), page 53, for more information.]

6. How is the concept of impedence matching involved in the design of a violin? What constitutes the transformer?

7. In Section 15.2, it was mentioned that the natural frequency of a vibrating string is altered if the string supports are not perfectly rigid. Should instruments be designed with perfectly rigid supports?

PROJECT

Find the frequency of the wolf note of a cello. Does it occur at one of the normally played notes? Can you excite the wolf note in a viola? How would you describe its sound?

NOTES

[1]H. Meinel, "Regarding the Sound Quality of Violins and a Scientific Basis for Violin Construction," *Journal of the Acoustical Society of America* 29 (1957):817.

[2]C. Hutchens, "The Physics of Violins," *Scientific American* (November 1962):78.

[3]R. Jones, "Musical String Vibrations," *Physics Teacher* (March 1977):145.

[4]J. C. Schelling, "The Physics of the Bowed String," *Scientific American* (January 1974):87.

[5]A. H. Benade, *Fundamentals of Musical Acoustics* (New York: Oxford University Press, 1976), chap. 23.

[6]E. Janssen, N. E. Molin, and H. Sundin, "Resonances of a Violin Body Studied by Hologram Interferometry and Acoustical Methods," *Physica Scripta* 2 (1970):243.

[7]I. M. Firth and J. M. Buchanan, "The Wolf in the Cello," *Journal of the Acoustical Society of America* 53 (1973):457.

Chapter Seventeen

STRINGED INSTRUMENTS

A wide variety of instruments that employ strings are used today. These vary from the violin to the piano. Almost all of them have certain features in common, including activated strings connected through a bridge to a soundboard, which in turn forms a part of a resonant case. The sizes, shapes, and methods of exciting the strings vary widely. Several selected examples will be discussed in this chapter, with the intent of covering most of the major categories of stringed instruments.

17.1 The violin family

Of the many members of the violin family once in existence, only the *violin, viola,* and *violoncello* (cello) are commonly used today. The latter instruments are basically larger forms of the violin design and have lower pitches. However, the sizes are not scaled up enough from the violin to match the lower tuning of the strings. As a result, they are not simply scaled-up violins but have distinct "voices" of their own.

The **bass viol**, or **string bass**, evolved from the historically different viol family. However, the design of the bass viol of today is close to the violin design, with only minor differences such as sloped shoulders. See Figure 17 – 1. The size of the bass viol can vary but is limited mainly by the difficulty in simultaneously fingering and bowing. With appropriately composed music,

FIGURE 17-1
The bass viol is the largest stringed instrument classified today as belonging to the violin family. Note the sloped shoulders.

the strings complement each other quite nicely in groups (such as the string quartet) or as a section of an orchestra.

The range of sizes of the bowed stringed instruments appears to leave several size gaps, particularly between the viola and cello. But with the advent of acoustical understanding, there has been a resurgence of interest in redesigning and extending the family to fill these gaps.[1] In particular, the Catgut Acoustical Society* has supported the development of a new family of eight instruments, based on the careful acoustical study of the characteristics of fine violins done by Carleen Hutchins and her co-workers. See Figure 17-2. This new family was conceived with the intent of retaining the characteristic voice

*The Catgut Acoustical Society is a worldwide association of scientists, musicians, and other enthusiasts dedicated to applying the science of acoustics to the understanding and the construction of instruments of the violin family. Their semiannual newsletter is a major resource in the acoustics of the violin family.

FIGURE 17 – 2 **Eight instruments of a new violin family. (Courtesy of Catgut Acoustical Society, Inc.).**

of the violin, while transposing its pitch over the full orchestral range. See Table 17 – 1. It is not the intent to replace the string quartet but rather to develop an alternate approach, requiring new or rewritten music for these particular instruments. Performances of the string octet have been well received, a tribute to the effort put forth by Carleen Hutchins and her associates.

17.2 The guitar

Many plucked stringed instruments evolved from the ancient cithara, including the lute, zither, and dulcimer. The Spanish guitar as we now know it was developed in Spain around the thirteenth century and resulted from the inter-

TABLE 17-1: The Eight Members of Carleen Hutchin's String Octet

NAME	SOUND BOX LENGTH (cm)	OPEN-STRING TUNING	
Treble violin	25	$G_4\ D_5\ A_5\ E_6$	Played under chin
Soprano violin	30	$C_4\ G_4\ D_5\ A_5$	
Mezzo violin[a]	36	$G_3\ D_4\ A_4\ E_5$	
Alto violin[b]	50	$C_3\ G_3\ D_4\ A_4$	Optional
Tenor violin	66	$G_2\ D_3\ A_3\ E_4$	Played vertically
Baritone violin[c]	86	$C_2\ G_2\ D_3\ A_3$	
Small bass	104	$A_1\ D_2\ G_2\ C_3$	
Large bass[d]	130	$E_1\ A_1\ D_2\ G_2$	

[a]Essentially a traditional violin.
[b]Equivalent to a viola.
[c]Equivalent to a cello.
[d]Equivalent to a bass viol.

action of the Spanish and Moorish cultures.[2] The modern Spanish guitar, shown in Figure 17-3, consists of a fretted fingerboard attached to a sound box with a single round hole in the flat front surface. The bridge is fastened directly to the top plate (soundboard) and supports the tension of all six

FIGURE 17-3
A Ramirez classical guitar. (Courtesy of Sherry-Brener, Inc.)

strings. The frets permit a specific note to be accurately produced but restrict the ability to continuously alter the pitch. The bass response of the guitar is determined primarily by the air and top plate resonance modes of the sound box.[3] See Figure 17–4. The larger the sound box, the lower are the resonant frequencies. For classical guitar performance, a relatively modern development, gut or nylon strings are used. High harmonics are poorly excited when picked by the fingers and quickly damped in the strings, resulting in a mellow timbre.

A flamenco guitar, used in the folk music of the Spanish gypsy culture, is usually smaller than the classical guitar, resulting in brighter tones. A plastic guard plate is used next to the strings to protect the wood against the drum taps that turns the flamenco guitar into a partly percussion instrument.

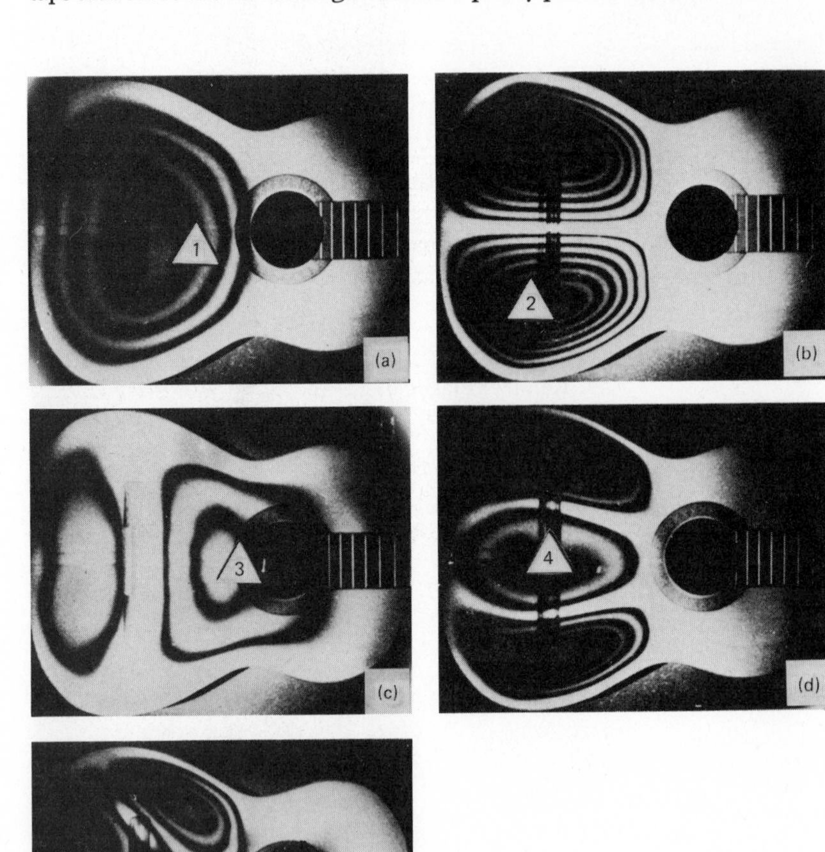

FIGURE 17–4 Time-averaged holographic interferograms showing the lowest five resonant modes of the top plate of the guitar. Triangles represent the point at which the plate was caused to vibrate by an electromagnetic driver. Vibration frequencies are (a) 185 Hz, (b) 287 Hz, (c) 460 Hz, (d) 508 Hz, (e) 645 Hz. (Courtesy of N.-E. Molin, H. Sundin, and E. Jansson)

Although the Spanish guitar is often used in America for folk ballads and entertainment in small groups, its low sound output makes it inadequate in accompanying most other instruments. The 12-string guitar is commonly used for greater sound output. It uses six pairs of steel strings, four of the pairs tuned an octave apart. The body is relatively large, enhancing the bass response. The result is a loud sound with high harmonic content, adapting it to the blues and jazz styles as part of performing combos. See Figure 17–5.

FIGURE 17–5 A Martin 12-string guitar. (Courtesy of CF Martin Organization)

17.3 The banjo

Modern banjos typically have four strings stretched across a small light bridge. The bridge rests on the head, which consists of a piece of skin or plastic stretched tightly over an open wooden hoop. The light and flexible head responds well to the higher frequencies, producing a bright and harmonically rich sound, particularly when picked close to the bridge.

The banjo evolved from a West African skin-covered gourd instrument, which in turn probably evolved from the ancient lute. The banjo is nevertheless considered a uniquely American instrument, relating closely to the blues and jazz culture.[4] Additional modifications include the addition of a fifth drone string, which is used in southern mountain bluegrass music in the style of Earl Scruggs and others. More recent modifications include the addition of a back to complete a sound box. Although a brighter and louder sound is thereby produced, most of the sound output comes from the thin head. See Figure 17–6.

FIGURE 17–6
A five-string banjo.
(Courtesy of Norlin
Music, Inc.)

17.4 The sitar

Although the sitar is native to northern India, Indian instrumentalists have popularized it throughout the Western world. The sitar consists of a hollow fingerboard, about 1 m in length, ending in a gourd, which acts as a resonator box and soundboard. See Figure 17 – 7. The sitar uses 16 to 22 archlike frets, which can be easily moved, allowing intervals of any scale to be produced.

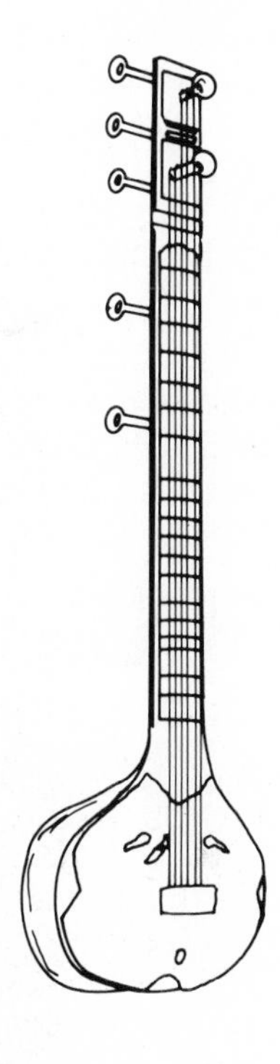

FIGURE 17 – 7
A plain sitar made of Tun-wood and having 7 strings, 19 brass frets, and rosewood pegs; it is 120 cm long. (Courtesy of James J. Peterson)

Seven strings are available on the modern sitar, with many additional sympathetic strings that run parallel to the main strings underneath the frets. The fingerboard is 7 – 8 cm wide, allowing the player to stretch the string laterally against each fret. This stretching increases the tension, causing the pitch of a plucked string to vary continuously (glissando) for as much as six semitones on a single fret.

17.5 The harp

The harp is the only plucked stringed instrument used in a modern symphony orchestra. Harplike open-stringed instruments date back into antiquity. Early harps were made with one string for each note. However, to include all the sharps and flats needed for most music would involve an excessive number of strings. This problem was solved during the eighteenth century by the addition of pedals that allow the harpist to vary the pitch of the strings by changing their length and tension. This pedal method culminated with the design of the modern double-action harp by Sebastian Erard in 1810. The harp is normally tuned to C-flat. Each of seven pedals can raise all the strings of a given key by one or two semitones. By pressing all pedals one step, the harp is tuned to C-natural. Two steps raise it to C-sharp. By proper selection of pedal positions, the harpist can play in any key. Pedaling during performance allows one to make chromatic alterations outside of key.

The modern harp, now a standing instrument, is shown in Figure 17–8. The body consists of a tapering sound box topped by the soundboard to which the bottom ends of the strings are attached. The pedestal contains the seven pedals. The pillar supports the upper structure and contains the rods that activate the pitch-changing mechanism. The neck is shaped to give the desired progressive increase in string length. Pegs along the neck terminate the strings and allow their tension to be adjusted. Below the pegs is the pitch-changing mechanism, which consists of two pins that push against the string and increase its tension.

FIGURE 17–8
A concert grand harp. The height is 188 cm. There are 47 strings with a range from C_1 to G_7. (Courtesy of Salvi Harps)

The harp normally uses 47 strings made of sheep gut or nylon. The lower strings are wound with steel to increase the linear density (mass per unit length). Including the chromatic capability, the double-action harp spans six and a half octaves. When plucked near the centers of the strings, only the first two of three harmonics are strongly excited. The use of fingers, rather than hard picks, serves to dampen the higher harmonics. The result is a tone well known for its purity and beauty.

17.6 The harpsichord

Early keyboard stringed instruments resulted from efforts to adapt a keyboard to known stringed instruments. For example, adaptation of a keyboard to a psaltry resulted in the *spinet,* the *virginal,* and a variety of other devices. Such instruments became quite popular from the fifteenth century onward.

The most versatile and popular of the plucked keyboard instruments was, and still is, the harpsichord.[5] See Figure 17–9. This instrument consists of many strings, each tuned to a different note of the musical scale, stretched across a soundboard to which they are coupled by a bridge. The action, the mechanism that plucks the strings, is shown in Figure 17–10. Pressing down

FIGURE 17–9
A modern concert harpsichord (Kurt Sperrkake) with two manuals, four registers, and lute stops. It has a keyboard range of five octaves and two notes, F to G. (Courtesy of Robert S. Taylor)

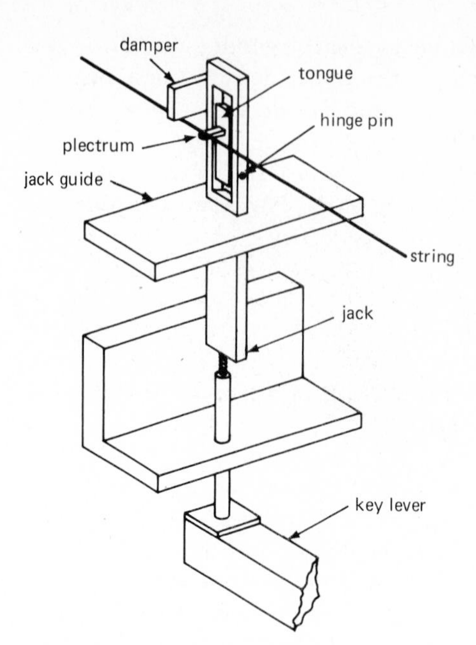

FIGURE 17-10 Harpsichord action. When the key is pressed, the other end of the key lever raises the jack and causes the plectrum to pluck the string.

on the key causes the jack to move upward in its guide. A hinged tongue rests on the jack, which holds a pointed plectrum. This plectrum plucks the string on the way up, but it avoids plucking on the way down because the tongue swings away.

It is often argued that the harpsichord (as well as the piano, guitar, etc.) should be classified as a percussion instrument, since the strings are excited percussively. Similarly, one could argue against classifying the flute and saxophone as woodwinds, since they are made of metal. Although the question of proper classification can lead to interesting discussions, it is ultimately not possible to sort the diverse array of musical instruments into a small number of clearly defined classes. The instruments in this text are grouped purely for convenience in discussing common features.

17.7 The pianoforte

If the harpsichord is a keyboard-activated psaltry, the piano must be considered as a keyboard-activated dulcimer. Early in the 1700s this new keyboard instrument (although the idea of hammering strings was not new at the time) was developed by combining the characteristics of the harpsichord, clavicord, and hammered dulcimer. Although early pianos were little more than the adaptation of hammers to a harpsichord, the piano evolved over the following century into a unique instrument.[6] See Figure 17-11. The major advantage of the piano was that, although the player had little control over the loudness of the harpsichord, the pianoforte could be played with wide dynamic (loudness) variations.

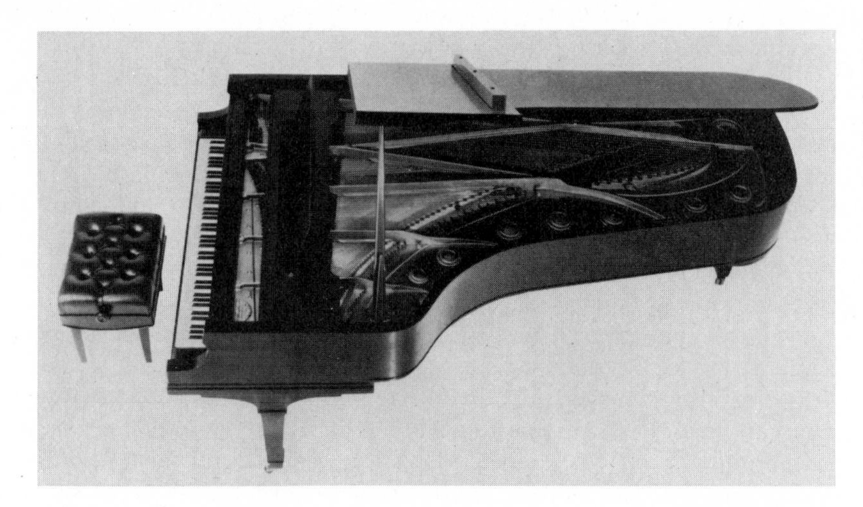

FIGURE 17-11
Top view of a concert grand piano. (Courtesy of Baldwin Piano and Organ Company)

17.8 The concert grand piano

The quality of the tone of a piano seems to improve with size, up to a limit. In the harpsichord, hammers are unable to transfer sufficient energy to the small, light strings. But if the size and tension of the strings are increased, a good energy transfer can be achieved. However, this situation requires a strong steel frame to support the strings.

The fundamental frequency of each string in a piano is given by Equation (11g), which for convenience is repeated here.

$$f = \frac{1}{2L}\sqrt{\frac{F}{m}} \qquad [17a]$$

where L is its length in meters ($\lambda = 2L$), F is the tension in newtons, and m is the linear density in kilograms/meter. The major parameter that is varied is the length. The tension varies little from string to string over most of the piano, and it is the parameter that is carefully adjusted to tune the string to the proper pitch. See Figure 17-11.

Below about C_3, the length of a string would become unreasonably long if the other parameters were held constant. The tension could be decreased, but a slack string does not receive sufficient energy from the hammer and does not sustain the high harmonics well, sounding dead. Increasing the diameter seems the only remaining way to produce low notes. However, if the diameter of a steel wire is increased sufficiently, the wire becomes somewhat rigid, vibrating to some extent as if it were a rod. The overtones of a flexible string under tension are harmonic, whereas the overtones of a vibrating rod are not.

Thus a string that is under tension, and slightly rigid, produces overtones that are slightly sharp, giving a timbre described as "tubby." Since piano strings are never perfectly flexible, this is always a problem. In order that higher notes be consonant with the overtones of lower notes, the musical intervals on a piano are normally "stretched" somewhat. Thus higher octaves are tuned increasingly sharp and lower octaves flat.

This problem in inharmonicity may be minimized by keeping the tension as high as possible, by using strings that are as long as possible, and by increasing the flexibility by replacing a solid wire with a thin flexible wire with a metal winding around it. The improvement with string length is one reason that larger pianos usually sound better than smaller ones. Alternatively, a certain amount of inharmonicity seems to be desirable to give the piano its characteristic sound.[7]

Since sound output is less for the higher strings, most pianos use only a single string for the lowest notes, two strings from F_1 to A_1 (on the grand piano), and three above B_1. The multiple strings in each group are normally tuned to be almost the same frequency, although they never remain so for long. If they are seriously mistuned, audible beats occur, which result in the honky-tonk, barroom piano sound. The lower strings are usually crossed over the upper to conserve space. The strings are secured over a *hitch pin* on the far end and tied around the adjustable *tuning pin* on the near end. See Figure 17–12. The vibrating length of the string is defined by a bar *(capo)*, or a block with holes *(agraffe)*, on the near end. The far end is defined by the *bridge*, which transfers the string vibration to a large *soundboard*, about 1 cm thick. Sound is transferred to the air by the vibation of the soundboard and modified by the piano *case*. All these components must be built of quality materials.

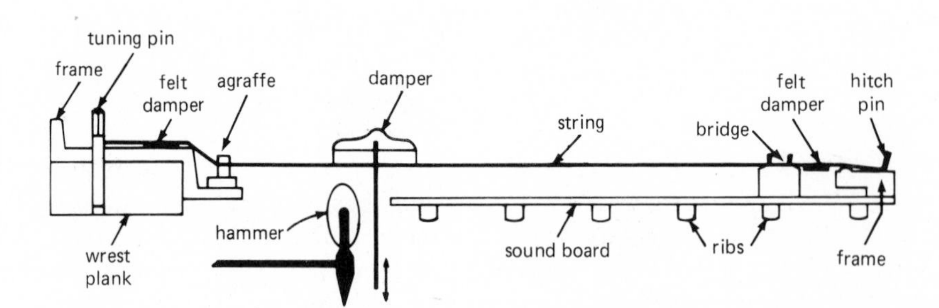

FIGURE 17–12
Sound production components of a grand piano.

17.9 The piano action

When a key is pressed in a piano, the hammer must strike the string. In doing so, several conditions must be met:

1. The thrust on the key must control the strength of the hammer blow.
2. The hammer must then immediately withdraw to avoid damping.
3. Withdrawal must be independent of subsequent key position.

4. The hammer must then be checked, to avoid its bouncing back to the string.

5. The hammer must respond to immediate and continuing repetition of the blow.

6. Repetition must be possible even if the key has not returned to the normal position.

7. The right pedal must raise and lower each damper.

8. The sostenuto (middle) pedal must hold up those dampers already raised.

All these features must be exercised flawlessly and as noiselessly as possible.

The action of a grand piano has undergone a century of development to arrive at the present configuration.[6] The action shown in Figure 17–13 meets the listed requirements almost perfectly. The major features are the *double action,* the *escapement mechanism,* and the *check.* When a key is pushed down-

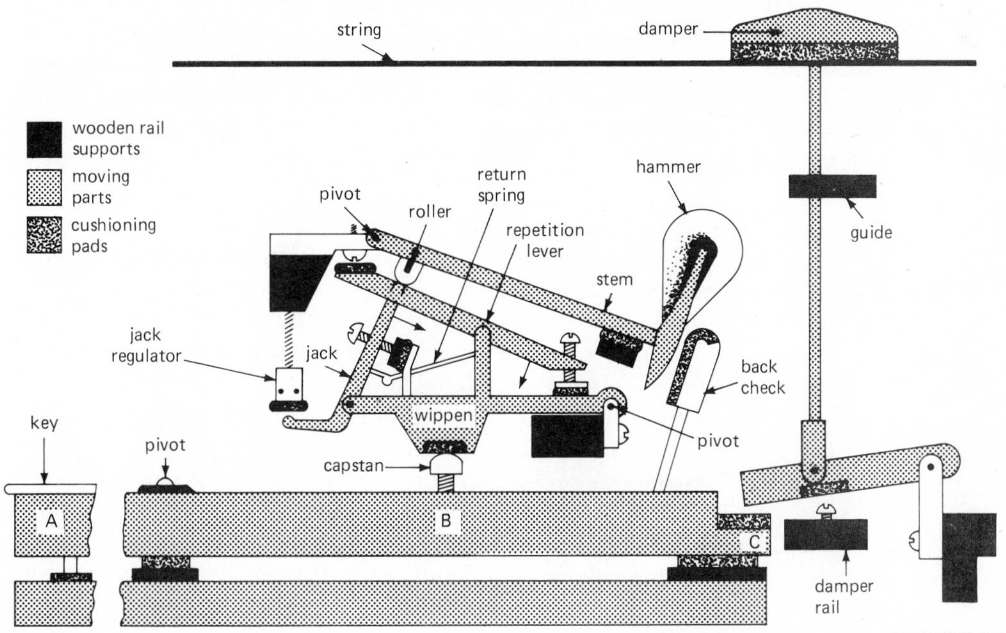

FIGURE 17–13
Action mechanism of a grand piano.

ward, the main lever AC is pushed downward at A and upward at B and C. This lifts the *capstan* and pushes up the left end of the *wippen,* as well as the *jack* and *roller.* This lifts the *hammer* toward the string. Just before the hammer reaches the string, the lower end of the jack strikes the *jack regulator.* This causes the jack to rotate counterclockwise so that its upper end comes out from under the roller and no longer pushes on it.

When the piano key is pressed slowly, the hammer will rise about 5 mm from the string and then fall back. When the key is pressed quickly, the hammer's momentum will keep it moving after the upper end of the jack comes out from under the roller, so the hammer will hit the string. At that moment it is completely free from all contact with the lifting mechanism. When the hammer bounces back from the string it is caught by the back check, thereby preventing it from bouncing up to hit the string again. The instant the depressed key begins to be lifted, the roller rests on the *repetition lever,* which is pulled by a small spring in the direction of the arrow. This permits the upper end of the jack, also pulled by a spring in the direction of its arrow, to slip back quickly under the roller. This mechanism therefore resets itself when the key return has moved less than halfway to its normal rest position.

When the key is depressed normally, the damper rises off the string, allowing it to vibrate until the key is released, whereupon the damper drops back and stops the string's motion. The dampers on all strings may be raised off the keys independently by means of a damper rail running lengthwise to the piano and attached to a pedal operated by the foot.

Since the hammer is free of the key at the time of impact, the pianist cannot control the way in which the string is struck apart from the strength of the blow. The so-called touch characteristics, which are often attributed to various pianists, must then be due simply to how hard the keys are pressed and a sense of timing.

17.10 Piano characteristics

When a note is struck, a variety of factors are involved in creating the desired sound.[7, 8] The hammers are covered with a special wool felt of the appropriate width. If the felt is too wide or soft, the higher harmonics will be damped before the hammer is withdrawn. If the felt is too hard or narrow, the overtone content will be too high. Harmonic content is also controlled by having the hammers strike one-seventh to one-ninth of the string length from the capo.

Following the sounding of a note, the loudness and timbre rapidly change as a result of the rapid decay of the fundamental and higher modes, and the effects of vibration coupling between strings of the double and triple string groups.[9] These effects are particularly noticeable when the sustaining (right) or the sostenuto (middle) pedal are used to sustain the note. The soft (left) pedal is used to shift the entire keyboard system sideways, causing most of the hammers to miss one of the strings.

The piano represents the culmination of a long endeavor to achieve excellence and versatility in keyboard stringed instruments. Its use in the orchestra, in the concerto, as a solo instrument, and as a standard instrument in the home attests to its success.

QUESTIONS

1. Compare the advantages and disadvantages of frets.

2. Describe how string vibration is possible, and explain why this is musically desirable.

3. Describe the evolution of piano design. Study some outside sources.

4. The data in Table 11 – 1 indicates that the upper partials of a piano string are sharp with respect to the desired harmonics. Why?

5. The **clavicord** is a historic keyboard instrument that produces a note by having a metal *tangent* strike the string, causing the string to vibrate between the nut and the point of tangent contact. Explain why such a mechanism produces so little sound. (See Section 16.6.)

6. The piano is tuned increasingly sharp in the upper octaves. Explain how this might compensate for the "tubby" sound of a piano string.

PROJECT

Construct a simple stringed instrument, such as a cigar box guitar or a washtub bass. Experiment with various materials for the strings, the top plate, and other components.

NOTES

[1]C. M. Hutchins, "Founding a Family of Fiddles," *Physics Today* (February 1967):23; and Carleen Hutchins, "The Physics of Violins," *Scientific American* (November 1962):78.

[2]H. Turnbull, *The Guitar* (New York: Scribner, 1974).

[3]E. V. Jansson, "A Study of Acoustical and Hologram Interferometric Measurements of the Top Plate Vibrations of a Guitar," *Acoustica* 25 (1971):95.

[4]For an interesting history and construction notes for the banjo and plucked dulcimer, see E. Wigginton, ed., *Foxfire 3* [New York: Doubleday (Anchor Books), 1975].

[5]W. J. Zuckermann, *The Modern Harpsichord* (New York: October House, 1969).

[6]W. L. Sumner, *The Pianoforte* (London: Macdonald & Co., 1966).

[7]E. D. Blackham, "Physics of the Piano," *Scientific American* (December 1965):88.

[8]A. Benade, *Fundamentals of Musical Acoustics* (New York: Oxford University Press, 1976), chap. 17.

[9]G. Weinreich, "The Coupled Motions of Piano Strings," *Scientific American* (January 1979):118.

Chapter Eighteen

WIND INSTRUMENTS

A variety of instruments produce musical tones by energizing vibrating air columns directly from air emitted from the player's respiratory system. Although they differ from each other in many ways, they are generally noted to fall into one of two distinct classes, **woodwinds** and **brasses**, as indicated in Table 18–1. This division is not perfect, but it is significant that each set of characteristics appears to be correlated in distinguishing a specific instrumental type. Moreover, it permits a framework in which we can study the wind instruments.

TABLE 18–1: Classification of Wind Instruments

	WOODWINDS	BRASSES
Excitation	Reeds and edge tones	Vibrating lips (lip valve)
Material	Wood, plastic, metal	Metal (occasionally plastic)
Bore[a]	Cylindrical, conical	Conical or cylindrical-conical
Pipe scale[b]	Large	Small
Length variation	Opening holes	Tube extension
Examples	Edge tone: flute	Conical: tuba
	Reed: clarinet	Conical-cylindrical: trumpet

[a]Bore geometry is usually somewhat irregular.
[b]Ratio of diameter to length; see Section 12.8.

When the flutist blows across the embouchure hole on the flute, edge tones are initially produced. (See Section 12.6.) The small gusts on the inside of the flute produce waves which then travel down the flute toward the open end. See Figure 18–1. The abrupt end represents an impedance mismatch, reflecting most of the waves back into the instrument. Standing waves are therefore established in the pipe. Since both ends are open to the air, a pressure node (displacement antinode) is established near each end, causing it to resonate as an open-open pipe, with the tube length equal to one-half the fundamental wavelength. (See Figure 12–2.)

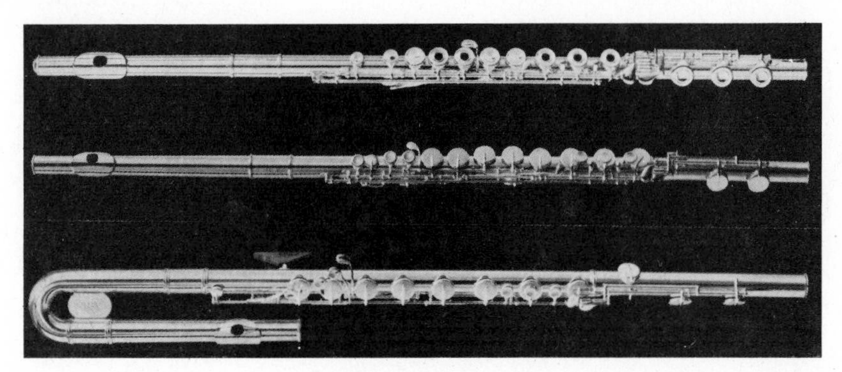

FIGURE 18–1
A conservatory flute, an alto flute in G, and an Ogilive model bass flute in C. (Courtesy of Artley, Inc.)

Although the node position at the playing end is near the embouchure hole, the short cavity behind the hole acts as part of the effective length, especially at the higher frequencies. As a result, the flute can be fine-tuned by adjusting the cork plug that sets the cavity length. See Figure 18–2. If one or more of the tone holes are opened, the far node occurs near the position of

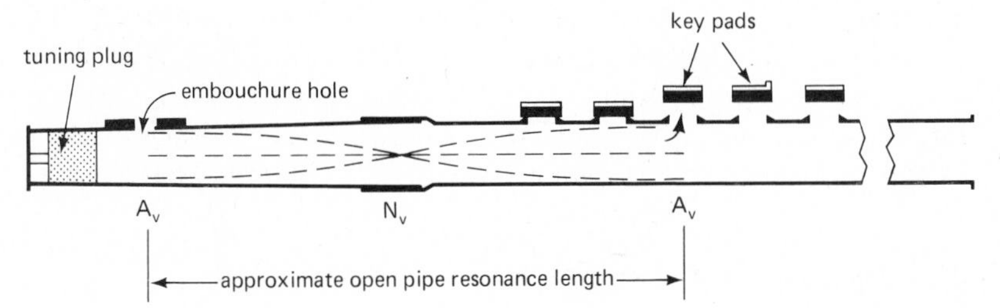

FIGURE 18–2
Cross-sectional diagram of a flute showing open covers and a corresponding displacement wave.

the opened hole that is closest to the embouchure hole. The resonant length is then approximately the distance from the embouchure hole to the first open tone hole, as shown in Figure 18–3(a).* More detailed measurements and calculations show that the effective resonating length is affected slightly by the size of the open tone hole. Even the presence of closed tone holes in the resonating pipe increases the pipe volume slightly and will extend the effective length slightly.[1]

The length of pipe beyond the first open hole is normally not part of the resonating column. However, if the opened hole is followed by several closed holes, a second section of pipe is available for resonance, which can be excited if its resonant frequency corresponds to one of the resonant harmonics of the primary resonant section. This will occur when the lengths of the primary and secondary sections are approximately the ratios of simple integers. See Figure 18–3(b). Certain high notes are excited by opening specific combinations of holes, a technique known as **cross fingering**.

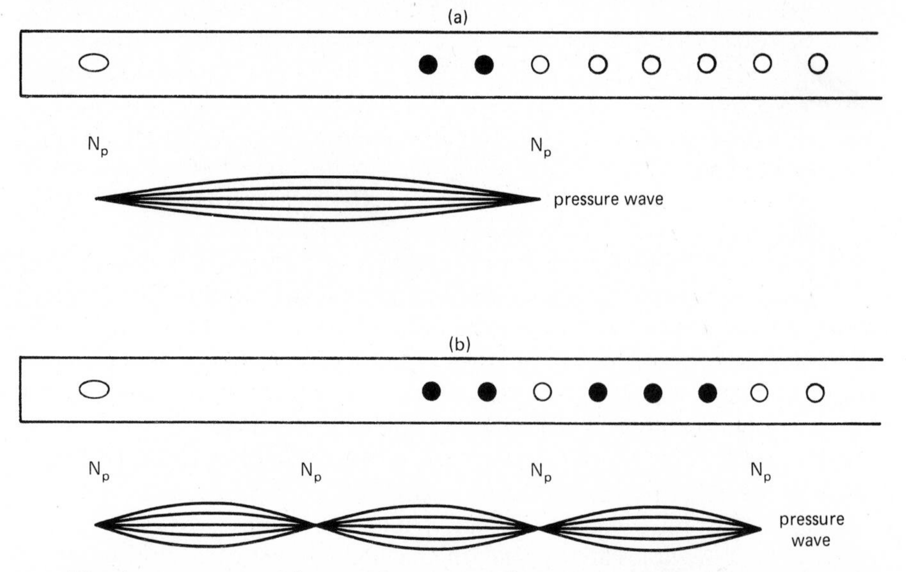

FIGURE 18–3
(a) Normal fingering on a woodwind instrument usually produces the fundamental mode. (b) Cross fingering selects appropriate higher notes.

The lowest octave (**first register**) on a flute is excited by blowing gently across the embouchure hole and opening the necessary tone holes. The holes are positioned to produce the notes of the tempered scale. As the standing wave becomes established in the air column, the airstream at the embouchure hole loses its independent behavior. The series of vortices that characterize a self-induced edge tone is now replaced by an airstream, which moves synchronously in and out of the pipe. This has the effect of feeding air into the column at just the time it is needed to sustain its oscillation. These sharp gusts of

*Pressure waves are illustrated instead of displacement waves in Figure 18–3 because of the convenience in identifying pressure nodes. Pressure nodes occur at displacement antinodes (Section 4.3).

air are capable of creating upper partials in addition to the fundamental. However, the large pipe scale (bore-to-length ratio) favors the resonant response to the lower vibrations, giving the flute its pure clear timbre.[2]

By increasing the blowing pressure, the sound amplitude is increased in the upper partials.[3] By varying the blowing pressure by 10 percent at about 5 Hz, a timbre vibrato is produced, a rapid increase and decrease of the upper partials.

The **second register** of the flute is created by moving the lower lip closer over the embouchure hole and blowing faster. This causes the transit time across the embouchure hole to synchronize with the second harmonic of the resonating pipe. In this second register, the length of the flute between the tone hole and first open hole (including end corrections) corresponds to a full wavelength of the standing wave. (See Figure 12–2.) Each hole position produces a note one octave higher than the corresponding note in the first register.

The increased lip coverage in the higher registers has the effect of lengthening the effective vibrating length and thereby flatting the notes. This is corrected for by a slight bore contraction in the flute toward the embouchure hole. This is one of the many design corrections that are made to wind instruments to keep them on key.[4, 5] By controlled overblowing, the flutist can force the edge tone to match any of the lower resonant harmonic frequencies of the pipe, and specific notes are selected by opening the appropriate holes. Slight tuning errors can be corrected by controlled lip coverage. As a result, the flute can be played over a wide range of notes.

The flute family* contains a number of musical instruments. The **recorder** is a simple keyless instrument of limited capability, which is therefore relatively easy to play and is popular among those interested in early musical styles. See Figure 18–4. In contrast, the modern Boehm concert flute represents a carefully arranged tone and keying design, resulting in high facility and versatility.[6] The **piccolo**, an instrument about half as long as the flute, is an octave higher in pitch.

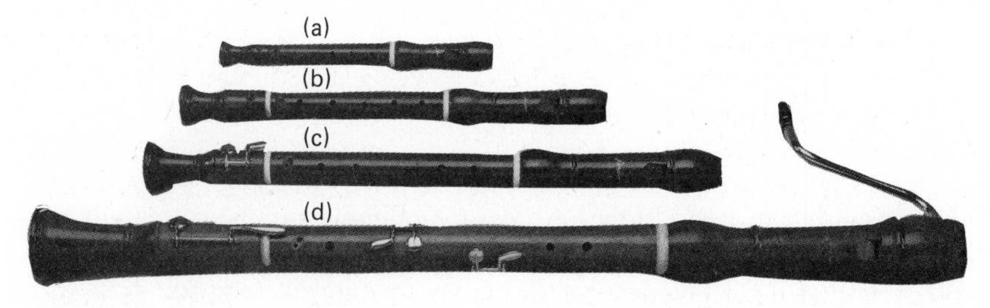

FIGURE 18–4
Four Schreiber recorders with different ranges: (a) soprano, (b) alto, (c) tenor, and (d) bass. (Courtesy of Buffet Crampon & Cie)

*The largest collection of flutes in the world was made by D. C. Miller of "ether drift" fame and is to be found in Washington, D.C., in the Library of Congress.

18.2 Reeds

If air is caused to move across a thin flat surface, the pressure drops below the static air pressure. If the air is stationary on the other side of the surface, the pressure there will not drop. In other words, a fast-moving stream through the opening creates a low-pressure region on top of the reed. If the surface is flexible, the imbalance in pressure will force the surface into the moving stream. This **Bernoulli effect** is responsible in part for a variety of phenomena, such as the curve of a thrown baseball or the lift of an airplane wing.

To illustrate the Bernoulli effect, we will consider the clarinet reed. The clarinet reed (see Figure 18–5) usually consists of a blade of cane, thinned to about 0.1 mm at the tip. This reed is clamped against the mouthpiece of a clarinet, covering all but a thin slot through which air can be forced. The player places his lips across the mouthpiece, pressing against the reed near its center, and blows into the opening. As a result of the vocal tract blowing pressure, as well as the Bernoulli effect, the reed will be forced toward the slightly curved lay of the mouthpiece, slowing or stopping the air flow.[4] When the mouthpiece is attached to the body of the instrument and a standing wave is started (by the first puff through the opening), the motion of the reed becomes controlled mainly by the difference between the vocal tract blowing pressure and the pressure in the mouthpiece. When the standing wave in the clarinet causes a drop in mouthpiece pressure, the blowing pressure and the Bernoulli effect cause the reed to close ("**beating reed**" behavior). When the mouthpiece pressure rises to a maximum, the reed is forced open.

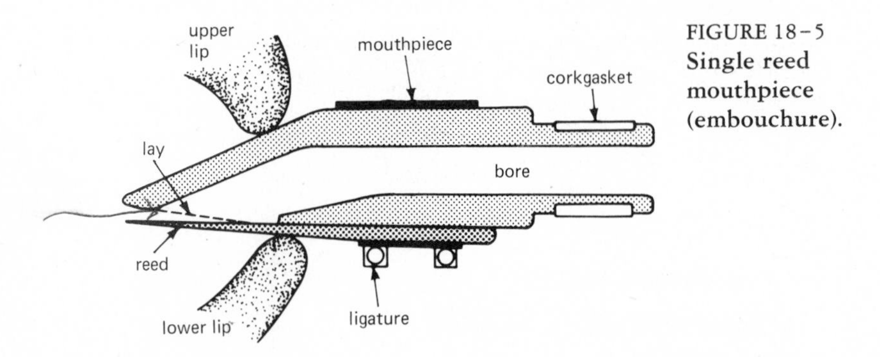

FIGURE 18–5
Single reed mouthpiece (embouchure).

Provided that the blowing pressure remains sufficiently larger than the maximum mouthpiece pressure, puffs of air will continually enter the mouthpiece at just the right time of the cycle to energize the standing wave. At lower blowing pressures, the reed can vibrate without complete closure ("free reed" behavior), but less of the higher harmonics are produced. Blowing is easier and better controlled if several of the resonances of the instrument match the higher harmonics generated by the reed as closely as possible.

A major part of the instrument design, including the bore shape, tone hole size, and placement, are arranged to accomplish this.[5]

Although the natural vibrating frequency of the reed is typically 2000–3000 Hz, the resonating air column will cause it to lock into synchronism at frequencies far lower. The player still has some control over the reed vibration. This control is exercised by controlling the air speed, by placement and pressure of the lips, and by prior selection and treatment of the reed. The reed should have proper size, shape, flexibility, and damping for proper tuning. (Selection of a good reed usually also involves an element of luck.)

Instruments such as the oboe, bassoon, and English horn use double reeds. In the accordion and harmonica, reeds receive little feedback from the surrounding enclosure. Such instruments have a pitch determined by the reed alone.

18.3 The clarinet

The clarinet is an instrument with a length close to that of a flute (about 60 cm). See Figure 18–6. Its bore is close to being constant, but it does expand slightly, especially close to the open end. When one or more tone holes are

FIGURE 18–6 Three wood clarinets: (a) B♭ with 17 keys, (b) B♭ with 20 keys, and (c) E♭ to A♭ alto clarinet. (Courtesy of Buffet Crampon & Cie)

opened, the resonant length is determined by the distance between the mouthpiece and the first open hole. Corrections must be made for the effects of the reed and mouthpiece, variations in bore shape, the size of the open hole, and the presence of intermediate closed holes.

At low frequencies, the standing wave terminates just beyond the first open tone hole. The sound wave is emitted primarily through this hole. Its frequency is generally unaffected by whether the following holes are open or closed. At somewhat higher resonant frequencies, the standing wave extends further. See Figure 18–7(a). As a result, the next holes begin to emit some of

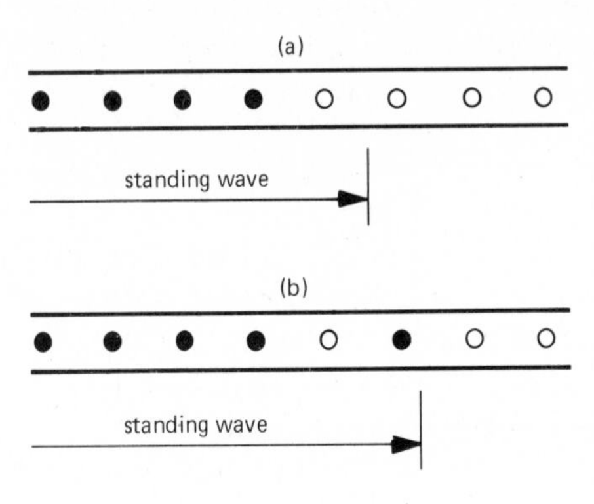

FIGURE 18–7
(a) The effective endpoint for standing waves in a clarinet is normally beyond the first open hole by an amount which depends upon frequency, hole size, and spacing. (b) Forked-fingering will often extend the endpoint, lowering the pitch somewhat.

the sound. Covering one or two subsequent holes now can cause a small drop in frequency. Shifting pitch in this way is called **fork-fingering.** See diagram (b). At still higher frequencies, the wave can run right through the array of open holes. The cutoff frequency (above which this happens) is determined mostly by the hole sizes and spacing. The instrument will not support the resonance of any partials above this cutoff frequency, typically 1500 Hz (about G_6) for the clarinet, 1400 Hz (about F_6) for the oboe, and 500 Hz (about B_4) for the bassoon. The cutoff frequency therefore strongly affects the tone color of the instrument.[5]

Unlike the flute mouthpiece, the reed mouthpiece acts to form a pressure antinode in its proximity rather than a node. As a result, the clarinet—typical of any cylindrical reed instrument—resonates as a pipe closed at one end and open at the other. In the fundamental resonant mode, the length is about $\frac{1}{4}\lambda$. The fundamental frequency is therefore about an octave lower than it would be in an open-open pipe of the same length, such as the flute. The actual frequency also involves the variations in bore, the open-end correction, and the effect of the vibrating reed mechanism. Since the even harmonics of a closed-open pipe are missing, only the odd upper harmonics will be accentuated by resonance in the clarinet, giving it its characteristic hollow or "woody" timbre.

Because of the large pipe scale (bore-diameter-to-length ratio) and reed characteristics, the clarinet will normally resonate in its fundamental mode, called the **chalumeau register.*** The next mode, called the **clarion register,** has three times the frequency, corresponding to an octave and a fifth. Therefore, there must be enough tone holes on the clarinet to span 18 semitones in the lowest resonant mode before forcing the clarinet into its next highest mode. To be specific, the C clarinet plays an E_3 in its chalumeau register with no stops opened. The clarion register starts with B_4. Therefore, the holes must span the notes from E_3 to $B^{\flat}_4$ in the fundamental mode. (For details, see Figure 19–15.)

18.4 The register hole

Although the flute can be forced into higher registers by overblowing, a reed instrument such as a clarinet needs some extra help. This is accomplished by opening a small hole, located a bit less than a third of the distance from the mouthpiece to the bell. See Figure 18–8. This has the effect of creating a pressure node at that point. As shown in diagram (c), the clarion register already has a node at about that point, whereas the chalumeau register does not. As a result, opening the register hole destroys the fundamental resonant mode, forcing the clarinet into the permitted clarion register.

If the register hole is too small, the fundamental will not be suppressed. If it is too large, it will cause a frequency shift for those notes whose pressure

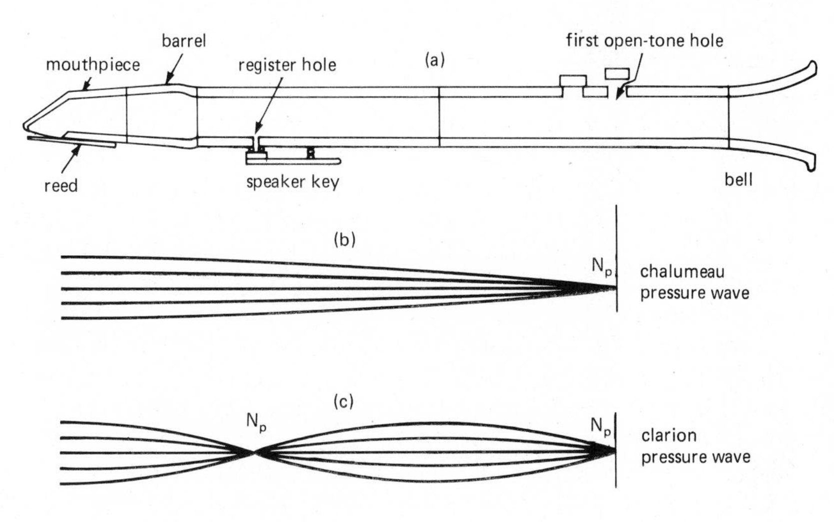

FIGURE 18–8
(a) Simplest clarinet (not to scale). (b) The fundamental (chalumeau) register produces a wave of wavelength approximately four times the distance from the embouchure to the first open tone hole. (c) Opening the register hole creates a partial pressure node at that point, forcing the instrument into the next (clarion) register.

*The lowest few tones in the fundamental register of the clarinet (called throat tones) were not included in the chalumeau, an ancient instrument.

nodes do not quite fall on the hole. The latter problem is handled in the saxophone and oboe by using two register holes. Although all notes in the clarinet register can be played by using a single register hole, other compromises in design (variations in diameter, etc.) are necessary to minimize frequency shifts.

By opening another hole further down on the clarinet, it can be forced into the next mode, the **altissimo register**, a major sixth higher in pitch. Al-

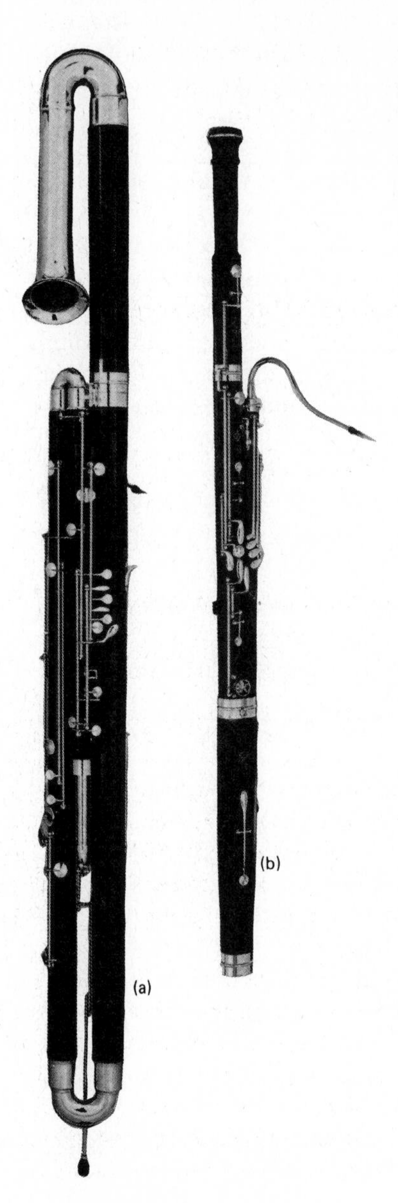

FIGURE 18–9
(a) C contrabassoon. (b) C bassoon. Both hand-polished maplewood. (Courtesy of Buffet Crampon & Cie)

though several notes can be played in this register, many of the higher notes are played by cross fingering (described in Section 18.1; for details, see Figure 19–15.)

18.5 Conical bores

As we have already discussed, the cylindrical bored pipe produces a fundamental frequency and a harmonic series of overtones. When one end is open and the other is closed, only odd harmonics are present. A bore that is not constant will generally produce an overtone series that is inharmonic. Such bores are only used if necessary for other acoustical reasons and if methods are available to correct for inharmonicity during play.

The conical bore, however, is a special case. The reed-driven conical instrument, such as the bassoon, operates as a closed-open resonator. See Figure 18–9. The overtone series is therefore such that the portion of wavelength corresponding to the standing wave is $\frac{1}{4}$, $\frac{3}{4}$, $\frac{5}{4}$, and so on. However, the slowly varying diameter distorts the wave shape so that the wave is stretched out in the narrow end and compressed in the widening end. See Figure 18–10. The

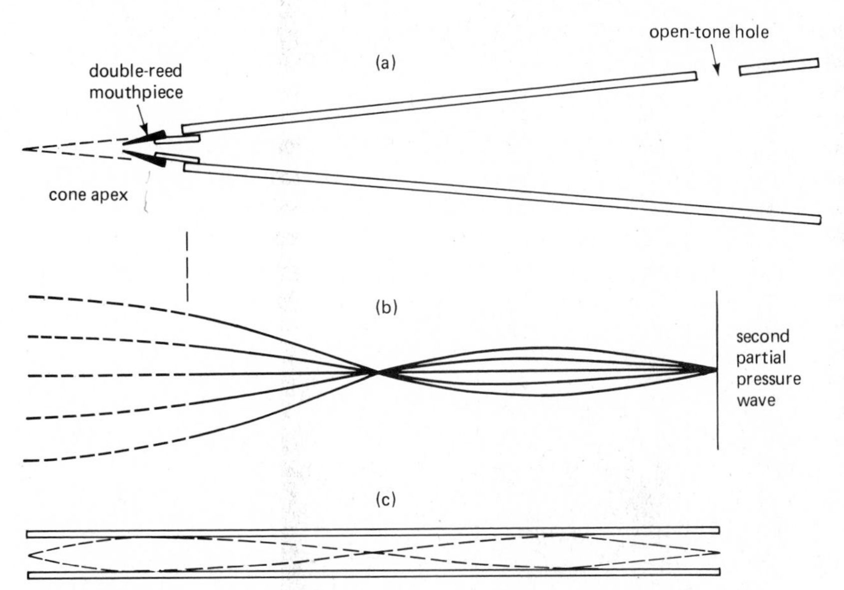

FIGURE 18–10
(a) Truncated cone (opening angle exaggerated) produces a nonsinusoidal standing wave. (b) Second partial, illustrating stretch of its wavelength and increase of its amplitude toward the apex. (c) Open-open uniform pipe, which would resonate at the same frequencies.

result is that the frequencies are identical with those in a open-open cylindrical pipe of the same length. In practice, the first 10 percent of the cone is truncated to allow for attachment of the mouthpiece. Since the standing wave is already stretched in this end, the fractional wavelength affected is small, although not necessarily insignificant.

18.6 Conical wind instruments

Reed instruments with perfectly conical bores will be expected to have a complete set of harmonic overtones, resulting in a richer timbre than the clarinet. The **oboe**, being fairly conical, does have a fuller tone than the clarinet. It is about an octave higher in its fundamental mode and overblows it by an octave. See Figure 18-11. The **bassoon** is a double-reed instrument considerably greater in length. Its bore opens somewhat erratically, forming an approximate cone.

The saxophones form another conical family of reed instruments. They have metal bodies rather than wood or plastic. Although the soprano saxophone is usually straight, the tenor and alto saxophone are bent to keep their unwieldy lengths within manageable bounds. See Figure 18-12. The upper bend is smooth and causes no significant problems. The lower bend, however, is sharp with respect to the diameter. Such sharp bends act as if the bore were wider and shorter at that point,[1] causing some minor tuning problems. The saxophone's conical shape is truncated at about 10-12 percent from its effective apex in order to attach the mouthpiece. Some soprano saxophones are

FIGURE 18-11
Three members of the oboe family. (a) C oboe, to low B♭ (African grenadilla wood). (b) C oboe, to low B♭, automatic F. (c) English horn, automatic F and F, A♭ to B♭ (Courtesy of Buffet Crampon & Cie)

truncated up to 16 percent. This, too, causes tuning problems, which must be corrected by other design (primarily mouthpiece-reed design) and playing modifications.

FIGURE 18-12
Three members of the saxophone family. (a) B♭ soprano saxophone. (b) B♭ tenor saxophone. (c) E♭ baritone saxophone. (Courtesy of Buffet Crampon & Cie)

18.7 The theater organ

Another use of reed excitation is in the reed organ pipe. See Figure 18-13. Unlike the exitation of the flue organ pipe (Section 12.7), the vibration in the reed organ pipe is controlled mainly by the air flow across a reed and by the tension applied by the tuning wire. The reed motion is affected little by the *feedback* of the standing wave in the resonator air column. Moreover, the timbre and loudness can be changed by adjusting the air speed over the reed, giving the reed organ *expression.* Periodic modulation of the airstream results in a *tremolo.*

Reed pipes (see Figure 18-14) are divided into two clases, *chorus* and *orchestral* pipes. The chorus pipe is an open-ended pipe with a length that resonates to the fundamental frequency of the reed, giving a clear pitch characteristic of that frequency. The air column will also resonante with whatever har-

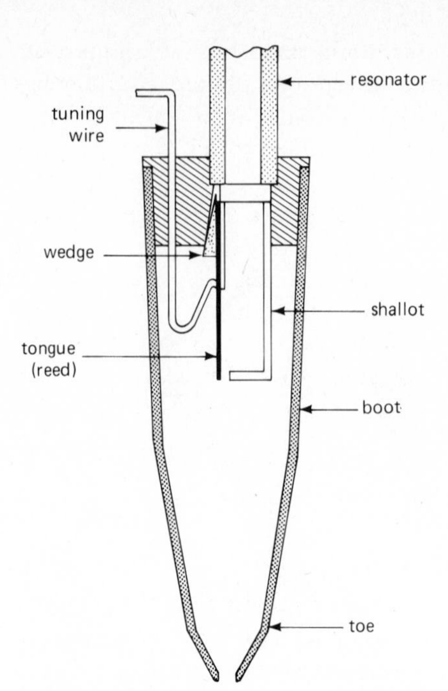

FIGURE 18-13
Tone-generating section of a lingual (reed) organ pipe.

monics are present in the vibrating reed. The result is a harmonically rich timbre. Chorus pipes are used to produce a strong solid tone.By appropriate design, they are made to simulate the tuba, trumpet, and other instruments.

The orchestral pipe has a length shorter than that necessary to resonate

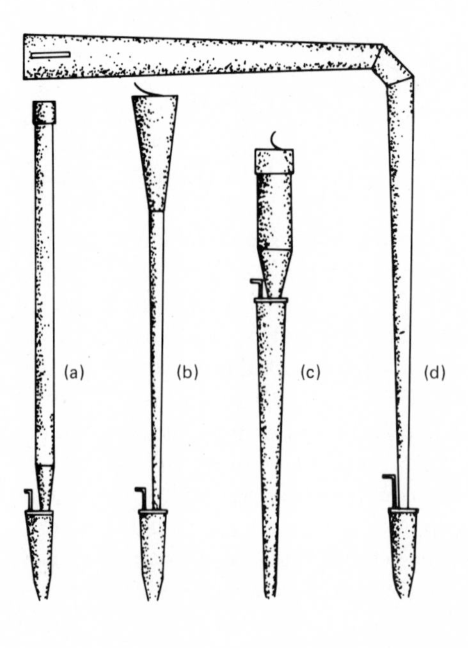

FIGURE 18-14
Lingual (reed) organ pipes: (a) clarinet, (b) oboe, (c) vox humana, (d) trombone.

with the reed fundamental, or the low overtones. As a result, the pipe resonates only to certain of the higher partials of the vibrating reed, giving it a buzzy sound of somewhat indefinite pitch. Orchestral reed pipes are often made to bear resemblance to clarinets, oboes, and even the human voice (vox humana).

Most theater organs use many sets (ranks) of reed and flue pipes of various shapes, pipe scales, and so forth to generate a variety of timbres. Whereas the diapason is the principal pipe for the church organ, the tibia—a large-scale, stopped wooden pipe—is basic to the theater organ. Moreover, we have seen that many pipe types are designed to simulate the sounds of various acoustical instruments. Although the resemblance is usually far from perfect, a sufficiently large collection of pipe types provides a very versatile instrument.

18.8 The ocarina

Most wind instruments involve setting up standing waves in long thin tubes. If the instrument is short and fat, it acts more like a cavity (Helmholtz) resonator. (See Section 6.2.) The vibrating mass consists mostly of the air rushing in and out of the opened hole or holes. The cushioning effect of the air in the cavity acts as the spring of the resonant system. The spring and mass functions are *separately identified* here, in contrast to the resonating tube in which these characteristics are *distributed* along the tube.

The ocarina operates as such a cavity resonator. See Figure 18–15. Here, the overall shape and the location of the holes have only a minor effect on the pitch. Rather, the pitch is determined by the total volume of the cavity and the total area of the open holes.

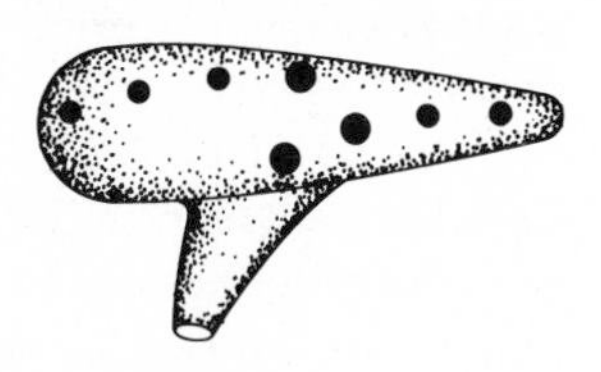

FIGURE 18–15 The ocarina, a tunable cavity resonator.

18.9 Materials

Musicians have long debated the merits of various materials in the construction of their instruments. Thin-walled wind instruments in many cases have wall vibrations that affect the standing waves, usually in detrimental ways. This is most noticeable in certain flutes and metal organ pipes. Such problems are usually avoided if the bodies of wind instruments are made of materials that are strong, rigid, smooth, and nonporous and are not easily warped, dented, or corroded. These properties may, in principle, be met by plastic as well as by ebony.

In practice, selection of materials is often made for their ability to be worked into the desired shape. It is often these shape differences that cause variations in tone quality, rather than the materials themselves. For example, problems with the tone of a plastic clarinet might be traced to turbulences set up at the sharp edges of the tone holes, whereas a wooden clarinet may be constructed with more rounded edges.

Provided that the material can be worked in accordance with the criteria above, the quality of performance will be due primarily to the overall geometry of the instrument, the tone-generating mechanism, and the ability of the instrumentalist. In spite of these facts, many performers as well as instrument manufacturers insist that superior performance requires the use of special materials, especially rare and exotic woods and expensive metals such as silver and gold.

18.10 The energy chain

As the clarinet player blows across the reed, the opening and closing of the reed will break the airstream into puffs, which generate the wave inside the body. Once the standing wave is established, it is essential that the reed motion synchronize with the standing wave in such a phase that the wave is augmented rather than damped.

The sound heard from the clarinet results from a portion of the wave escaping from the instrument. This sound leaks from the bell if the tone holes are all closed. Otherwise, it is emitted primarily from the first opened hole.

It is commonly believed that a wind instrument acts like a megaphone, channeling sound energy from the mouthpiece to the open air. Actually, almost all the energy is trapped as a standing wave in the vibrating pipe, and it is eventually transformed into heat by friction against the inner wall. Only one or two percent of the input energy escapes the instrument as sound.

In this chapter, we have selected certain common instruments to illustrate the main features of the flute and reed instruments. Many interesting but subtle features have been omitted or glossed over in order to limit discussion to basic concepts. The reader who is interested in details of specific instruments or instrument components is referred to any of several books and articles on the subject (see the Bibliography).

QUESTIONS

1. Why does opening a register hole cause a wind instrument to sound a higher harmonic?

2. Although the flute and clarinet have about the same length, the fundamental pitch of the flute is about an octave higher than the clarinet. Why?

3. Why do woodwinds normally sound the fundamental tone?

4. Why do cylindrical and conical bore woodwinds have different timbres?

5. Why is the saxophone classified as a woodwind and not as a brass instrument?

6. What does the energy chain of the wind instruments have in common with the energy chain of a stringed instrument? (See Section 16.11.)

7. How does the use of the register hole relate to the method of playing in harmonics? (See Section 16.6.)

PROBLEMS

1. Calculate the approximate length of a B♭ clarinet, which plays D_3 in the chalumeau register with all holes stopped.

2. If a clarinet plays A_3 in the chalumeau register, what note would be played in the clarion register with the same stops opened? What note would be played in the altissimo register?

PROJECTS

1. Measure the length of several common cylindrical woodwind instruments (e.g., flute and clarinet) and compare your results with the length calculated from the known fundamental pitch with all holes closed. Try to account for any differences found.

2. Try adapting a flute head joint to a clarinet as a replacement for its normal mouthpiece and barrel joint. Does the combined instrument sound like a flute or a clarinet? Explain your results.

NOTES

[1]C. J. Nederveen, *Acoustical Aspects of Woodwind Instruments* (Amsterdam: Frits Knuf, 1969).

[2]N. H. Fletcher, "Acoustical Correlates of Flute Performance Technique," *Journal of the Acoustical Society of America* 57 (1975):233.

[3]W. L. Klein and H. J. Gerritsen, "Comparison Between a Musical and a Mathematical Description of Tone Quality in a Boehm Flute," *American Journal of Physics* 43 (1975):736.

[4]A. H. Benade, "The Physics of Woodwinds," *Scientific American* (October 1960):144.

[5]A. H. Benade, *Fundamentals of Musical Acoustics* (New York: Oxford University Press, 1976), chap. 21.

[6]T. Boehm, *The Flute and Flute Playing in Acoustical, Technical, and Artistic Aspects* (reprint ed., New York: Dover Press, 1964).

Chapter Nineteen

BRASSES AND THE HUMAN VOICE

Early in human history it was found that blowing through tightened lips into a hollow animal horn would create a raucous honk. Under the right conditions, however, a clear musical note could be produced. The modern brass instrument is the descendent of such a horn, improved by trial and error, correction and compromise, to become the still imperfect instrument that it is today. We will look at some of the characteristics of the brass instruments—and the human voice—in this chapter.

19.1 Modern horn design

The brass wind instrument has become fairly well understood acoustically in recent years.[1] In order to appreciate the major features of the modern design, it is helpful to discuss the design as if it evolved from acoustical considerations.

Consider the four pipe shapes shown in Figure 19–1. Since the mouthpiece end acts as a closed end, similar to a reed instrument, the horn acts as a closed-open pipe, as in diagram (a). To produce the desired bright timbre, all harmonics should be present. This suggests using a conical bore rather than a cylindrical pipe, as in diagram (b). However, a conical taper is difficult in the central section where the valve-operated tube elongation occurs, and it is impossible in the case of a slide trombone. This leads to a central straight section, as in diagram (d).

The mouthpiece must be designed to fit comfortably against the lips. The flared end, or bell, must be carefully shaped to get adequate sound transmission to the open air, yet reflect enough of the acoustical wave to establish a standing wave in the horn. These considerations must be taken together in order to appreciate the flaring shape of the modern brass instruments.

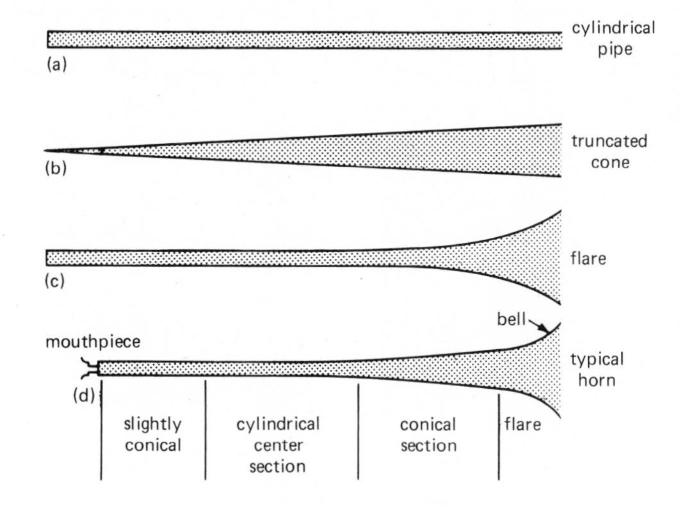

FIGURE 19–1 **Common horn shapes, such as (a) the cylindrical pipe, (b) the truncated cone, and (c) the flared horn, are (d) combined in the construction of a typical brass instrument.**

19.2 The mouthpiece

To excite a musical note in the horn, the musician presses his lips against the mouthpiece and expels air from his vocal tract into the horn, causing his lips to vibrate. The tone produced depends on the lip separation, lip tension, placement on the mouthpiece, and mouthpiece shape. The combination of these attributes is called the **embouchure**. These features, along with the control over the airflow, represent the player's ability to control the input to the instrument. See Figure 19–2.

The **mouthpiece** consists of a hollow cavity followed by a **mouthpipe**, a short tube somewhat smaller than the following section of the horn. The abrupt mismatch of the mouthpiece to the horn has the general effect of somewhat decoupling the mouthpiece acoustically. The mouthpiece has a reso-

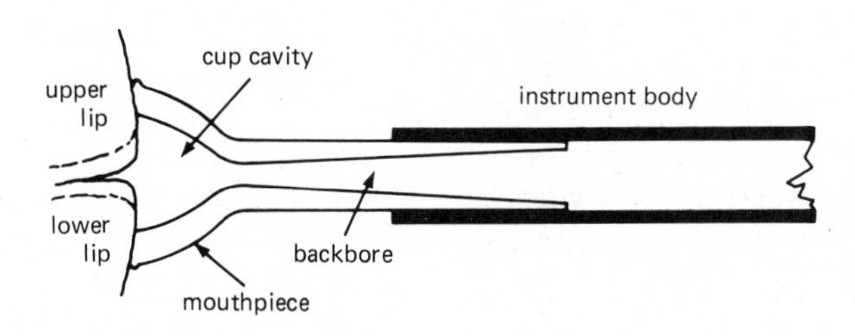

FIGURE 19–2 **Embouchure involves proper relation of player's lips and mouthpiece, which is coupled to the rest of the instrument through a narrow backbore.**

nance frequency of its own, which enhances the response of a significant portion of the response of the horn as a whole. In addition, the effective length of the mouthpiece varies with frequency in a somewhat complex way, depending upon the volume of the cavity and its own natural frequency.[2]

As the lips vibrate, the opening is rapidly varied, causing the air to be injected in puffs. If the lip frequency matches one of the resonant modes of the instrument, a standing wave will be established. The lips are close to a pressure antinode, so that at the part of the cycle in which the pressure increases, the lips are forced apart. Since the blowing pressure exceeds the horn pressure, air is expelled from the player's airway at that part of the cycle that will add to the standing wave. This action provides the acoustinal *feedback* that locks the lip motion to the standing wave, and it causes the player to continue to add energy to the instrument at its resonant frequency.

If the higher resonant frequencies of the horn are harmonically related to the fundamental, then the higher overtones of the lip vibration cause their own standing waves, which serve to further lock in the lip vibration. As a result, the playing frequency is stabilized, resulting in a clean tone. If the higher resonances are somewhat mismatched, the lip vibration frequency may be a compromise, resulting in an overall poorer tone.

19.3 The bell

If a horn were terminated with a smooth gradual flare, it would act as an impedance-matching transformer, such as a megaphone or a loudspeaker (Section 5.6). See Figure 19–3(a). In this case, the traveling wave would move from the mouthpiece down the horn and out. Alternatively, if there were no flare, the match to the air would be poor, and most of the wave would be reflected.

The bell of a horn, Figure 19–3(b), actually serves both functions. For

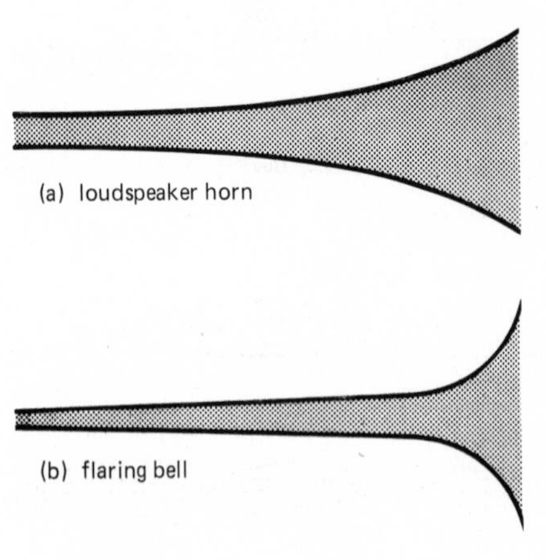

FIGURE 19–3
The loudspeaker horn (a) provides good energy transfer beyond the opening, whereas the abrupt flare of the instrument bell (b) reflects most of the sound wave energy.

notes in the playable range, the bell flare is abrupt compared with the wavelength, and the waves are reflected, forming a resonant standing wave. At sufficiently high frequencies, the flare of the horn is gradual compared to the wavelength, and such waves are efficiently expelled from the bell.

The vast majority of resonant wave excitation at the playing frequencies is trapped inside as a standing wave and dissipated almost entirely by friction with the interior wall. The audible sound from the horn resonance is due to only a small fraction that is "leaked" through the bell into the open air. It seems truly amazing that the opening end of a horn actually acts as a very efficient barrier to most traveling sound waves.

For a cylindrical pipe, the standing waves terminate at a pressure node slightly outside the open end. (See Section 12.3.) For a bell opening, the position of the node seems somewhat indefinite. Acoustical analysis[1] shows that the effective node in an open bell end depends upon wavelength, moving deeper into the bell with lower frequencies. See Figure 19–4. In effect, the waves are reflecting at that point at which the rate of flare is rapid compared with the wavelength. By proper shaping of the overall flare, the effective lengths of the various resonant modes can be adjusted to bring them fairly close to a harmonic relationship.

As the higher modes terminate farther out in the bell, they tend to radiate better, giving the flared horn its characteristically bright sound. Above a certain cutoff frequency, however, the waves radiate almost completely. This lack of reflection precludes the buildup of standing waves and prevents the player from locking onto the desired frequency.

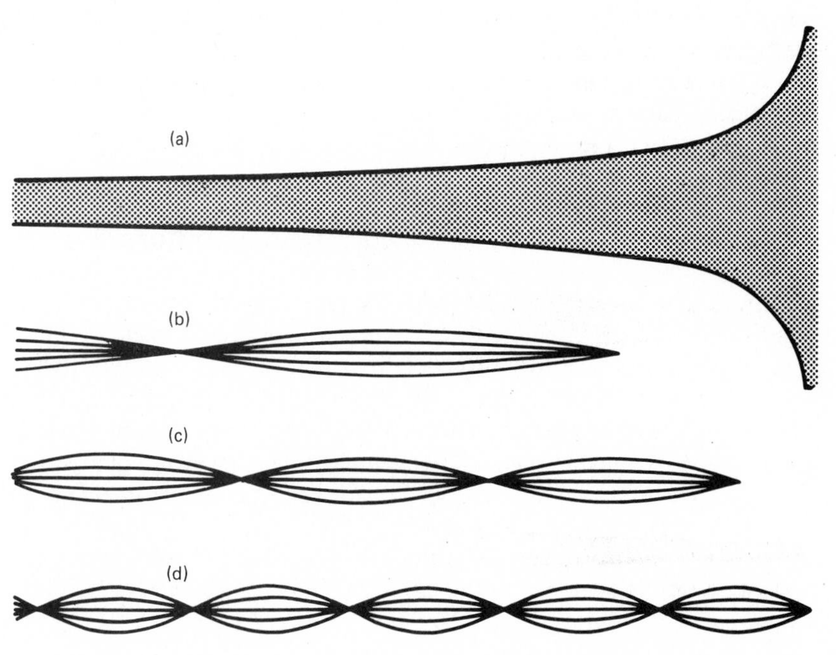

FIGURE 19–4
(a) The horn of an abruptly flared instrument and standing pressure wave patterns for (b) a low frequency note, (c) a middle frequency note, and (d) a high frequency note.

The somewhat indefinite position of the node in the bell means that the horn will still resonate fairly well if the wavelength is somewhat altered, corresponding to a small shift in nodal position. This allows the player to alter the resonant note somewhat by control of the embouchure, a technique known as **lipping-in.** Unfortunately, if it is easy to alter a note in this way, it is correspondingly more difficult to keep on key. The **French horn** and the **cornet,** for example, tend to suffer from this problem more than a trumpet. The nodal position in the bell can also be changed by alteration of the effective bell shape, such as by sticking the hand properly into the bell.[1]

The flared bell of the horn also affects the directionality of the emitted sound. Diffraction normally causes sound waves that leave small openings to spread widely (see Section 7.1). For a horn, high-frequency sound will have a wavelength smaller than the bell opening, resulting in the forward projection of the highs. This results in aiding the bright sound of a horn, provided that it happens to be pointed at the listener.

19.4 The pedal tone

As a result of the factors discussed so far, the typical brass instrument will have several resonant frequencies that fall approximately along a harmonic series. The fundamental frequency, whose higher harmonics correspond to horn resonances, is called the **pedal tone.** Because of the compromises in horn design, this note, and sometimes the next harmonic or two, will not correspond to natural resonances of the horn. For example, the lowest resonant frequency of the trumpet is several semitones lower than the pedal tone. Curiously enough, this pedal tone can usually be played anyway, although with some difficulty. Such a nonresonant note is called a **privileged note,** and it is made possible because the higher harmonics of the pedal lip vibration are locked in by the higher resonances in the horn.

19.5 The alpenhorn and bugle

The Swiss alpenhorn (alphorn) is a fine example of a straight conical horn. See Figure 19–5. Approximately four meters in length, it is made of wood and comes apart near the middle for easy transportation in the Swiss Alps. The horn has a typical mouthpiece at the small end and a slightly upturned bell. The haunting tone of the alpenhorn so impressed Brahms that he incorporated an alphorn theme in the fourth movement of his first symphony.

It was early discovered that the acoustics of a wind instrument are practically unchanged if the instrument is bent, provided that the bends are not too abrupt and that there are no abrupt changes in diameter. The valveless (natural) bugle is a simple 1.4-m, conical-cylindrical brass instrument coiled

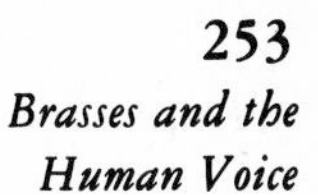

FIGURE 19-5
Alpenhorn. (Courtesy of Howard and Barlett Mel)

into a manageable size, as shown in Figure 19-6. The effective pedal tone is C_3, consistent with the harmonic series*:

C_3,	C_4,	G_4,	C_5,	E_5,	G_5,	($A\sharp_5$),	C_6,	D_6,	E_6
1	2	3	4	5	6	7	8	9	10

normal playing range (3–6)

FIGURE 19-6
The valveless (natural) bugle is an example of a conical-cylindrical brass instrument. (Courtesy of Jerald P. White)

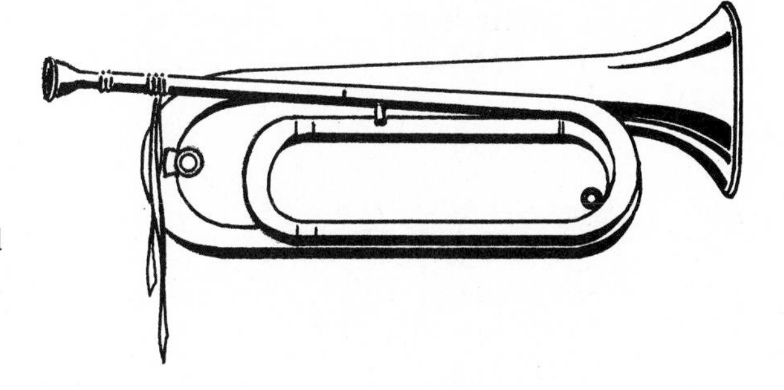

Of these, the C_3, and to a slight extent C_4, does not correspond to resonant frequencies. In addition, the narrow pipe scale of the bugle, as well as most

*Although bugle music is written in the key of C, the modern B♭ bugle is pitched a whole tone lower.

brass instruments, facilitates excitation of the higher harmonics but inhibits the lower ones.

At the upper end of the bugle scale, the seventh harmonic is dissonant. The harmonics 8 to 10 and many of the higher ones are valid diatonic notes (see Problem 2) but are so close together in frequency that considerable embouchure control is required for the player to select the desired note. As a result, standard bugle tunes such as *taps* and *reveille* are written to require only harmonics 3, 4, 5, and 6.

19.6 The trombone

The limited compass of the natural bugle presents no problems to the U.S. Cavalry or the Scouts. However, the serious musician will want to be able to play all semitones within the playing range. This can be accomplished by providing a means of selectively extending the central cylindrical section of the horn. See Figure 19–7. The slide trombone lengthens the central section by extending a U-shaped section which slides over close-fitting pipes underneath.

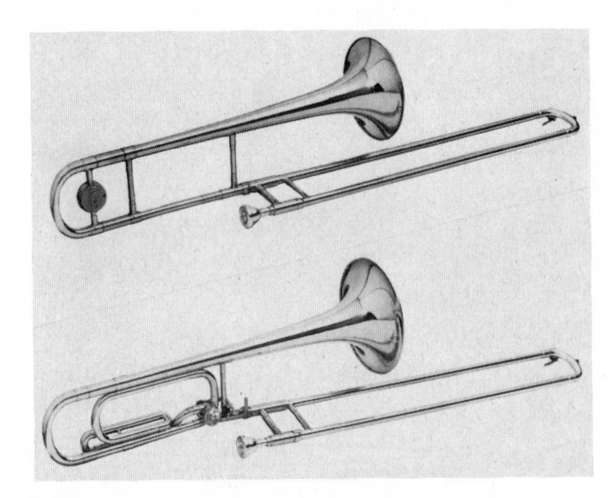

FIGURE 19–7
Photos of two B$^\flat$ trombones. (Courtesy of C. G. Conn, Ltd.)

Consider the slide brought to its shortest position and sounded in the third harmonic. The second harmonic is a musical fifth lower, seven semitones. To span this gap in the chromatic scale, six additional lengths are required. All these are used when the trombone is played in the third harmonic. See Figure 19–8. A decrease in a semitone requires a decrease in frequency of about 6 percent, requiring an additional 6 percent extension of the total pipe length. Since the total length increases from one note to the next lower, each further 6 percent increase requires increasingly larger extensions, as shown in Figure 19–8. The trombone player learns where those positions are by practice. If a note is slightly missed, it can be quickly corrected by a smooth glide to the proper pitch. The player can easily vary the frequency to achieve a vibrato effect.

With seven notes, the player can span the gap from the third harmonic

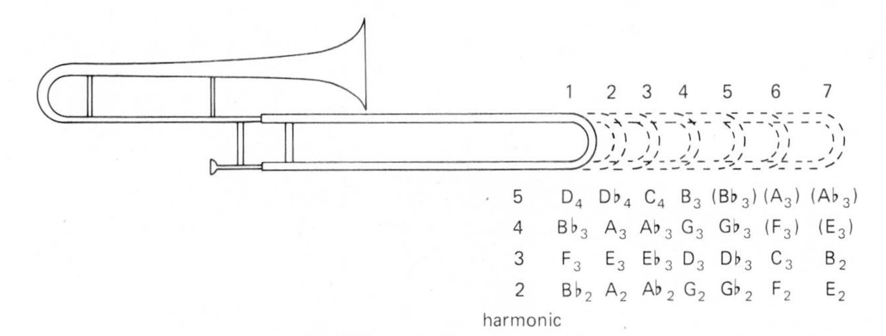

1 2 3 4 5 6 7
5 D_4 Db_4 C_4 B_3 (Bb_3) (A_3) (Ab_3)
4 Bb_3 A_3 Ab_3 G_3 Gb_3 (F_3) (E_3)
3 F_3 E_3 Eb_3 D_3 Db_3 C_3 B_2
2 Bb_2 A_2 Ab_2 G_2 Gb_2 F_2 E_2
harmonic

FIGURE 19–8
Playing positions of the slide trombone.

(register) to the second and can play low notes well down into the second register. Since the gaps between higher registers are closer together, they can be amply filled by the available length positions. Changing registers is accomplished by properly changing the embouchure and blowing pressure.

19.7 The trumpet

The trombone is the only major wind instrument that permits smooth and continuous progression from one note to the next (portamento). The others increase the central length by valve action. Some bugles, for example, incorporate one or two valved extensions. The **valved cornet**, the **flugel horn**, and the **trumpet** are instruments similar to the bugle, but they have three (or more) valves. See Figure 19–9. Pressing the second valve, the central one,

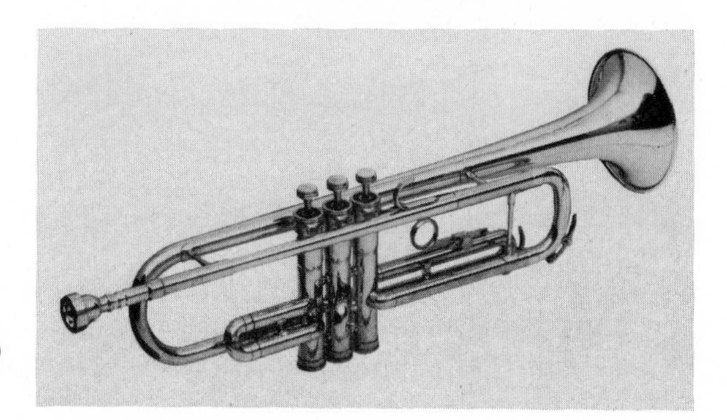

FIGURE 19–9
The trumpet, showing its three valves for lengthening and shortening the total length of the vibrating air column. (Courtesy of C. G. Conn, Ltd.)

adds a section of pipe called a crook into the central section of the pipe. See Figure 19–10. This increases the basic length by about 6 percent, lowering the pitch about a semitone. Pressing the first valve increases the length about 12 percent, lowering the pitch about a whole tone. To drop it a semitone further now requires a length of additional tubing a bit more than that provided by adding crook one. Pressing valves one and two together to try to reach a

FIGURE 19–10
The cornet and flugel horns each have three valves for lengthening or shortening their vibrating air columns. (Courtesy of C. G. Conn, Ltd.)

note three semitones lower will therefore result in a note slightly sharp. The player must therefore lip this note down. The third valve could be made of a longer length to bring in this note. However, it is made even longer yet so that when used in combination with valves one and two to reach even lower notes, the combined length discrepancies do not get out of the range where they can be corrected by lipping in.

These considerations are summarized in Figure 19–11 (a). It can be seen here that none of the notes are perfectly in tune. Neither are they very far off.

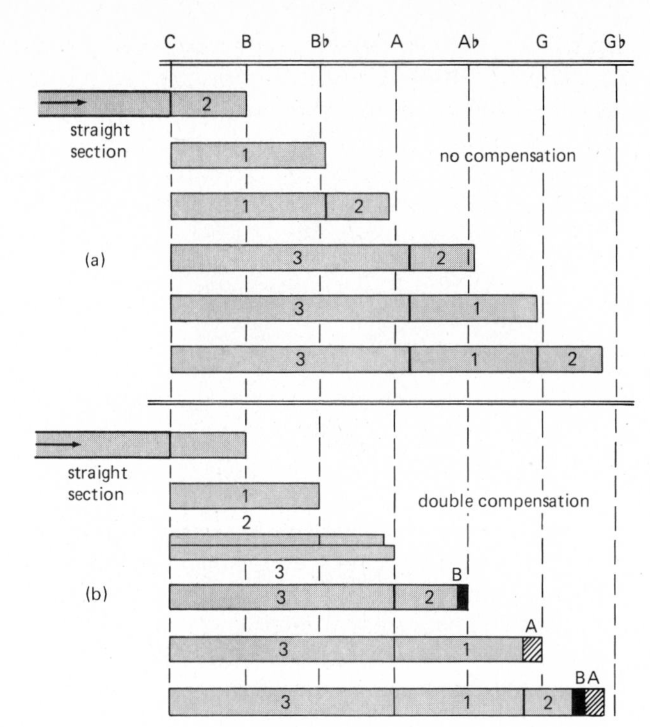

FIGURE 19–11 **Valved extension of a trumpet and a tuba (represented here by a uniform pipe for convenience) cannot be compounded perfectly. Compensation crooks, or other means of extension, are necessary for proper intonation.**

19.8 Compensation crooks

In an effort to overcome the problem of variable crook lengths, many instruments have incorporated additional valved crooks. For example, the double French horn uses a total of eight valved crooks. See Figure 19–12. Alternatively, many instruments have a movable slide to compensate for the discrepancies due to combined valves. Some brasses, such as the tuba, may incorporate extra crooks when two or more valves are simultaneously depressed. The plumbing system itself is complex as well as clever. However, the results are easily depicted, as shown in the lower portion of Figure 19–11. When valves 1 and 3 are pressed, extra crook A is automatically added to crooks 1 and 3. When valves 2 and 3 are depressed, extra crook B is added to crooks 2 and 3. When all valves are depressed, all five crooks are added. Only in this latter case will the total still be slightly short.

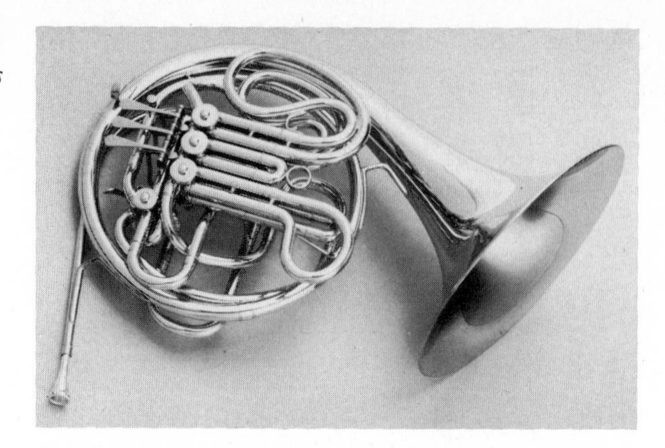

FIGURE 19–12
A double F to B♭ French horn. (Courtesy of C. G. Conn, Ltd.)

19.9 Other brasses

The trumpet and trombone family of brasses have a sizable amount of cylindrical pipe in the center of the instrument and an overall flare toward the end. A second family, including the **alto horn, baritone horn,** and, to some extent, **tuba** and **sousaphone,** is closer to an overall conical shape. See Figures 19–13 and 19–14. These horns differ from each other in a variety of subtle ways, giving each its characteristic timbre. The playing range is determined by the overall length, varying from E_3 to $B\flat_5$ (and above) for the B♭ trumpet to E_1 to $B\flat_3$ (and above) for the $B\flat_0$ tuba. The operation of most of these instruments is basically the same as that already discussed. The reader who is interested in the nuances of specific brass instruments will find further information in the references.[1, 2]

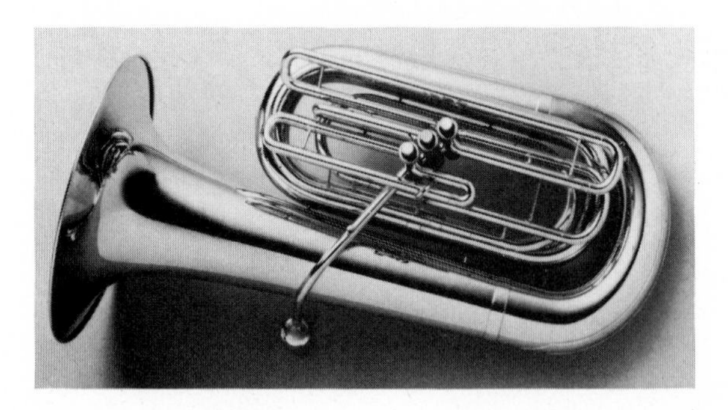

FIGURE 19–13
A tuba, used as a background instrument in a band. (Courtesy of C. G. Conn, Ltd.)

19.10 Brasses versus woodwinds

Early in Chapter 18 it was noted that woodwinds and brasses varied in several respects. In summary, it is interesting to compare the playing range of the bugle, the trumpet, and the clarinet. See Figure 19–15. In both cases, the trumpet and clarinet notes are selected by varying the harmonic (register) and

FIGURE 19–14
A fiberglass sousaphone, used largely in marching bands. (Courtesy of Norlin Industries, Inc.)

the tube length. The clarinet is played primarily in the first two registers, whereas the trumpet is played primarily in registers 2, 3, 4, 5, 6, and 8. As a result, more variations in length are required for the clarinet than the trumpet. This distinction between favored registers is due in part to the smaller average pipe scale (bore diameter) of the trumpet. The favoring of higher partials in the trumpet is a further result of the narrow bore, as well as the involvement of the bell as the sound emission point.

Finally, it should be noted that each instrument has its unique history of development and its unique place in the musical world. It is hoped that the acoustical evaluation of these instruments will result in a deeper appreciation of the idiosyncrasies with which the musician must cope to coax music from his or her instrument.

19.11 The human voice

Most sounds of the human voice[3,4] originate in the vibrations of the **vocal cords** in the *larynx.* See Figure 19–16. As the air from the lungs passes these stringlike membranes, the *Bernoulli effect* (Section 18.2) sets them vibrating. The frequency of vibration depends mainly on the muscular tension applied to the cords. When singing, these tensions are altered, just as stringed instruments are tuned by changing the string tension. A good singing voice can normally span about two octaves by varying vocal cord tension.

Each time the cords part, a sharp gust of air is emitted through the open-

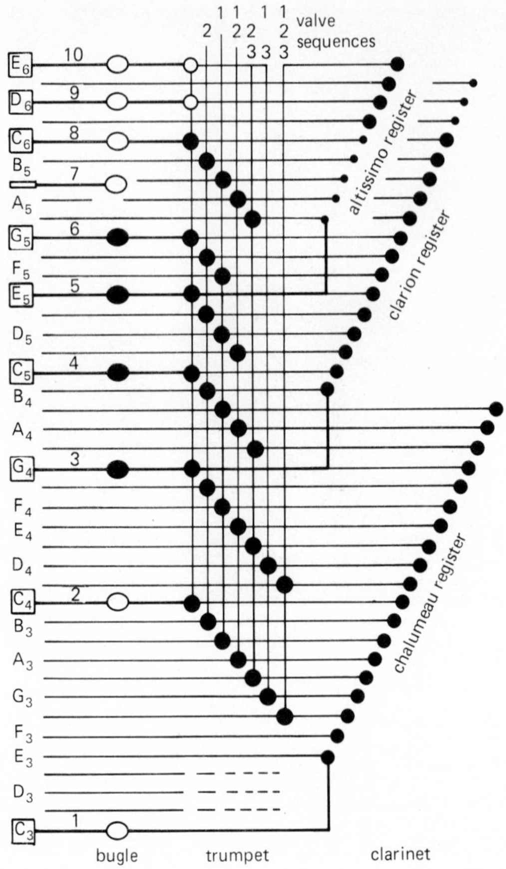

FIGURE 19–15 **Comparison of the playing sequences of the C bugle, the C trumpet, and the C clarinet.**

ing (glottis) into the oral cavity. A series of such sharp gusts, as shown in Figure 19–17(a), is characterized by a series of harmonics of only slowly decreasing amplitude. Such an overtone-rich sound produces a **bright tone.** If the air from the lungs is passed through the larynx cords more gently, the vocal cords do not completely close, resulting in a train of waves more closely resembling a sine wave, as shown in Figure 19–17(b). Such a wave is dominated by the fundamental frequency, with a rapidly decreasing set of overtones. The resulting sound is generally described as a **dark tone.**

The vocal tract acts as a resonator; it is similar in many ways to a reed organ pipe in that it acts as if it were closed at the larynx and open at the mouth. See Figure 19–18. If it were a cylinder of length L it would resonate at frequencies whose wavelengths are $\frac{1}{4}\lambda$, $\frac{3}{4}\lambda$, $\frac{5}{4}\lambda$, and so on (see Section 12.2). However, the soft walls of the vocal cavities absorb the sound waves much more than do the hard walls of the reed instruments. This damping of the sound wave broadens the resonant frequencies, so that resonance occurs over

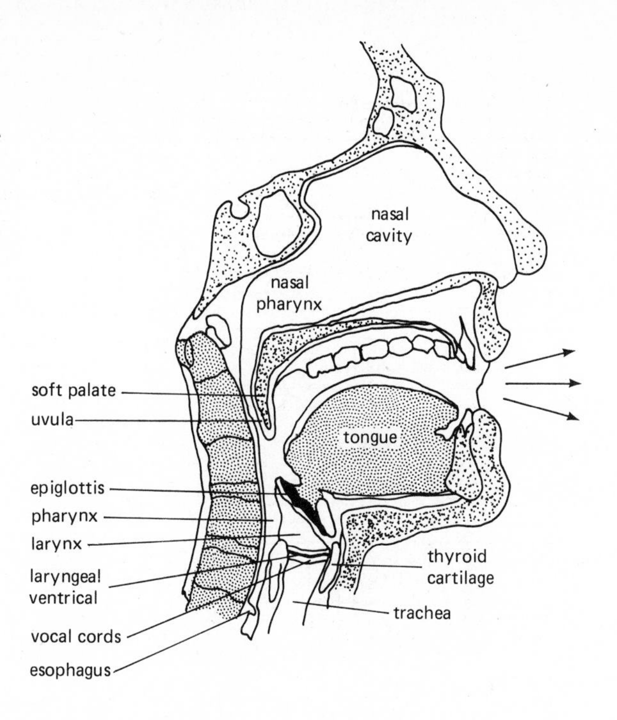

FIGURE 19–16
Cross-sectional diagram showing the location of the vocal chords in relation to other related features of the human head.

several broad frequency ranges instead of at narrowly defined frequencies. See Figure 19–19(b). These broad resonances are called **formants.**

The vocal tract, however, is hardly a perfect cylinder. As a result, the overall response curve and formant frequencies are altered in accordance with the true shaping of the vocal tract. In contrast to what occurs in most instru-

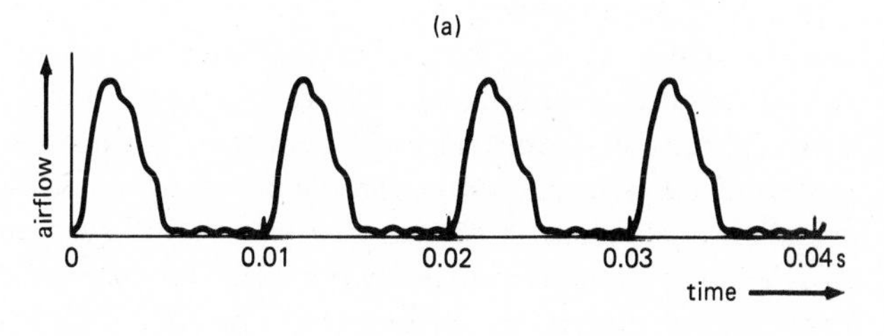

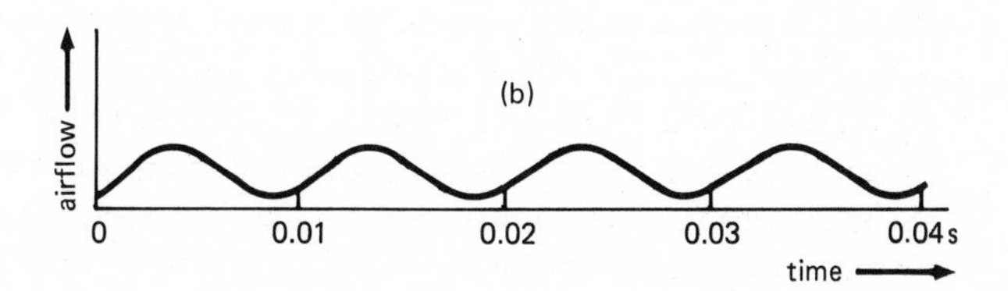

FIGURE 19–17
Typical wave forms for the larynx at (a) high intensity and (b) low intensity.

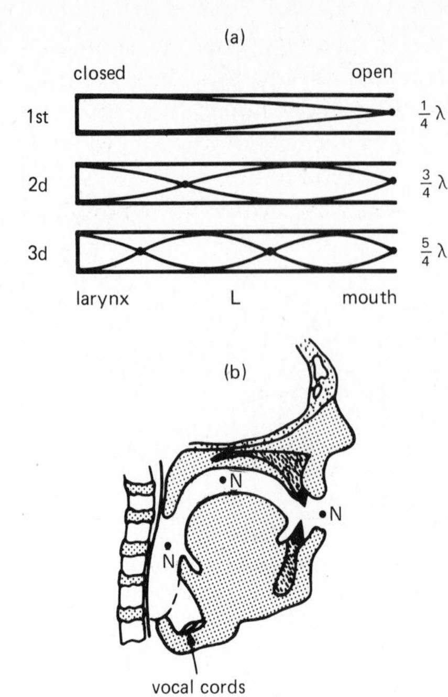

FIGURE 19–18
(a) Diagrams of pressure differentials in the first, second, and third formants, showing nodes and loops in straight air columns.
(b) Diagram of the vocal tract showing the larynx as a short tube at the base of which are the vocal cords. Appropriate nodes N are shown for the third formant resonance.

ments, however, the tract resonances have little feedback control over the vocal cords.

A **whisper** is caused by allowing air to move past stationary vocal cords, causing a hiss rather than a definite tone. Since a hiss is basically noise—a random combination of all frequencies—the resulting audible sound will be caused by those generated frequencies that resonate in the vocal tract at the formant frequencies. The rather indefinite pitch of the whisper is therefore characteristic of the formants.

If a tone is vocalized by causing the vocal cords to vibrate, any partials of the vocal cord frequency [Figure 19–19(a)] that fall within the formants will produce a strong resonant response, as shown in Figure 19–19(c). If different frequencies are generated by the vocal cords, but the vocal cavity remains unchanged, different partials will be accentuated. The quality of a voiced sound is determined by the size and shape of the larynx, throat (pharynx), mouth, and nasal cavities. The singer has only partial control over these variables, although voice training can extend the degree of control. Studies indicate that the jaw angle and lips control mainly the first formant, whereas the position and size of the constriction between the tongue and mouth roof determines mainly the second formant.[5] Control of the velum (soft palate) can open the nasal cavity, adding another resonance at around 1000 Hz.

The adult male generally can vary his first formant between 200–700 Hz and the second formant between 700–2500 Hz. The third formant will be higher yet. Male opera singers often develop an extra **singing formant** in the range 3000–3500 Hz by reshaping the larynx-pharynx transition so that the larynx cavity resonates somewhat independently of the rest of the vocal tract.[5]

The female voice generally runs 15–20 percent higher than the male, and the child's voice is higher yet. In the high ranges, the vocalized partials in the region of the formants may be few, as shown in Figure 19–19(e). If these partials fall in the regions of low response, the singer's voice will be subdued.

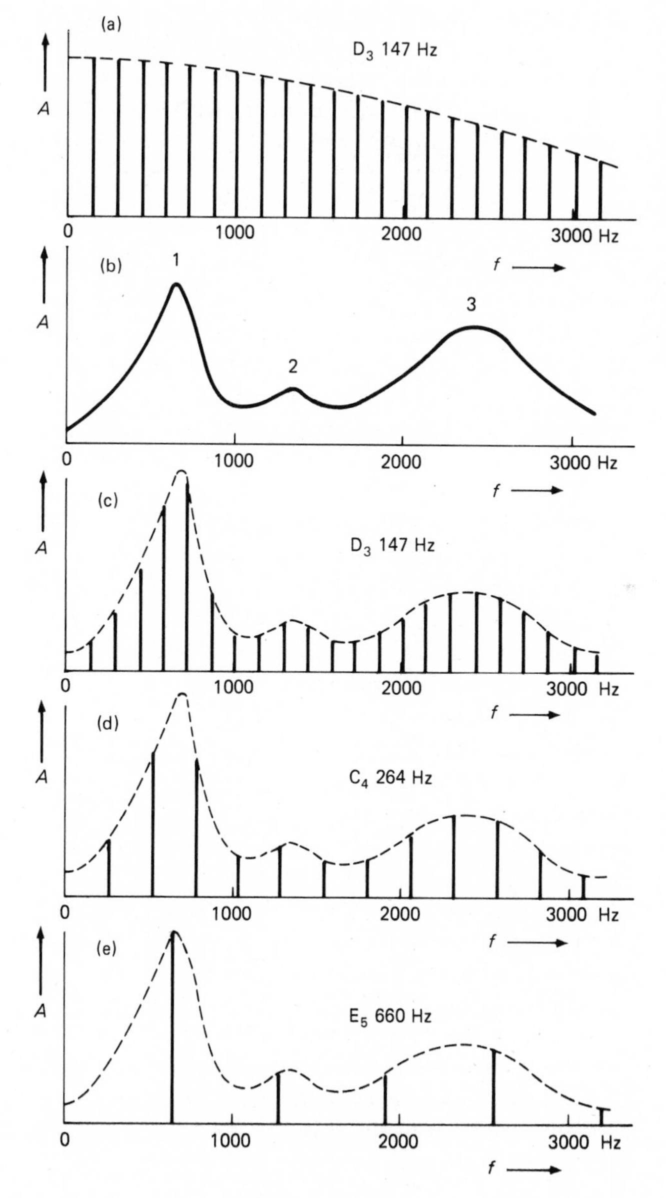

FIGURE 19–19 Partials of a vocalized D_3 (a) are modified by the formants of the vocal tract (b) to produce the resultant spectrum (c). Higher notes in the same vocal tract, (d, e), are also shown.

Through training, however, the soprano can often reshape her vocal tract to adjust one or more of the formants to match partials of a sung note.[5]

An interesting experiment may be performed with the human voice to show the effect of gas density on the quality and pitch of the emerging sounds. If the vocal tract is filled with a gas lighter than air, the speed of sound will increase [see Equation (12e)] and change the voice quality considerably. A demonstration can be performed by exhaling completely and then inhaling pure helium gas to fill the lungs. Upon speaking or singing loudly, the experimenter as well as all who hear him will be astonished at the peculiar high-pitched voice, which must be heard to be appreciated. The peculiarities arise from the fact that the fundamental pitch, due to the frequency of the vocal cords, remains practically normal, while the formants produced by the mouth, throat, and nasal cavities are raised by about two and a half octaves.

19.12 Vowels

A single well-defined sound is called a **phoneme.** A phoneme may be represented as a single letter of the alphabet, such as "s." It may also be represented by a combination, such as "sh." Some letters, such as "a" actually represent

TABLE 19–1: The English Vowels[a]

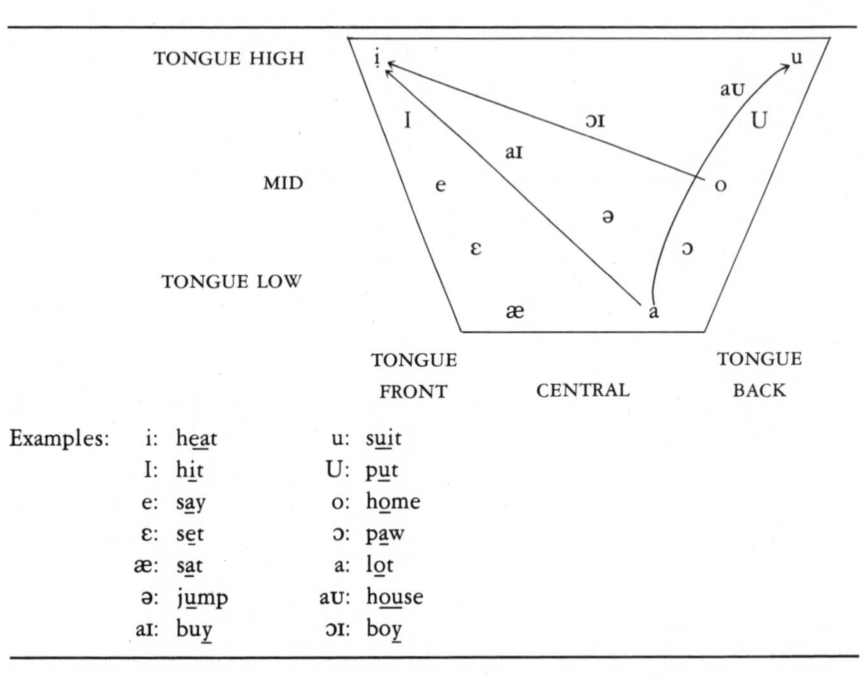

Examples:				
	i:	heat	u:	suit
	I:	hit	U:	put
	e:	say	o:	home
	ɛ:	set	ɔ:	paw
	æ:	sat	a:	lot
	ə:	jump	aU:	house
	aI:	buy	ɔI:	boy

[a]This system is used widely by American linguists and is an abridged and modified version of the International Phonetic Alphabet. Other assortments and symbols are readily found in the literature.

several possible phonemes. A phoneme may be voiced or unvoiced, depending upon whether or not the vocal cords are used to generate the phoneme.[6, 7]

Most of the sustained voiced phonemes are vowels. See Table 19–1. Although it is possible to excite an unlimited number of vowels, there are only about 14 (there is some dispute as to the proper number) recognized as distinct in American English. Three of these are shown in Figure 19–20. Of the 14 vowels, many are diphthongized vowels in that they slide more or less from one position to another.

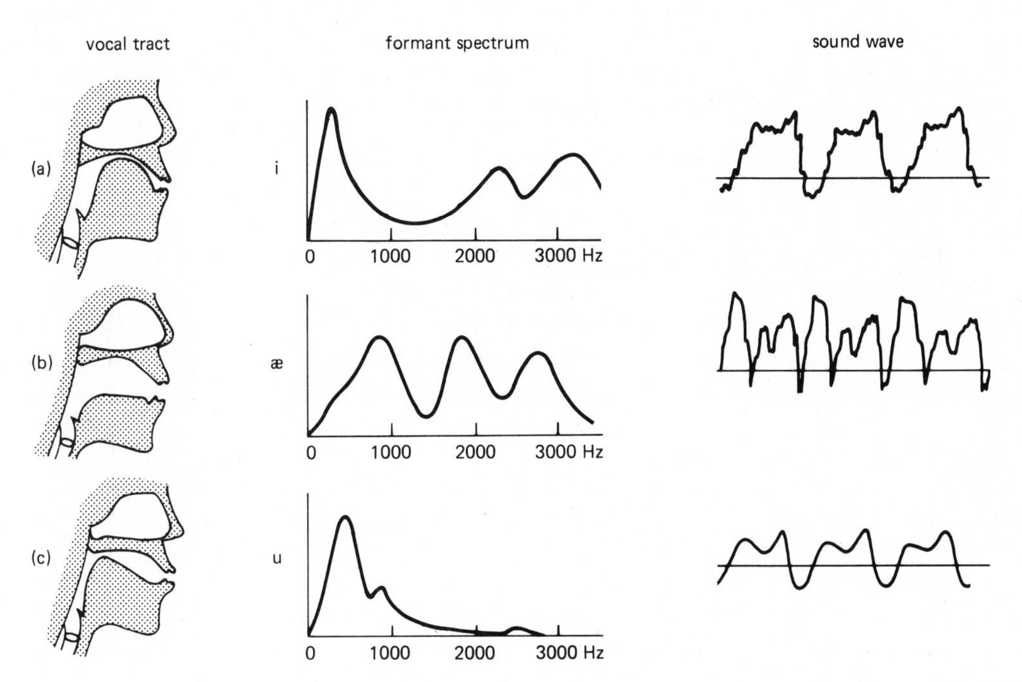

FIGURE 19–20 **Vocal tract shape, formant spectrum, and typical resonant sound wave form for the vowels: (a) i (as in h<u>ea</u>t), (b) æ (as in s<u>a</u>t), and (c) u (as in s<u>ui</u>t).**

Vocalized vowels have waveforms unlike those of other musical instruments. The fundamental mode is present and determines the overall repetition rate of the pattern. However, the pattern is dominated by the higher partials, particularly those corresponding to the formants.

19.13 Semivowels and consonants

Although linguists can debate endlessly on the number of consonants in American English, we list 24 generally regarded as distinct. These can be divided into the general classes shown in Table 19–2. This is the most widely used arrangement, although there are others.

The phonemes l, w, r, and y are voiced by the vocal cords, similar to the

TABLE 19–2: The English Consonants[a]

	BILABIAL	LABIODENTAL	DENTAL	ALVEOLAR	PALATAL	VELAR	GLOTTAL
PLOSIVES	p b	—	—	t d	—	k g	—
FRICATIVES	—	f v	θ ð	s z	ʃ ʒ	—	h
AFFRICATES	—	—	—	—	tʃ dʒ	—	—
NASALS	m	—	—	n	—	ŋ	—
LATERAL	—	—	—	l	—	—	—
SEMIVOWELS	w	—	—	r	y	—	—

[a]The second member of each pair is voiced.
Examples: θ: thin ð: then
ʃ: ship ʒ: azur
tʃ: chain dʒ: jump
ŋ: sing

vowels. The nasal consonants m, n, and ŋ (ng) are voiced using the nasal cavities for resonance, with the sound being emitted from the nose rather than the mouth. See Figure 19–21(a). **Humming** a tune consists of vocalizing purely nasal sounds.

The **fricatives** and **affricatives** are produced by forcing the air through a very small opening, producing turbulent noise, as shown in Figure 19–21(b). These consonants, such as f, θ (th), s, ʃ (sh), and so on, are sounded when the vortices produced by the turbulence produces sound waves. Although their pitch is indefinite, the range of frequencies produced differs for each point of articulation. For example, "s" is formed between the tongue and upper teeth, producing a much higher pitch range than "f," which is produced between the teeth and lower lip. The shape of the vocal tract will also affect the fricative sound. Early onset of vocalization will transform f to v, s to z, θ (th as in *thin*) to ð (th as in *then*), and so on.

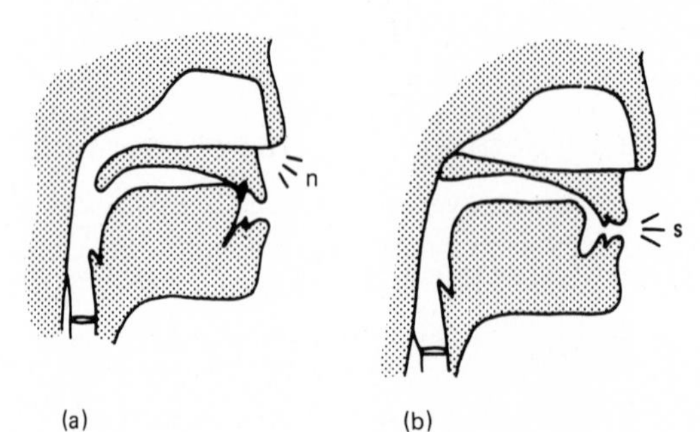

FIGURE 19–21 Vocal tract shapes for the production of (a) the nasal n and (b) the fricative s.

The **plosives (stops)** are consonants that start with a brief silence, followed by sudden aspiration and the immediate onset of a vowel. The plosives b, p, d, t, g, k are initiated by sudden openings formed in different points of the vocal tract. The first sound emitted is an aspiration followed almost immediately by a vowel. Both are characterized by the initial shape of the vocal cavity. The cavity then rapidly changes to sound the vowel that is to follow the plosive consonant. The sound produced by the plosive is, then, primarily a rapid transitory vowel. The plosives b, d, g involve a slightly earlier onset of vocalization than the respective plosives p, t, k.

The spoken voice consists of an appropriately arranged series of phonemes. Such a sentence is best presented as an **acoustical spectrogram,** a frequency analysis of the voice showing the changes occurring as the sentence is pronounced. See Figure 19–22.

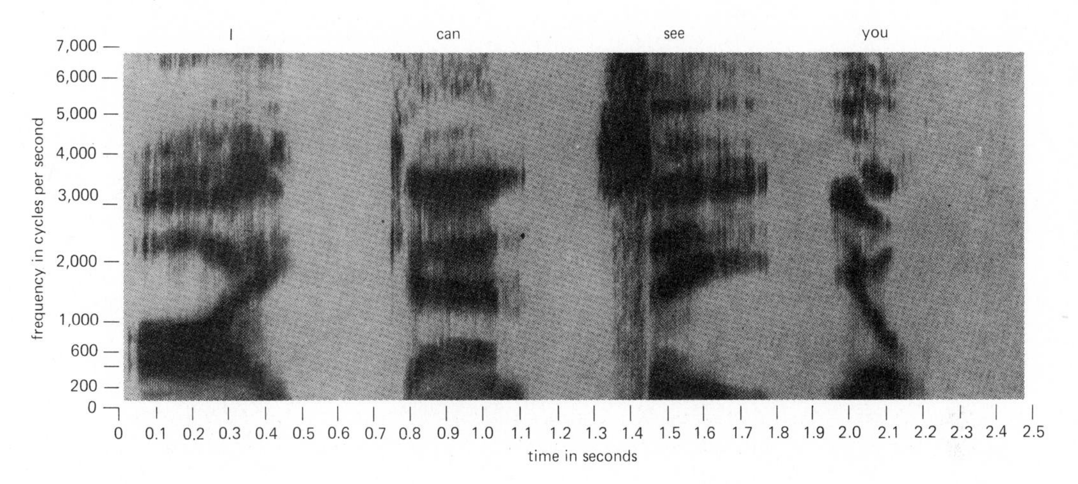

FIGURE 19–22
A voice spectrogram of the vocalized words: "I can see you." Note the shifting formant positions (dark bands) in "I" and "you", the abrupt plosive "c," and the broad noise spectrum of "s." (Courtesy of Bell Laboratories)

19.14 The singing voice

From the vocalist's point of view, it is the vowels that carry the sound. Vowels are generated by the vocal cords and usually sustained as long as desired. When the drill sergeant shouts "hor-hur-har," he knows that he has pronounced all the sounds in "forward march" that would be audible on the parade ground. The "h" is not actually a vowel but results from the vocal cords being initially open as the air is expelled from the lungs.

Vocalists are usually encouraged to use consonants as well as vowels for clarity of diction. In spite of these admonitions, most singers usually appear to be less concerned with clear diction than with the variable vocalization made possible by sustained vowels. Moreover, the modification of the formant response curve for better musical effect, as discussed in Section 19.11, may result in an alteration in the vowel. The resulting slurring or alteration of words may not be objectionable to the listener if the overall sound is pleasing.

There is an immense flexibility in the voice as a musical instrument, and studies are still being done to better understand it.

QUESTIONS

1. What is the difference between a pipe and a horn?

2. Why does a megaphone act to facilitate the transfer of sound energy, whereas a trumpet bell acts to reflect it?

3. What is an "acoustical spectrogram"?

4. Describe the role of voicing and formants on the sound of a sustained sung tone.

PROBLEMS

1. Calculate the length of a conical horn that has a fundamental frequency of C_2.

2. The fourth octave of the bugle, or the valveless trumpet, is called the *diatonic octave* because most of the notes of the just diatonic scale correspond to natural harmonics in this range (harmonics 8 through 16). Show which of the diatonic notes in this octave correspond to natural harmonics.

PROJECT

Measure the length of a brass instrument. Calculate its pedal frequency and several higher partials. Compare the results with the actual playing pitches.

NOTES

[1]A. H. Benade, "The Physics of Brasses," *Scientific American* (July 1973):24.

[2]A. H. Benade, *Fundamentals of Musical Acoustics* (New York: Oxford University Press, 1976), chap. 19.

[3]I. Lehiste, ed., *Readings in Acoustic Phonetics* (Cambridge, Mass.: M.I.T. Press, 1967).

[4]P. Denes and E. Pinson, *The Speech Chain* (Garden City, N.Y.: Anchor Books, 1973).

[5]J. Sundberg, "The Acoustics of the Singing Voice," *Scientific American* (March 1977):82.

[6]H. Fletcher, *Speech and Hearing in Communication* (New York: Van Nostrand, 1953).

[7]C. K. Thomas, *Phonetics of American English* (New York: Ronald Press, 1958).

Chapter Twenty

PERCUSSION INSTRUMENTS

By hammering, scratching, stroking, scraping, rattling, and plucking, almost any object can be induced to emit a sound. Many of these percussion instruments have a definite pitch. Those that have an indefinite pitch are still useful to carry rhythm, dramatize passages and notes, or create special effects. We will look at some typical percussion instruments in this chapter.

20.1 Percussive notes

A musical note is generally characterized by its pitch, timbre, and loudness. It is also characterized by its transient response, the *attack, sustain,* and *decay.* (See Section 14.12.)

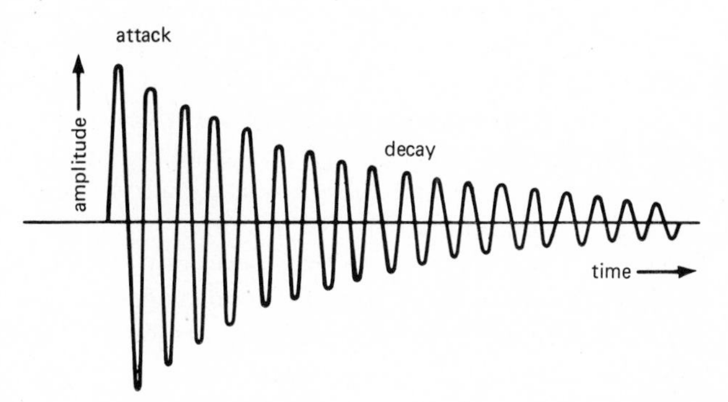

FIGURE 20–1
A percussive note is characterized by a sudden attack followed by a decay.

A violin note is dominated by the sustain, while the attack and decay are of minor importance. A percussive note, on the other hand, has a sharp attack, followed immediately by a decay of variable duration. See Figure 20–1. If a xylophone bar is struck with a soft mallet, the fundamental frequency dominates the response, and a clear pitch is heard. If a tambourine is shaken, a large number of vibrating modes are excited, and no pitch is discerned.

20.2 Types of percussion instruments

A variety of stringed instruments emit percussive notes, including the harpsichord, piano, and even the guitar. However, their structure and tonal character relates them more closely to the stringed instruments. There are also instruments such as sirens, bird whistles, and others that create a series of percussive bursts of wind pulses. These instruments are difficult to classify as wind or percussion, and they are usually used only for special effects.

There are a variety of ways in which percussion instruments can be classified. These groupings may be based upon the way in which the sound is produced, the nature of the sound itself, the groupings in the orchestra, and so forth. For the purposes of this study, we will limit our consideration of percussion instruments to three general classes:

1. **Tuned bars, tubes, and plates,** such as vibraphones, chimes, and steel drums.
2. **Whole-body vibrators of indefinite pitch,** such as triangles, gongs, and cymbals.
3. **Vibrating membrane instruments,** such as drums and tympani.

20.3 Tuned bar instruments

The **xylophone,** like the orchestra bells (Section 13.2), is composed of a series of flat bars of varying lengths. When struck with a mallet, the bars are set vibrating in their fundamental mode. See Figures 20–2 and 13–3. By mounting the bars horizontally on soft felt padding at their nodal points, the fundamental vibration is not rapidly absorbed by the mounting structure. A variety of more complex motions are also excited, creating higher partials. (See Figure 13–4.) A characteristic timbre is created, depending upon where the bar is struck as well as upon the nature of the beater.

Unlike metal bars of the orchestra bells, the rosewood or synthetic bars of a xylophone internally absorb the vibration energy rapidly, especially at the higher vibration modes. The resulting sound from wooden bars is therefore crisp, resulting from the rapid decay of vibration. The higher partials of a flat bar are not harmonic with the fundamental pitch. However, by properly hollowing the underside of each bar, as shown in Figure 20–3, the second

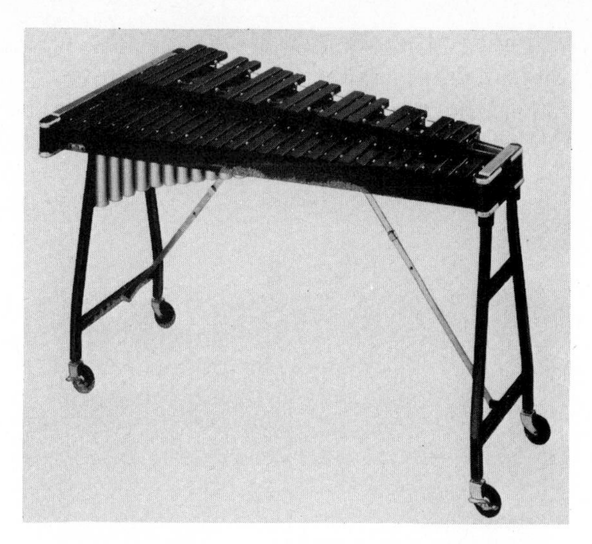

FIGURE 20-2
The xylophone contains bars of different lengths that produce the notes of a musical scale. (Courtesy of Ludwig Drum Company)

mode is raised from $2.76f$ to $3f$. The second partial is then consonant, a twelfth above the fundamental.

Hollow tubes are usually suspended below each xylophone bar, each closed at the lower end, and tuned to resonate with the corresponding bar. This tends to enhance the loudness of the fundamental pitch and any consonant odd harmonics.

20.4 Marimba and vibraphone

The marimba is similar to the xylophone, except that it uses larger bars, a softer beater, wider resonator tubes, and extends its keyboard to lower notes. See

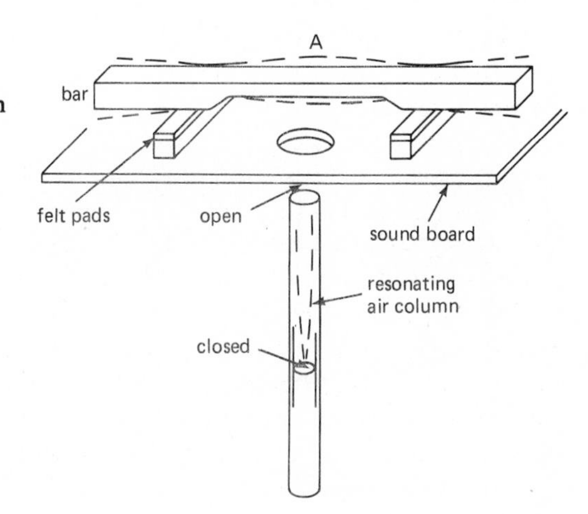

FIGURE 20-3
Details of a xylophone bar with a resonating pipe suspended below.

Figure 20–4. The underside of each bar is hollowed more deeply than the xylophone bars, raising the second mode from $2.76f$ to $4f$, two octaves above the fundamental. The effect of these features is a mellow, boomy sound. Although the pipes in the marimba are cut to lengths that provide symmetry to the instrument as a whole, they are closed at the bottom for the lower frequencies and some distance up the tubes for the higher frequencies, so that all the air columns resonate to the fundamentals of the corresponding bars directly above them.

FIGURE 20–4
A marimba made of rosewood bars above and resonating metal pipes below. (Courtesy of Ludwig Drum Company)

The **vibraphone** employs aluminum bars, hollowed as are marimba bars. Stops are located at the top of the resonant tubes. By means of a motor-driven system, these stops can be opened and closed at a regular rate, typically 2 Hz

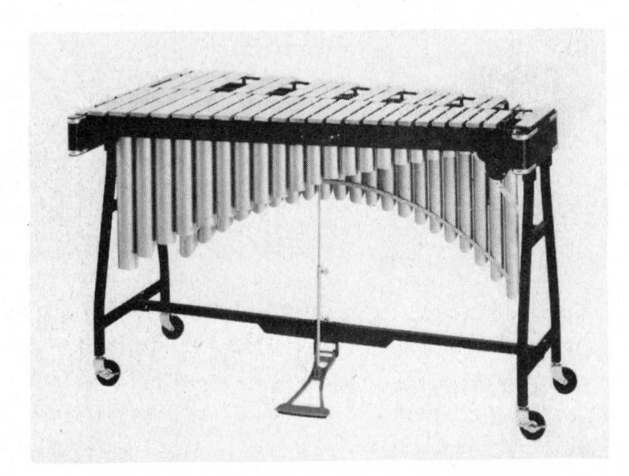

FIGURE 20–5
Vibraphone, or vibrachord, often called vibes. (Courtesy of Ludwig Drum Company)

(slow), 4 Hz (moderate), or 6 Hz (fast). See Figure 20–5. Each time the tube is closed at the top, the resonance ceases, causing a drop in the fundamental loudness. Since the nonresonant overtones are less affected by the stops, the timbre of the note will change. Although the fundamental frequency does not change, its suppression causes a subtle rise in pitch. As a result of activating the motor drive, a *timbre vibrato* is produced, accompanied by a *tremolo* (loudness variation) as well as a slight *vibrato.* See Figure 20–6.

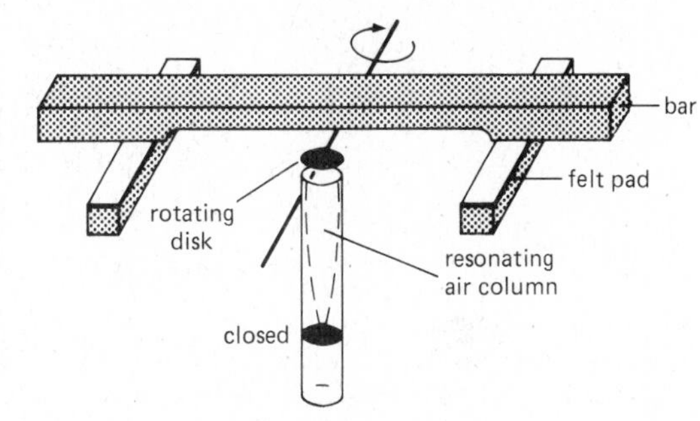

FIGURE 20–6
Single vibraphone bar showing the rotating stop over the open end of its resonating pipe below.

20.5 Tuned tubes, rods, and plates

Orchestral bells and the bell lyra, shown in Figures 13–5 and 13–6, are good examples of musical instruments that use straight bars that are uniform in cross section. They are made of metal and struck with a hard mallet. These features accentuate the high inharmonic overtones and give the instruments their bell-like tones.

Chimes (tubular bells) are conceptually similar to the bars of a xylophone. See Figure 20–7. Chimes are loosely suspended from a single point. When struck near the top, they vibrate in transverse modes similar to xylophone bars, although the pitch is determined by several of the higher partials.[1] The plug in the upper end of each chime tube modifies the weight distribution and is designed to adjust the relative mode frequencies so as to result in a timbre similar to carillon bells (Section 13.5).

The **steel drums** of the Caribbean Islands are good examples of tuned vibrating plates. The steel drum consists of a number of individual convex, circular sections, hammered into the head of a steel oil drum. Cutting a shallow groove around each section serves to isolate it from the rest, so that each acts as a separate vibrating diaphragm. Each section is tuned by varying the thickness through careful hammering. See Figure 20–8.

The **musical saw** is an example of a single vibrating plate that is tuned by flexing the saw. Originally, joiners handsaws were used, but modern musical saws are designed strictly to be bowed or struck to produce a musical note. A wide *glissando* can be produced by flexing the saw while bowing or while a struck note is decaying.

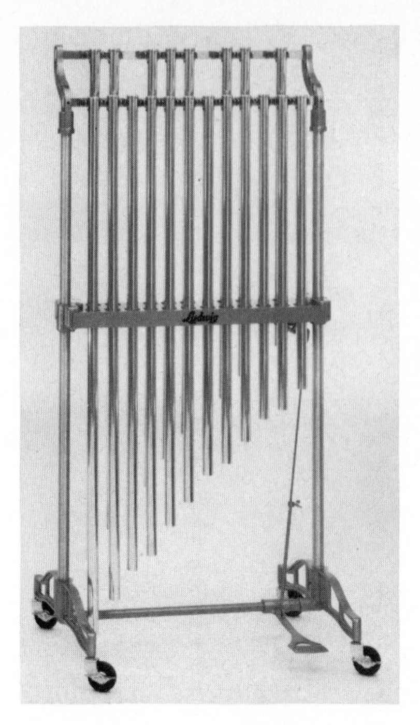

FIGURE 20-7
Chimes are hollow tubular metal bars, each with a plug in the upper end. (Courtesy of Ludwig Drum Company)

20.6 Indefinite pitch vibrators

When a pair of **cymbals** are struck together, so many vibrating modes are excited that no one frequency is dominant. See Figure 20-9. However, a dominant pitch range is evident, if compared with larger or smaller struck plates. The **triangle** will generate predominately higher pitches than a cymbal. A **gong** or **tam-tam** will generate a lower range. See Figure 20-10. All these instruments are considered to have indefinite pitch.

A series of instruments of common shape, such as wooden **temple blocks,** are often used as percussion instruments. See Figure 20-11. If a single temple block is struck, its pitch is difficult to discern. If each is struck in

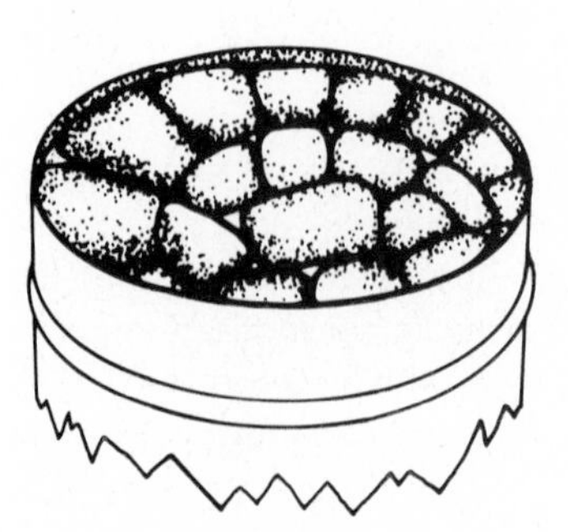

FIGURE 20-8
A steel drum, fabricated by hammering out isolated, diaphragms on one end of a 50-gallon oil drum.

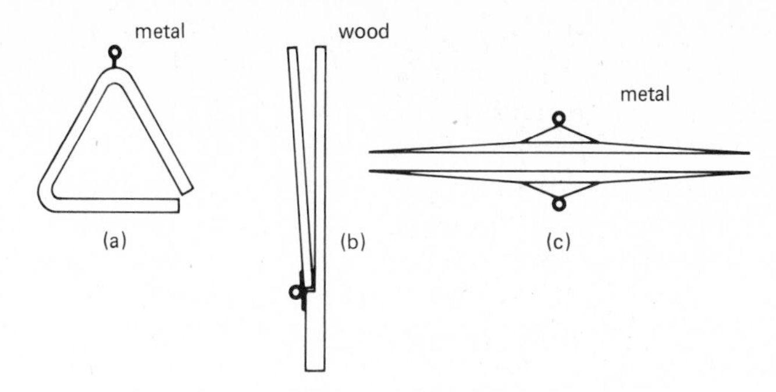

FIGURE 20–9
Vibrating bodies of indefinite pitch are illustrated by (a) the triangle, (b) the whip, or slapstick, and (c) clash cymbals.

order of decreasing size, a sense of progressively increasing pitch is clearly evident, although one may still be unable to identify the pitches themselves.

A wide range of clicks, clanks, and clunks can be created by striking steel plates, anvils, and wooden blocks of an almost endless variety. As a result of their indefinite pitch, these instruments blend well with any orchestral notes.

FIGURE 20–10
A tam-tam.

20.7 Shaking, stroking, and scraping

A cymbal or drum can be scraped with wire brushes, giving a considerably different effect than that achieved by the usual striking. The result is a contin-

FIGURE 20–11
Temple blocks.

uous series of many soft percussions, giving a sense of a sustained sound rather than a transient sound.

Metallic sounds of indefinite pitch can be generated by shaking **sleigh bells** or scraping a **washboard.** However, most instruments used for generating such sounds are made of wood, such as the **guiro.** See Figure 20–12. The clicking of **castanets** is identified with Spanish music, and the shaking of various **rattles** is used extensively in Latin American music. Such instruments are used primarily for rhythm. Their indefinite pitch neither enhances nor detracts from the melodic component of the music.

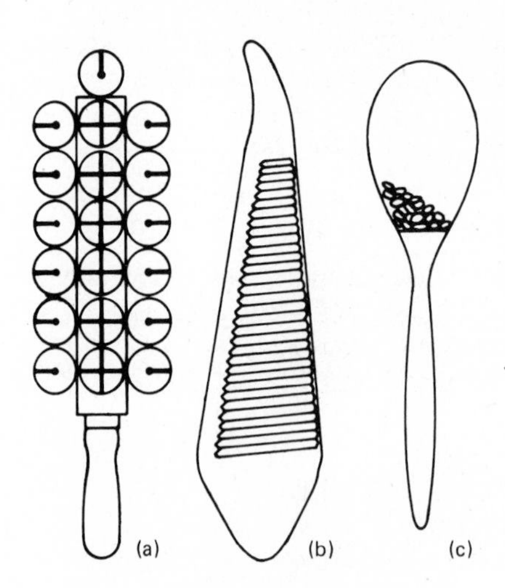

FIGURE 20–12 Sounds of indefinite pitch are produced by (a) jingling sleigh bells, (b) scraping a guiro, or (c) rattling maracas.

Rattles and scrapers often have some sort of enclosure that enchances a certain range of pitches. For example, **gourds** and **maracas** are often used; these instruments are filled with dried seeds or shot. Each shake results in a succession of light taps, enhanced by air vibration within the enclosure.

20.8 Drums

The drum consists of a membrane of skin or plastic stretched tightly across a cylindrical shell. The membrane, or "head," can be excited in a variety of modes by striking or scraping it. See Figure 20–13. The sound is generated directly by the vibrating membrane itself. The lowest frequency is generated when the head moves up and down as a whole, the center moving most. The next partial is produced when the head moves in halves, as shown in Figure 20–14(a). Successively higher partials are produced by more complex modes, both segmented and circular. See Figure 13–15.

FIGURE 20-13
Typical drum section of a concert band. From left to right: snare drum, tom-tom, bass drum, tom-tom. Also shown are three sizes of cymbal plates. (Courtesy of Ludwig Drum Company)

20.9 Tuned membranes

The sound of most drums is characterized by a number of partials, with no simple harmonic relationships. As a result, one cannot generally recognize a particular pitch. To produce a drum with a noticeable pitch, it is necessary to modify the membrane vibration modes so as to bring several of them into a harmonic relationship. In the case of the **tabla**, from India, this is done in part by adding a layer of flexible gum to the center of the membrane.[1] If properly done, this results in a membrane of varying thickness whose lower partials are approximately harmonic. This treatment is similar in effect to the shaping of xylophone and marimba bars.

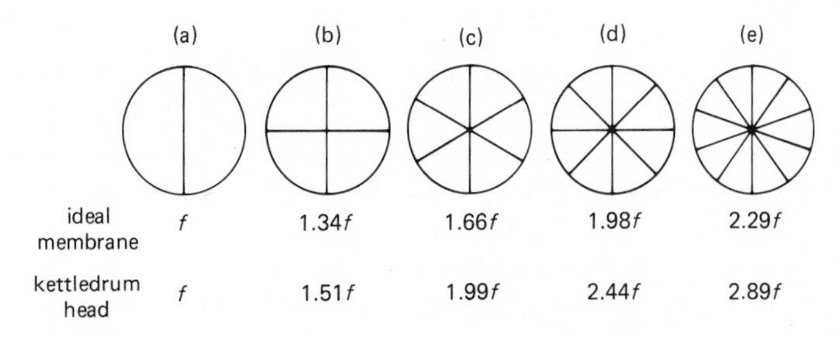

FIGURE 20-14
Nodal patterns of the circular membrane of a kettledrum head. (a) second mode, (b, c, d, e) higher modes.

The tympani (as well as the tabla, in part) employs a second process to bring the proper partials into harmonicity. This is done by controlling the acoustical environment of the membrane.

20.10 Tympani

Closure of the far end of a drum shell with a second flexible head will have only a modest effect on the timbre. However, closing it with a rigid surface, as is done in the kettledrum, will change its character decisively.

The kettledrum (a set is called tympani) consists of a calfskin or plastic membrane stretched across a hemispherical shell. See Figure 20–15. When the kettledrum is struck in the center, only a dull thud is heard. When struck near the rim, the mode in Figure 20–14(a) is excited, resulting in a clearly discerned pitch. This pitch can be tuned by varying the tension on the head with a foot pedal.

FIGURE 20–15 Tympani, a set of kettledrums. (Courtesy of Ludwig Drum Company)

The dominance of this mode [Figure 20–14(a)] takes some explanation. It is found that the fundamental mode of a relatively thin head is never strongly excited, even when struck near the center. By striking at the normal point, about one-fourth of the way in from the rim, the higher modes are selectively excited. The second mode is shown in Figure 20–14(b). As the left side pushes down on the enclosed air, the right side moves up to accommodate the displaced air. See Figure 20–16. As a result, the vibration of the head is accompanied by air sloshing back and forth inside. This sloshing of air

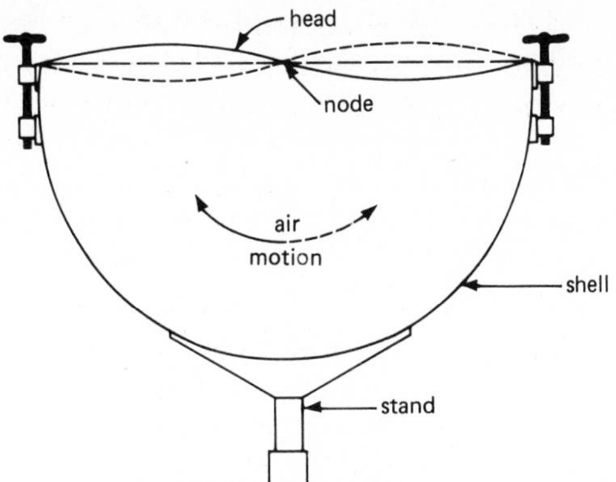

FIGURE 20–16
Cross section of a kettledrum showing the favored motion with a single nodal diameter.

tends to accentuate this vibration mode, as well as to lower the vibration frequency somewhat. The third mode is similarly affected by air motion.[2] Alternatively, the small amount of stiffness of the membrane will tend to raise the frequencies of the higher harmonics. The coupled air motion, as well as some stiffness of the membrane, serves to adjust the frequency ratio of these two modes to 2:3. Studies show that several of the higher modes are also brought into approximate harmonicity.[1, 2] The pitch of the kettledrum is, then, due to several approximately harmonic vibration modes.

20.11 Drum excitation

The degree to which the various modes of a drum are excited will be determined by the hardness of the beater, the position of strike, and the supporting structure. If the head is struck in the center, the fundamental mode will be preferentially excited, especially for drums with a small area and a thick membrane, such as a tom-tom. Drums with wide and thin heads have poorer response for excitation of the fundamental mode, but they respond much better in the higher overtones. Such drums are therefore struck off center, often near the rim, where the higher frequency modes have their antinodes. See Figure 20–14. Heads made of animal skins, especially thick ones, rapidly absorb the energy of motion, especially the higher frequency modes. Such heads therefore have a mellow sound. The brighter sound of a plastic head is due to the relative lack of absorption of the higher modes.

The main function of the drum shell is to support the head. If the head covers the end of a hollow log, the log will resonate to the low vibration modes of the head. Unlike the primitive log drum, most modern drums[3] have a diameter larger than the depth, and therefore will not resonate significantly. Moreover, the dominate drum head modes create air wavelengths that are much longer than the drum dimensions. As a result, the open drum shell has a relatively minor influence on the timbre of the sound produced. If a second head covers the far end of the shell, it also will vibrate. A **snare drum** has springs called *snares* stretched over the far side of the lower head. These rattle against the lower head when the upper head is struck.

Now that we have studied in some detail the strings, woodwinds, brasses, and percussions, we will compare their singing ranges with respect to each other, as well as with the instrument with the widest range of all, the piano. A chart showing the 88 keys of the grand piano is given at the bottom of Figure 20–17. A number of the principal instruments, classified into three groups, along with the human voice, are shown above. Their singing frequency ranges are blocked in by shaded areas, while their overtones are shown as open stripes extending to the right. Many of their audible overtones go beyond the chart, some of them reaching as high as 16,000 Hz. Space does not permit showing all instruments.

The discussion so far has been limited to acoustic instruments, those that transform vibrational motion directly into sound waves. In the next chapters we will investigate the advantages of using electronics to amplify, modify, and often produce musical sounds.

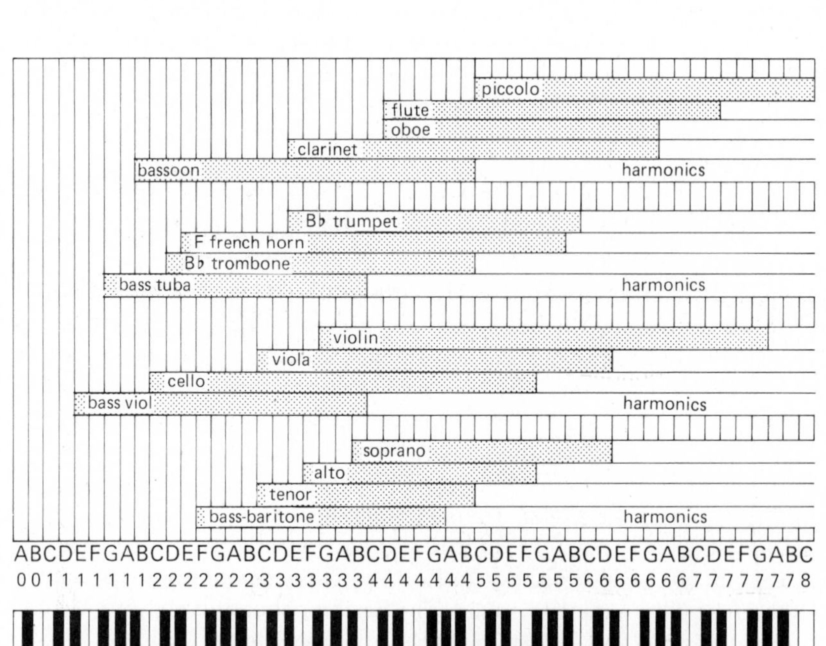

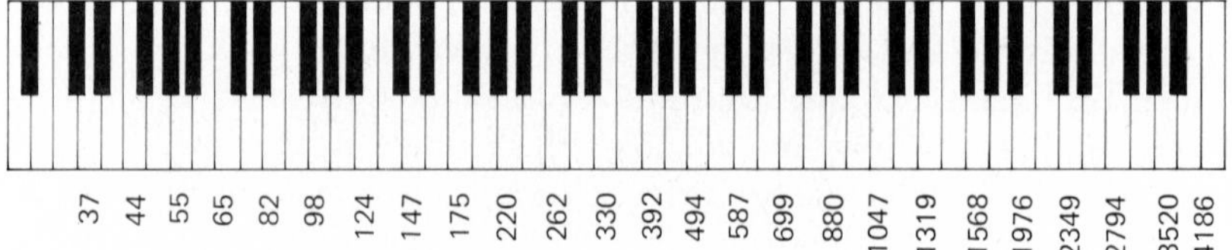

FIGURE 20–17
A chart showing the singing ranges of some of the well-known musical instruments of the band and orchestra, as well as the human voice. (From a chart developed for Illinois Bell Telephone Company by Bell Telephone Laboratories)

QUESTIONS

1. What is a percussive note?

2. What is the advantage of an instrument of indefinite pitch?

3. Why do drum heads made of animal hide and plastic have different timbres?

4. The piano is considered by some people to be a percussion instrument. What are the arguments for and against such a classification?

5. The drum has been used in many cultures for long-distance communication. Discuss the acoustical design appropriate for such use. Study some outside references.

6. The xylophone has a "brighter" sound than the marimba, due in part to the fact that the resonator tube responds to the second partial for the xylophone but not for the marimba. Why is this?

7. Placing a resonator below a vibrating bar increases the overall loudness but decreases the note's duration by at least 50 percent. Why might you expect this?

8. The orchestra bells, bell lyra, and upper registers of the xylophone, marimba, and vibes are not compensated by appropriately hollowing out the underside of the bars. Why might it be unimportant to compensate?

PROJECTS

1. Study the Chladni patterns of various drum heads. Poppy seeds or Christmas glitter may work better than sand.

2. Study the nodes of the tympani head. Determine the effect of creating nodes by pressing various points on the head. When the contact points cause little change in the sound quality, the main vibration mode probably already has a node at that point. Compare these with the nodes of Figure 20 – 13.

3. Construct a xylophone.[4]

NOTES

[1]T. D. Rossing, "Acoustics of Percussion Instruments: Part I," *Physics Teacher* 14 (December 1967):546; "Part II," *Physics Teacher* 15 (May 1977):278.

[2]A. H. Benade, *Fundamentals of Musical Acoustics* (New York: Oxford University Press, 1976), chap. 9.

[3]J. Blades, *Percussion Instruments and Their History* (London: Faber and Faber, 1970).

[4]T. F. Johnston, "How to Make a Tsonga Xylophone," *Music Educator's Journal* 63 (November 1976):38.

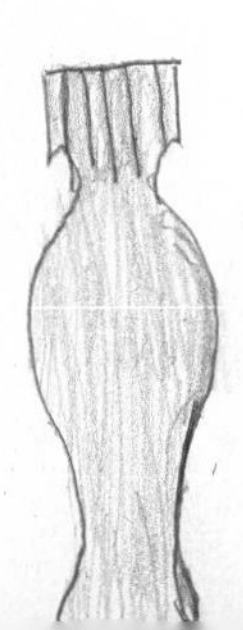

PART FOUR

Electronic Sound Systems

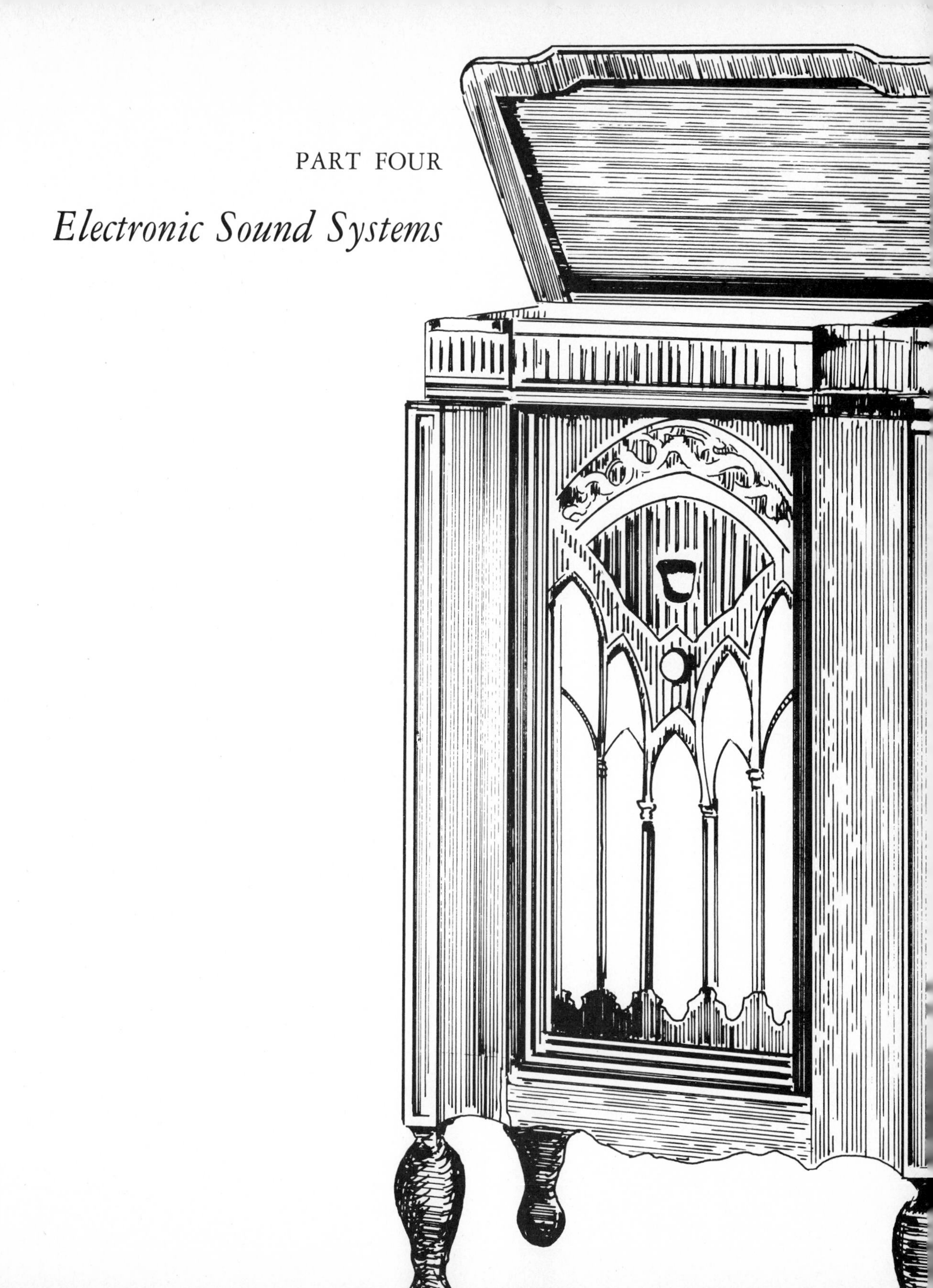

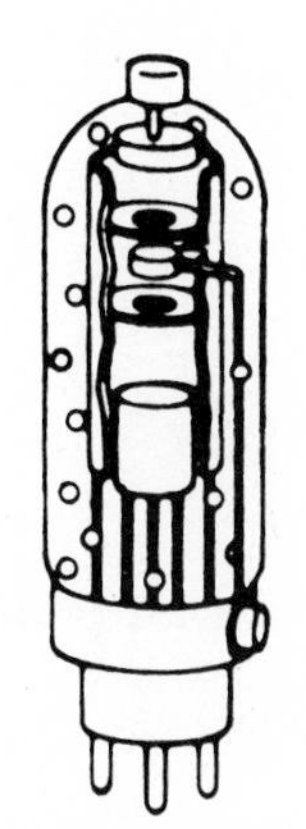

Chapter Twenty-one

HIGH-FIDELITY SOUND REPRODUCTION

Today, more people listen to music reproduced by electronic devices than by attending performances. Listening to music via an electronic reproduction system does not have the sense of immediacy of a concert hall. However, it is usually more convenient. Since recordings can be made under ideal studio conditions, the quality of reproduced music can be excellent, at least in principle.

21.1 Electronic sound systems

An electronic system is one that utilizes transistors, vacuum tubes, and so on to handle varying electric currents. A simple electronic sound system capable of increasing the original sound intensity level is shown in Figure 21 – 1. This system consists of a *microphone,* which converts the input acoustic energy to electric energy, an *amplifier,* which controls and boosts the electric energy, and a *loudspeaker,* which converts the electric energy back into acoustic energy.

The major advantage of an electronic system is in the ease with which the desired electric energy, or signal, can be controlled, modified, amplified, and distributed. Additional electronic components also allow for recording, playback, and radio transmission.[1-3] Offsetting these advantages are several inherent defects in an electronic sound system. These include depression or removal of some of the sound, addition of unwanted sound (noise and spurious pickup), and distortion. A system that minimizes these possible defects and permits faithful reproduction of the original acoustic energy is called a **high-fidelity system.** This is a much-abused term that should only refer to high-quality (and usually more expensive) systems capable of faithful reproduction. A **stereophonic system** is one in which two or more parallel sound channels are involved. Stereophonic sound systems are usually more or less high fidelity as well, although some are not. For simplicity, the following discussion will involve **monophonic** (single-channel) sound, leaving the discussion of stereophonic effects for Chapter 25.

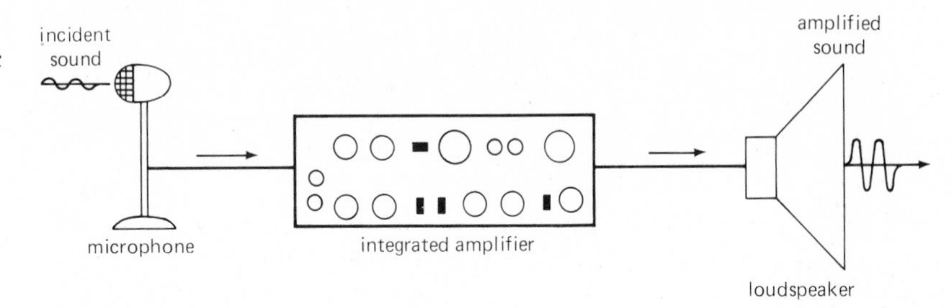

FIGURE 21 – 1
A simple electronic amplification system.

21.2 Transducers and amplifiers

A device that converts one type of energy to another is called a **transducer.** For example, the microphone and loudspeaker in Figure 21 – 1 are transducers. Generally, the fidelity of a transducer is not as good as that of the **amplifier,** which deals purely with electric energy. The amplifier can compensate for some of the defects of the transducer (equalization) but not for all of them.

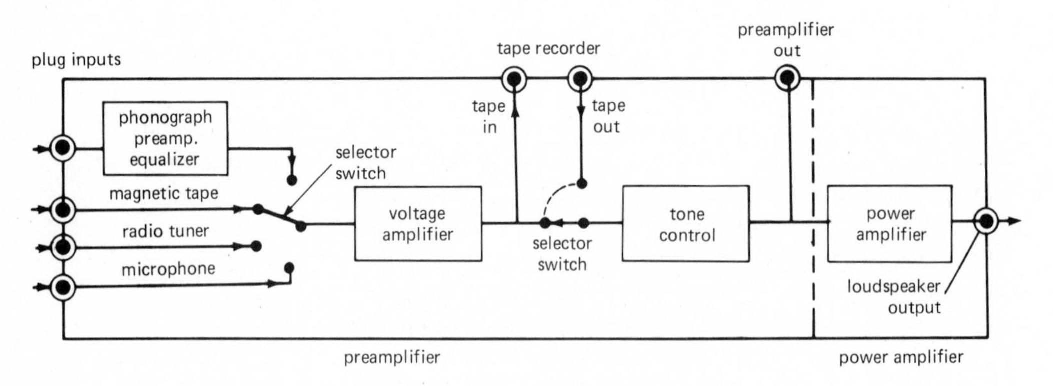

FIGURE 21 – 2
Block diagram of an integrated amplifier.

Transducers and amplifiers alike can be evaluated in terms of their performance, using commonly agreed upon specifications. Since the amplifier is the heart of a high-fidelity system, its performance will be discussed first.

An **integrated amplifier** is usually considered as two components, a **preamplifier** and a **power amplifier.** See Figure 21 – 2. The preamplifier allows for control of the sound, such as loudness and high- and low-frequency balancing. It also allows for selection and control of a variety of inputs, such as a microphone, tape deck, and turntable. The power amplifier simply boosts the output from the preamplifier to adequately drive the loudspeakers. These two components are usually housed in the same chassis or cabinet. Performance of an amplifier is determined by its *frequency response, amplitude response, signal-to-noise ratio,* and *power output.*

21.3 Electronic terminology

Within the amplifier, the wave variations take the form of varying **electric current** I, measured in amperes (abbreviated A). This current consists of many small charges flowing through the wires and components, mostly negatively charged electrons but sometimes positive charges too. The current is produced by an applied electric potential or **voltage** V, measured in volts (V). For the circuits normally used, the resulting current is proportional to the applied voltage. This relation can be illustrated by considering the flow of water between two reservoirs. See Figure 21 – 3. The current of water is analogous to the electric current, and the difference in height of the water surfaces is analogous to the voltage. If the height is raised or lowered, the current is raised or lowered accordingly.

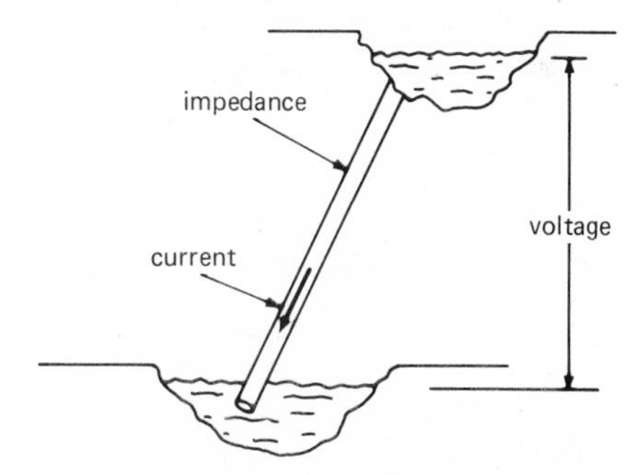

FIGURE 21 – 3
Electric quantities are analogous to the quantities describing the flow of water between two reservoirs.

The restriction imposed by the pipe is analogous to another electric quantity, the resistance, or **impedance.** Electric impedance Z is measured in ohms (Ω) and is related to the current and voltage by **Ohm's law:**

$$V = I \times Z$$
$$\text{volts} = \text{amperes} \times \text{ohms} \qquad [21a]$$

In contrast to the water analogy, the electric signals that correspond to acoustical phenomena involve sinusoidally **alternating current** (ac) and voltage, varying rapidly between positive and negative values. See Figure 21 – 4.

One final electric quantity is the power P, measured in watts (W). Power is given by the product of current and voltage:

$$I \times V = P$$
$$\text{amperes} \times \text{volts} = \text{watts} \qquad [21b]$$

By combining this power relation with Ohm's law, we see that

$$P = I^2 Z \quad \text{or} \quad P = \frac{V^2}{Z} \qquad [21c]$$

Therefore, the power produced is proportional to the square of the current or the voltage. For alternating current, the value of the power corresponds to an average over the variations.

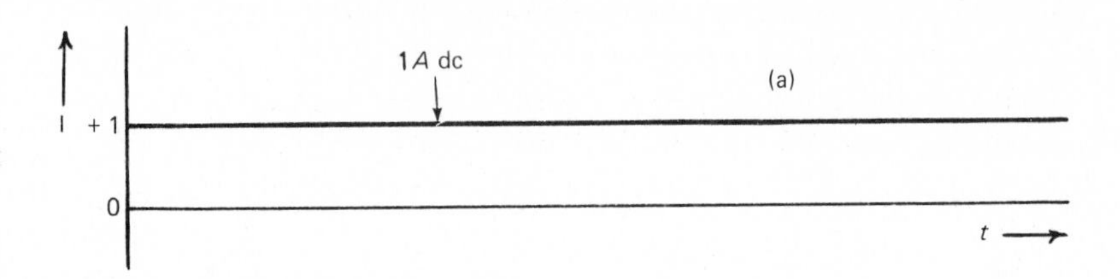

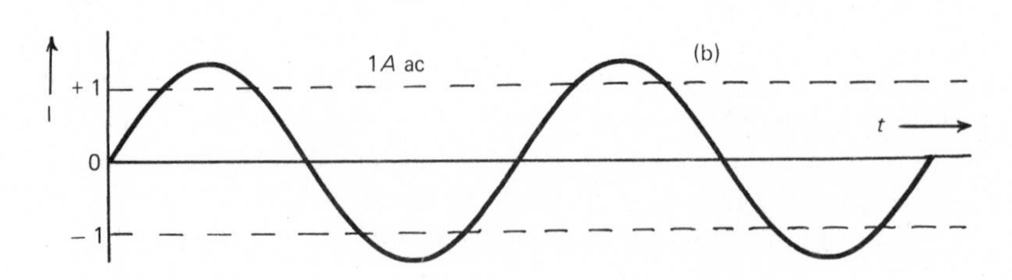

FIGURE 21–4
(a) Constant or direct current (dc) and (b) alternating current (ac), with effective average value of 1A.

Example

Suppose a current of 0.10 A is established in a 20-Ω resistor. Find the voltage across the resistor and the electrical power dissipated (transformed to heat).

Solution

Using Equation (21a) to find the voltage:

$$V = I \times Z$$
$$V = 0.10 \text{ A} \times 20\ \Omega$$
$$V = 2.0 \text{ V}$$

Using Equation (21b) to find the power:

$$P = I \times V$$
$$P = 0.10 \text{ A} \times 2.0 \text{ V}$$
$$P = 0.20 \text{ W}$$

21.4 Frequency response

The amplification of sound is dependent upon the frequency in a manner normally specified by a frequency response curve. See Figure 21-5. The frequency is plotted horizontally, using a logarithmic scale in order to adequately spread out the low-frequency end. The gain or amplification is plotted vertically, using a decibel scale. The decibel scale is itself a logarithmic representation of intensity. Since we will often refer to these plots, it is important that this meaning is fully understood.

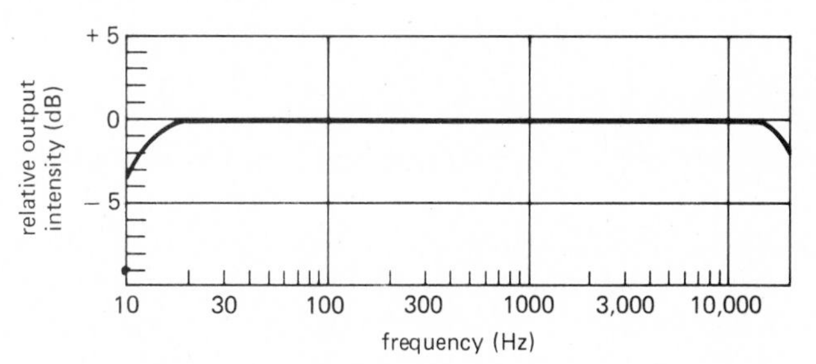

FIGURE 21-5
Frequency response curve illustrating the flat response of a high-fidelity amplifier.

A good amplifier should have a **uniform (flat) response** extending beyond the normal hearing range on both ends. An amplifier with a variation of less than 3 dB over the range 20-20,000 Hz is considered very good. In Figure 21-5 the sound intensity level of zero can be any desired value, so that plus and minus values represent variations from this level.

21.5 Tone control

Although an ideal amplifier should have a flat response over the audible range, there are a variety of reasons for modifying the frequency response. For example, suppose that the room acoustics are such that the low frequencies are somewhat suppressed. This deficiency can be compensated to some extent by boosting the bass frequency response. See Figure 21-6. Most amplifiers have both treble and bass controls to boost or attenuate (cut) the high and low frequencies, respectively.

Many listeners prefer to turn up the amplifier **gain control** (often called the **volume* control**) to produce a high sound level. This is usually done to allow the listener to hear the full frequency range, since sensitivity to high and low frequencies falls off significantly at low sound levels (**Fletcher-**

*Since the word *volume* is normally used to describe three-dimensional space, we will avoid using it in the present context in favor of the more technical word *gain*.

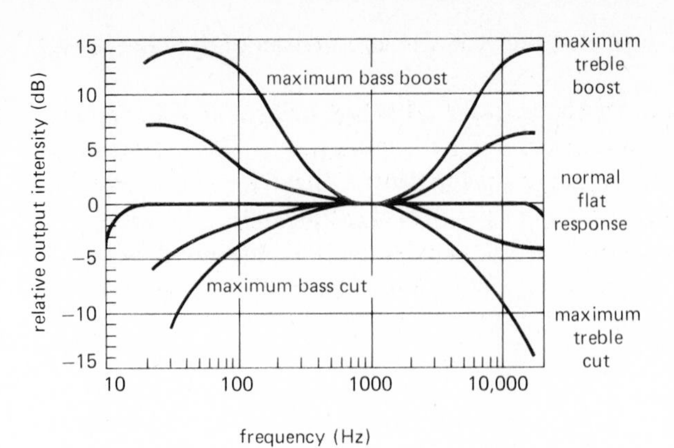

FIGURE 21 – 6
The frequency response curves for a typical high-fidelity amplifier with, and without, modification by tone controls.

Munsen effect). Many amplifiers have a circuit that compensates for the Fletcher-Munsen effect by boosting the high and low frequencies at low loudness levels. See Figure 21 – 7. This circuit is usually activated by a **loudness switch,** and it allows the listener to hear the full frequency range at low sound levels.

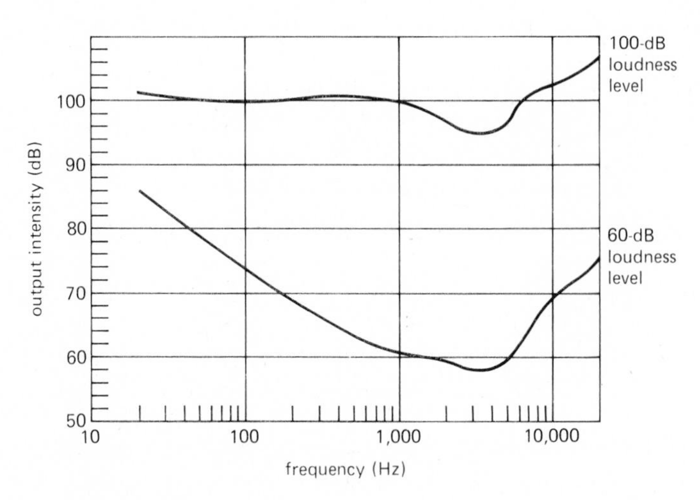

FIGURE 21 – 7
Frequency response of an amplifier, with ideal correction for normal hearing response, at various loudness levels.

21.6 Filters

It is often convenient to include circuits that limit the frequency response in a rather abrupt manner. A **high filter (low-pass filter)** is a circuit that passes all frequencies below the specified cutoff frequency and rejects those above. For example, a **scratch filter** that cuts off frequencies above 10,000 Hz is often used to eliminate record scratch, tape hiss, and other noise that dominates at these high frequencies. See Figure 21 – 8. However, desired sound above the cutoff frequency is also eliminated, resulting in a reduction of the high-fidelity response.

A **low filter (high-pass filter)** cuts off all frequencies below a specified cutoff frequency. For example, a **rumble filter** with a cutoff frequency at about 30 Hz is often used to eliminate unwanted low-frequency turntable noise. Again, desired bass frequency response is also reduced.

There are several circumstances under which it is desirable to limit the output to a fixed range between two cutoff frequencies. A circuit that does this is called a **bandpass filter**, and the frequency range that it permits to pass is called the **bandwidth.**

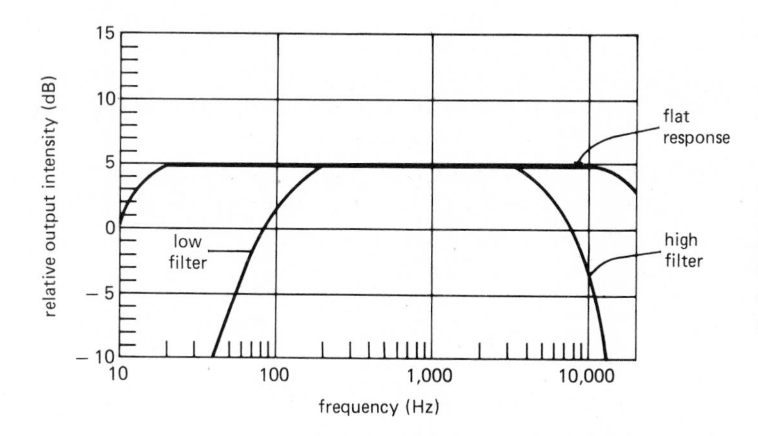

FIGURE 21 – 8
Frequency response curve showing the ways in which normal flat response can be modified with a low-frequency (rumble) filter and a high-frequency (scratch) filter.

21.7 Multichannel equalization

In earlier sections, we discussed dividing the audible frequency spectrum into two or three separately controlled bands (bass, midrange, and treble). However, circuits are made to divide the audible spectrum into 5 to 30 or more independently controlled bands.[4] This system is called a **multichannel equalizer.**

There are two good reasons for using the equalizer. The first is to compensate for uneven response in other parts of the system: the speakers (Section 22.10), room absorption (Section 28.6), and so on. For example, if drapes in the room absorbed strongly in the 5-kHz region, that portion of the frequency spectrum could be boosted in the equalizer. The second reason for use of the multichannel equalizer is to deliberately enhance or suppress certain sounds. We list here some typical uses, based on a 5-channel equalizer:

20 – 100 Hz: Boost for richness in very low bass sounds (drum, organ) or cut to eliminate hum or rumble.

100 – 500 Hz: Boost for clarity in upper bass sounds (cello, woodwinds, bass voice).

500–2000 Hz: Most effective for emphasizing or deemphasizing the human voice and the fundamental tones of most musical instruments.

2–8 kHz: Boost to clarify brass and stringed instruments and soprano voices.

8–20 kHz: Boost to clarify cymbals and brilliance of brasses and reeds, or cut to suppress tape noise and record scratch.

The equalizer is either built into an amplifier or added as a separate unit, usually between the preamplifier and power amplifier.

21.8 Amplitude response

The ratio of output to input of an amplifier is called its **gain**. For example, if an input of 4 V produces an output of 8 V, the amplifier has a voltage gain of 2. The gain should be constant for a given amplifier. For the example given, then, an input of 5 V should produce an output of 10 V, and −2 V in should produce −4 V out. With such an ideal amplifier, response is "linear" and is described by a straight line on an **output versus input graph**. See Figure 21–9. Such a device is called a **linear amplifier**.

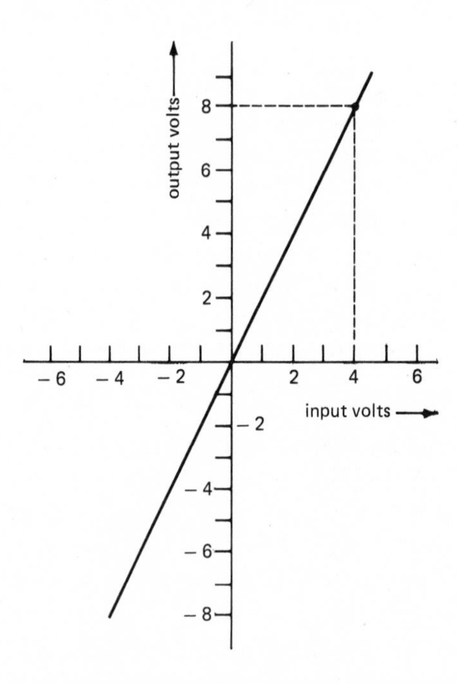

FIGURE 21–9
Voltage gain curve illustrating linear or proportional response of an amplifier with a gain of 2.

Any true amplifier, however, will have a maximum possible output, called its **saturation** level. Figure 21–10 describes a typical amplifier with a gain of 2. For an input not greater than 4 V (nor less than −4 V), the desired linear output is produced. From 4- to 7-V input, the output increases but is no

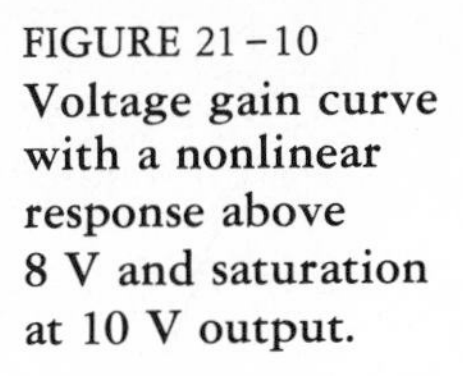

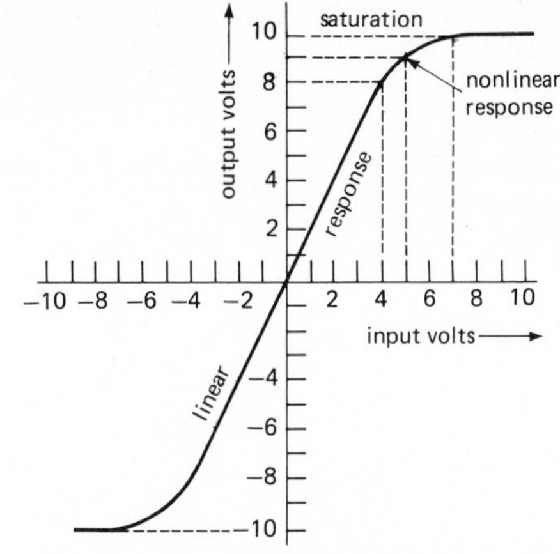

FIGURE 21 – 10
Voltage gain curve with a nonlinear response above 8 V and saturation at 10 V output.

longer proportional to the input. Finally, above 7-V input, the amplifier is saturated, and all output will be 10 V. It is therefore desirable to always keep the input voltage less than 4 V.

21.9 Harmonic distortion

Suppose the amplifier input consists of a pure sinusoidally varying voltage (sine wave) that never exceeds 4 V. The output will be identical in shape but twice the amplitude. See Figure 21 – 11(a). If a portion of the input reaches 5 V, however, the output waveform will no longer be similar in shape but will be rounded off at the peaks. See Figure 21 – 11(b). If the input exceeds 7 V, the output will be saturated during part of each cycle, producing an even more severely distorted flattop waveform. See Figure 21 – 11(c). Although the distorted output is no longer a pure sine wave, it can be considered to be a sum of pure sine waves (Fourier's theorem; see Section 8.2). The slight distortion indicated in Figure 21 – 11(b) would correspond to the original input waveform with a small additive of higher harmonics. The greater distortion induced in Figure 21 – 11(c) would correspond to a high intensity of added harmonics. The net effect of this **harmonic distortion** is to generate new harmonics that were not present initially, producing an output that is audibly different and generally displeasing.* The amount of harmonic distortion in a system is generally quoted. In a good system, it should be less than 0.5 percent of added harmonics.

21.10 Intermodulation distortion

Suppose there are two or more superimposed waves at the input of an amplifier. This could correspond to two different inputs, one input with noise or hum

*Also see Section 15.2 for a discussion of nonlinear effects as applied to the hearing process.

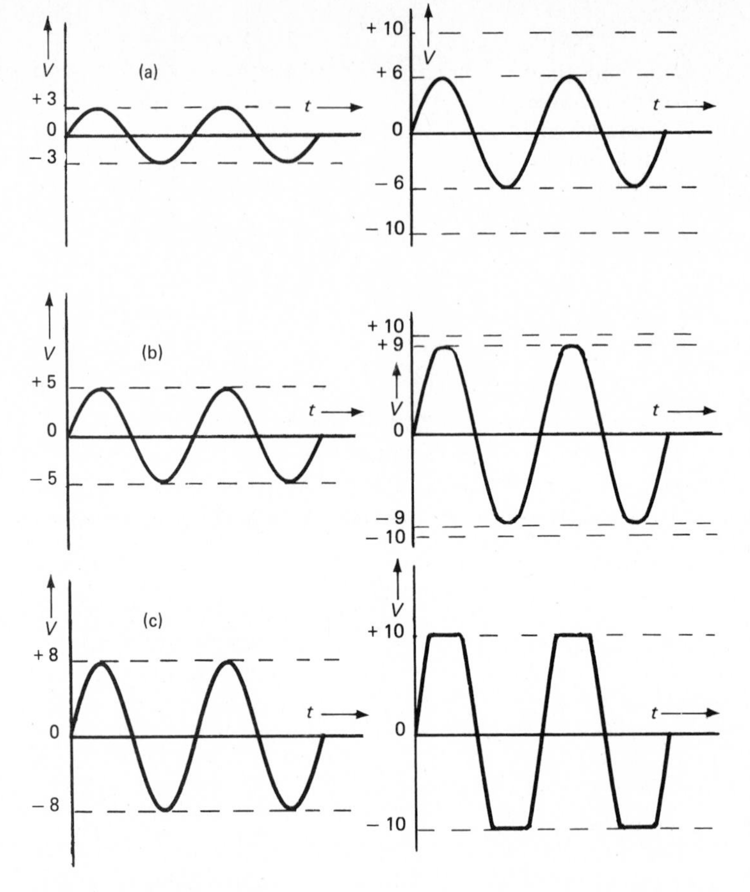

FIGURE 21 − 11
Graphs of input and output of an amplifier of gain ×2. Increasing the amplitude of the input signal can lead from (a) linear response to (b) nonlinear response and finally to (c) saturated response.

pickup, or just different frequencies from the desired single input. Normally, each waveform will be amplified as if it were independent of the others. Under some circumstances, however, one of the waveforms may cause the distortion of the others. There are a variety of ways in which this can happen. A high-frequency signal by itself would produce an undistorted output. With a low-frequency hum present, the two together would produce saturation, thereby distorting the original signal. See Figure 21 − 12. If such **intermodulation distortion** is significant in a supposedly high fidelity system, the listener may not recognize it as such but soon develops an irresistible urge to turn the system off.

21.11 Noise

A variety of unwanted sound sources may be present in an amplifier in addition to any noise that accompanies the input device. Among the most serious

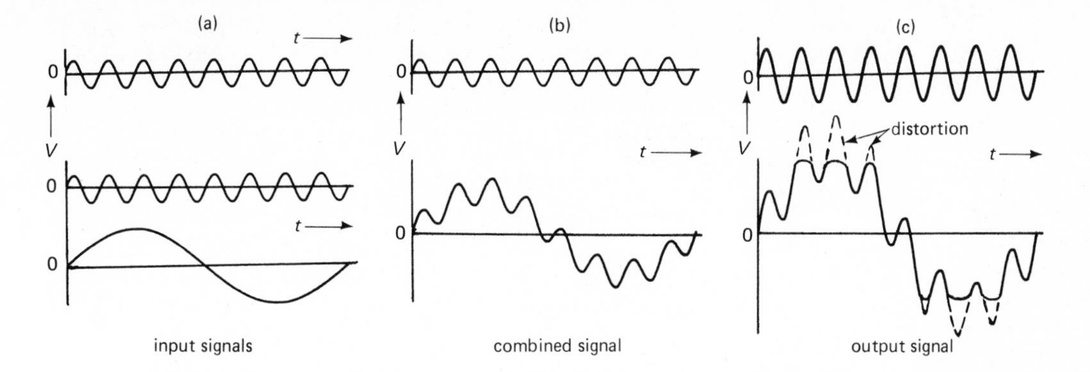

FIGURE 21–12 Upper trace: A single input signal (a) may be amplified linealy (c). Lower trace: Two or more input signals (a) may combine (b) to produce an amplified output (c) in which both signals are distorted.

of these is **hum**, caused from picking up the alternating voltage from the electric power service either through the amplifier power cord and supply or from nearby house wiring and appliances. This is usually heard as a 60-Hz tone along with one or more of its harmonics. Nearby motors, radio transmitters, and other electric devices emit electromagnetic waves that can be picked up as noise. Also, noise can be generated by temperature-induced electrical fluctuations in the amplifier itself. See Figure 21–13. The noise content in an amplifier (or any other component) is expressed as the **signal-to-noise ratio**, the ratio of desired signal to unwanted noise, usually expressed in decibels. Noise in an amplifier can usually be adequately suppressed by proper shielding and careful design.

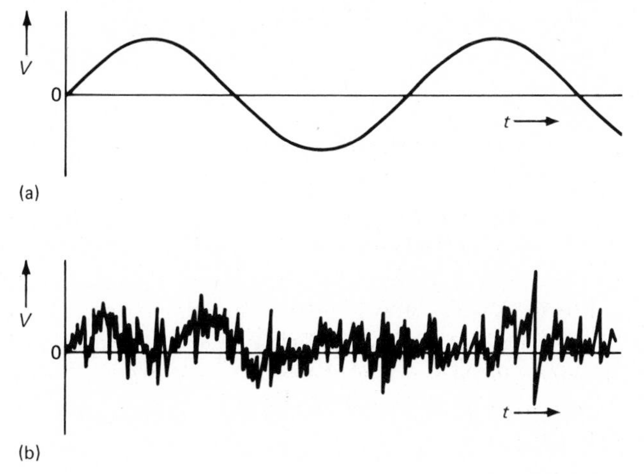

FIGURE 21–13 Unwanted pickup can consist of (a) 60-cycle (60-Hz) hum, (b) a random noise, or both.

21.12 Output

After the preamplifier has selected the desired input and shaped the frequency response, it is usually necessary to boost the power level enough to drive the loudspeaker (or loudspeakers). This is the function of the power amplifier. Although there are several conventions for specifying the capability of the amplifier, the most important of these is the **continuous power output**, expressed in watts. This is the amount of power that can be continuously fed to a speaker without exceeding a specified (usually small) amount of harmonic distortion and without damaging the amplifier.

The power requirements of an amplifier are determined by the loudspeaker that follows it. In order to ensure efficient power transfer, the output impedance should be matched to the speaker. This is usually 4, 8, or 16 Ω. It is therefore desirable to know the type of speaker to be used when deciding upon the power amplifier needs.

Generally, technology is available to produce as good an amplifier as effort and money require. As we will see in the following chapters, fidelity is usually limited by the transducers to which the amplifier is coupled.

QUESTIONS

1. What are transducers? Give two examples.
2. How does alternating current differ from direct current?
3. Why should the amplifier output impedance be matched to the impedance of a loudspeaker?
4. List five different reasons for modifying the frequency response of an amplifier.
5. How does harmonic distortion differ from intermodulation distortion?
6. What is meant by the linear response of an electronic component?
7. List systems with linear responses in the fields of social behavior, economics, and biology.

PROBLEMS

1. An 8-Ω speaker is fed by an ac current of 3 A. Find (a) the ac voltage applied to the speaker and (b) the power delivered to the speaker.
2. If an amplifier generates a current of 5 A in a speaker, the power delivered is 100 W. What is (a) the speaker impedance and (b) the voltage applied?
3. If 72 W is delivered to an 8-Ω speaker, what is (a) the current and (b) the voltage applied?
4. Tabulate the gain of the amplifier illustrated in Figure 21 – 10 for input voltages of 1, 3, 5, 7, and 9 V.

PROJECT

Locate a hi-fi system and study all the controls. Listen to a piece of music, and vary each control in turn. Compile a list, writing down what each control is supposed to do objectively and describing the perceptual result as you hear it.

NOTES

[1]K. R. Johnson and W. C. Walker, *The Science of High Fidelity* (Dubuque, Iowa: Kendall/Hunt, 1977).

[2]W. J. Strong and G. R. Plitnik, *Music, Speech, and High Fidelity* (Provo, Utah: Brigham Young University Press, 1977).

[3]*Understanding High Fidelity* (Moonachie, N.J.: Pioneer Electronic Corporation, 1975).

[4]J. Eargle, "Equalization in the Home," *Audio* 57 (November 1973): 54.

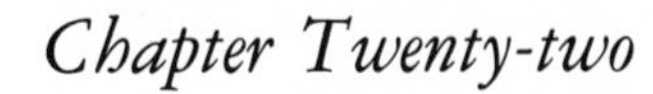

TRANSDUCERS

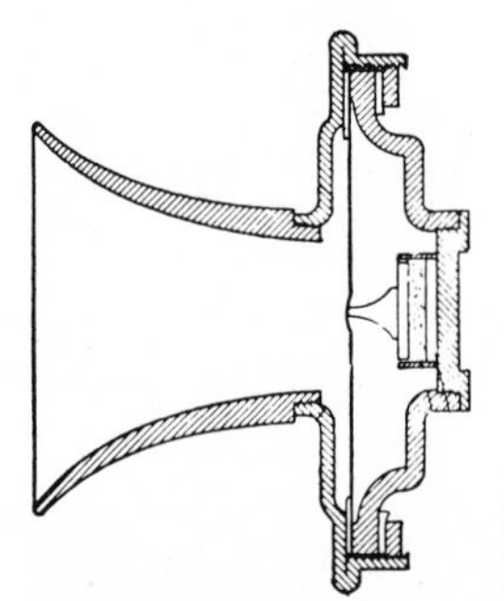

A wide variety of devices are available to convert sound waves from one energy form to another. We have seen in the last chapter that such devices are called *transducers* and are most useful if they convert to or from electric energy. For example, a loudspeaker converts electric energy to acoustic energy, and a phonograph cartridge converts mechanical energy to electric energy. In this chaper we will examine some common transducers.

22.1 Microphones

The microphone (see Figure 22 – 1) converts acoustic sound energy into electric energy. A microphone usually operates by allowing incident acoustic waves to strike some sort of a diaphragm, which is caused to vibrate at the frequency of the incident wave. The diaphragm in turn is attached to a device capable of generating electric current because of its motion.[1]

There are several types of microphones. A **dynamic microphone** contains a small coil of conducting wire that is caused to move through the field of a small magnet. See Figure 22 – 2(a). Here, the transducing action involves **electromagnetic induction**: the generation of electric voltages resulting from a changing relation between the conductor and a magnetic field.

A **crystal microphone** is based on the **piezoelectric effect**: the generation of electric voltages across certain materials by mechanical deformation. By connecting the diaphragm to a suitable crystal, such as Rochelle salt, pres-

FIGURE 22–1
A cardioid microphone. (Courtesy of Shure Brothers, Inc.)

sure variations at the diaphragm surface are transduced to electrical voltage variations at the microphone output. See Figure 22–2(b).

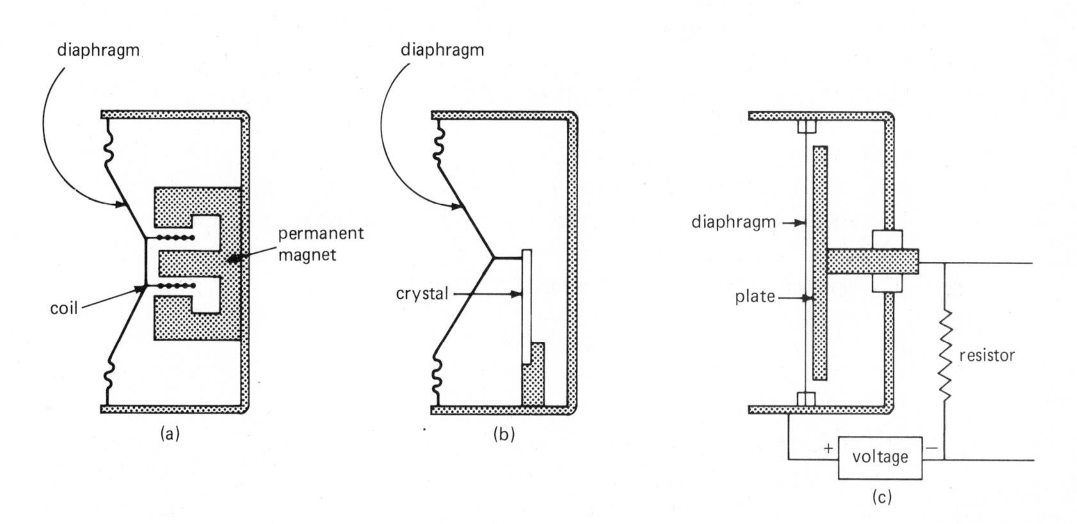

FIGURE 22–2
Schematic diagrams of (a) a dynamic microphone, (b) a crystal microphone, and (c) a condenser microphone.

A **condenser** (or **capacitor**) **microphone** consists of a diaphragm near, but not touching, a fixed plate. See Figure 22–2(c). When a fixed voltage is established between the plates, the electric charge on the plates will vary with the capacitance of the plates:

$$q = CV$$

where q is the charge on either place, C is the capacitance, and V is the voltage across the plates. The capacitance C depends in turn on the geometric arrangement of the plates, increasing as the plates move closer and decreasing as they move apart. (Capacitance can be thought of as a measure of the ability to store electric charge at a given voltage.) As the diaphragm vibrates, the capacitance of the plates varies accordingly. The charge then flows around the circuit in the form of a periodically varying electric current. This current, in turn, produces proportional voltage differences across the resistor *R*.

Although we have not covered all types of microphones used today, it is hoped that the examples above illustrate the principles involved.

The frequency response of a typical dynamic microphone is shown in Figure 22 – 3. Although an amplifier with such response would be considered poor, this response is quite good for a microphone. The variations in the frequency response are primarily a result of the various ways in which the microphone components resonate. The dramatic drop in response below 100 Hz and above 9 kHz illustrates the difficulty in getting any mechanical transducer to respond well over the entire audible frequency range. A mechanical device that responds well to a high-frequency wave will usually respond poorly to a low-frequency wave, and vice versa. This problem is even more severe with loudspeakers.

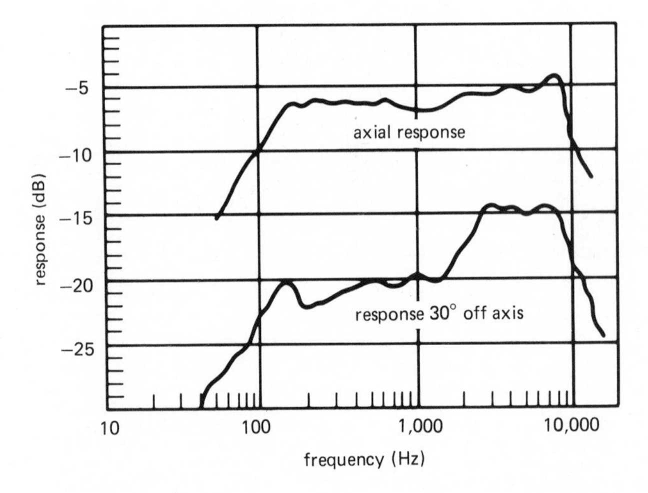

FIGURE 22 – 3
Frequency response curve for a typical cardioid microphone.

22.2 Directional response

Microphones and loudspeakers usually have a directional preference. This preference depends upon the specific transducer mechanism involved, its geometrical design, and the nature of the cage or enclosure that surrounds it.

A microphone that is designed to be equally sensitive to sound coming from all directions is called **omnidirectional.** Such a microphone would be useful, for example, at the center of an orchestra. Alternatively, it is often useful to have a highly directional or **beamed microphone.** Such a microphone would be useful for picking out music from a specific instrument or the voice of a specific speaker when surrounded by other sound sources. This is most easily accomplished by placing a small microphone at the focus of a concave reflector, as shown in Figure 22–4. All sound arriving from the beamed direction will be reflected from the concave surface, focused on the microphone, and picked up. Sound arriving from other angles will be focused elsewhere, such as point P. Such a system is usually very sensitive in the beamed direction and is capable of picking up normal conversations at 50 m, a field referee in a football game, and bird calls at 100 m.

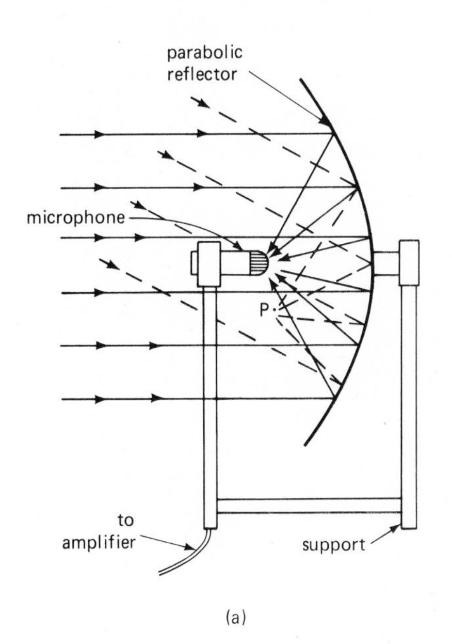

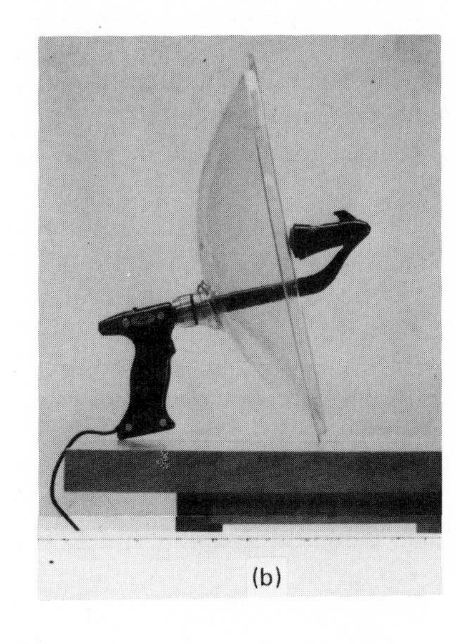

FIGURE 22–4
(a) A diagram of a beamed microphone system. (b) One type of beamed microphone.

Most ordinary microphones have a generally preferred direction. The degree of directionality is usually defined by a plot of response (in decibels) versus the angle that the incident sound makes with respect to the microphone axis. This is most conveniently done by plotting the angle about the circumference of a circle and plotting the response radially *(a polar plot).* See Figure 22–5. This example indicates maximum response at the forward angle (0°), lower response (−7 dB) to either side, and a minimum response (−23 dB) directly behind the microphone (180°). Since a heart-shaped or cardioid curve describes the response, such a microphone is called a **cardioid microphone**. The cardioid microphone is used by vocalists who wish to avoid sound pickup from their surroundings, including audiences. It also avoids **runaway acoustical feedback**, a situation in which amplified sound from the room's loudspeakers reenters the microphone, resulting in loud squealing sounds.

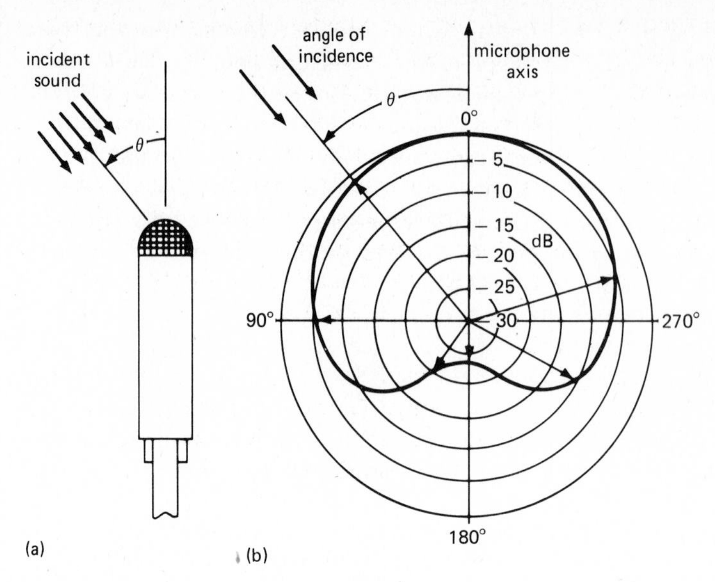

FIGURE 22–5
The response of a cardioid microphone (a) to the angle of sound incidence, represented by a polar plot (b).

22.3 Loudspeakers

Loudspeakers, sometimes referred to simply as speakers, are used to convert electric energy directly into sound energy. Since the quality of a high-fidelity system is usually limited by the speaker or speaker system, careful attention must be given to its selection and placement.

A typical **dynamic speaker**[2] is shown in Figure 22–6. A stiff paper cone

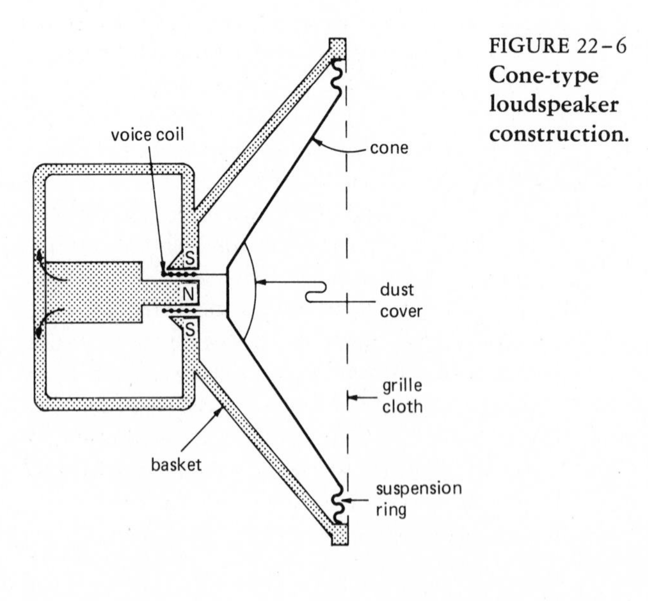

FIGURE 22–6
Cone-type loudspeaker construction.

is loosely supported at the rim, allowing it to vibrate freely in and out as a whole. Attached rigidly to the center of the cone is the *voice coil,* usually a hollow lightweight spool on which a small wire coil is wound. Such a voice coil surrounds a permanent magnet that is fixed to the metal frame but does not touch the voice coil. When electric current from the amplifier flows through the coil, the current interacts with the magnetic field, causing a force to be exerted on the coil. Therefore, when an audio signal is applied to the voice coil, it will move back and forth along its axis in accordance with the sound signal variations. These motions are transferred to the cone, which is then caused to vibrate in and out, moving the adjacent air mass and thereby producing acoustic waves.

Comparison of this speaker design and the dynamic microphone design (Figure 22–1) shows them to be conceptually identical. For this reason, such speakers can be used as microphones. In fact, many intercommunication systems use a single transducer as both microphone and speaker.

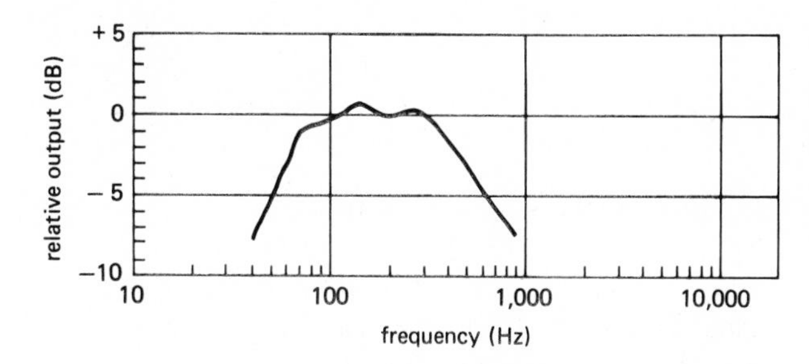

FIGURE 22–7
Frequency response of a typical low-frequency, cone-type loudspeaker.

An ideal loudspeaker should be capable of faithfully reproducing the entire audio spectrum. This means that all areas of the cone should respond accurately and immediately to the many different frequencies that must be reproduced. The frequency response of a typical large (30-cm-diameter) speaker is given in Figure 22–7. It is immediately seen that the frequency response falls off steeply above 400 Hz. The cone simply has too much mass (inertia) to follow the high-frequency changes with the same fidelity with which it reproduces the lows. In fact, the outer area of the cone is unable to keep pace with the high-frequency motion of the cone near the voice coil. This tells us that for good high-frequency response, it is necessary to use a small low-mass cone. However, such a small speaker will not be able to move much air and will therefore have poor low-frequency response. Both problems are somewhat alleviated by distributing the driving force over the moving surface, rather than leaving it concentrated at a cone apex. For example, **electrostatic speakers** use electric voltages to move a large diaphragm,[3] whereas the **Heil driver** uses current-carrying ribbons in the presence of a magnetic field to drive a corrugated surface.[4]

22.4 Multispeaker systems

The popular solution to the frequency response problem has been to use two, three, and sometimes more, speakers to span the audible range. A large dynamic-type **woofer** is used to span the low frequencies, typically 20–500 Hz. A **midrange** speaker spans the 500–5,000-Hz range, and small **tweeters** are used to cover 5,000–20,000 Hz. The **crossover frequencies** at 500 and 5,000 Hz are only typical values; they may vary with each speaker system. See Figure 22–8. In order that adequate sound may be produced without overdriving a speaker to distortion, a given speaker system may consist of more than one of each type of speaker. For example, the desired response may require two woofers, one midrange, and nine tweeters. See Figure 22–9.

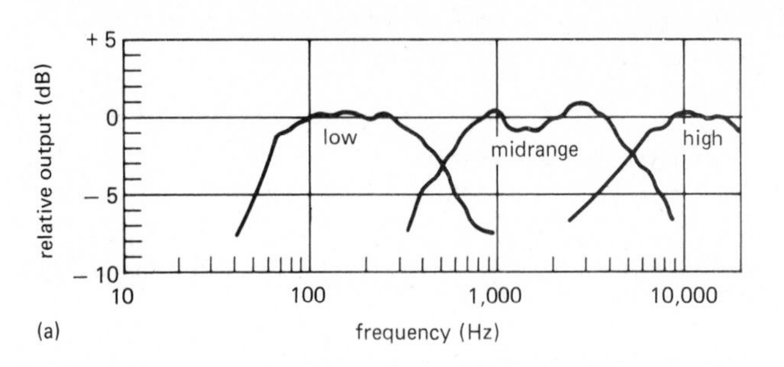

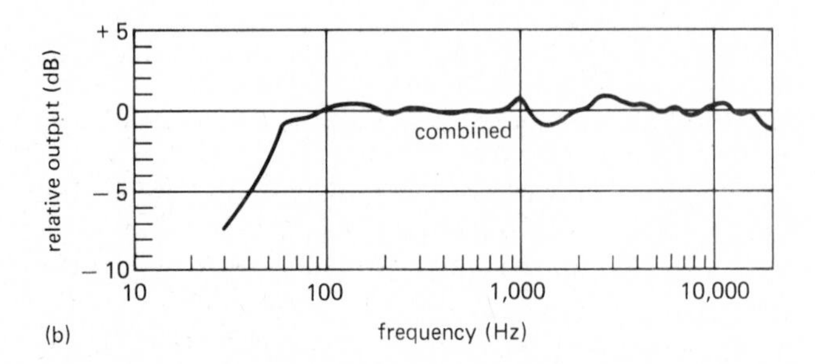

FIGURE 22–8 **(a) Frequency response of a low-frequency, cone-type speaker (woofer), a midrange speaker, and a high-frequency speaker (tweeter). (b) Frequency response of the combined system.**

Although the output from the power amplifier could be sent to all the speakers in a system directly, it is usually routed to the speakers through a **crossover network**. This network isolates the speakers from each other. It consists of a series of high-pass and low-pass filters (see Section 21.6) that feed each speaker only the range of frequencies to which it can efficiently respond. See Figure 22–10. Otherwise, some electric energy may be wasted on a speaker that could not effectively convert it to sound, and this could cause intermodulation distortion of the emitted sound[5] (see Section 21.10).

Although the speakers in a system can be isolated from each other, they should be connected so that their cones move in unison. In this case the speakers are said to be connected in phase, resulting in more efficient output.

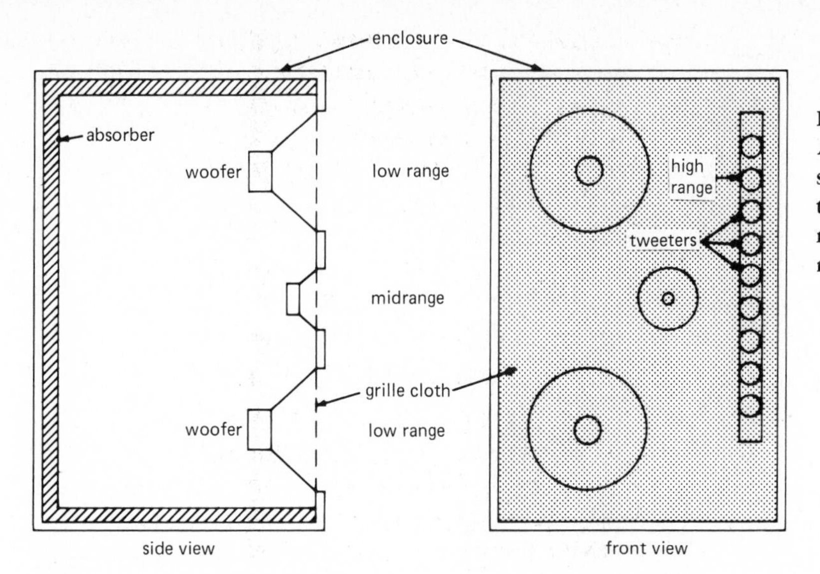

FIGURE 22–9
A loudspeaker system composed of two woofers, a midrange, and nine tweeters.

Also, it should be noted in particular that the diffraction of waves by the speaker apertures is a principal element in determining the angular spread of the sound from any speaker or set of speakers. (See Figures 7–3 and 7–4.)

22.5 Horn-type speakers

Many specialized speakers have been developed in recent years to meet special needs. The horn speaker has long been used in public address systems

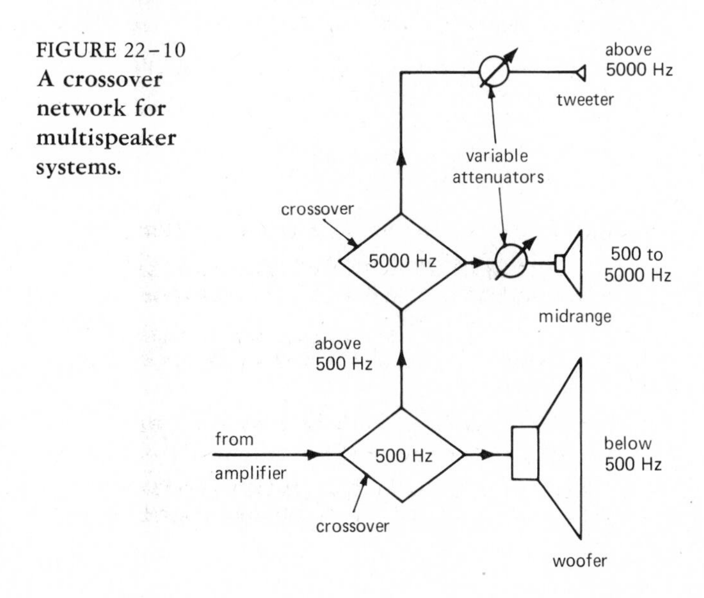

FIGURE 22–10
A crossover network for multispeaker systems.

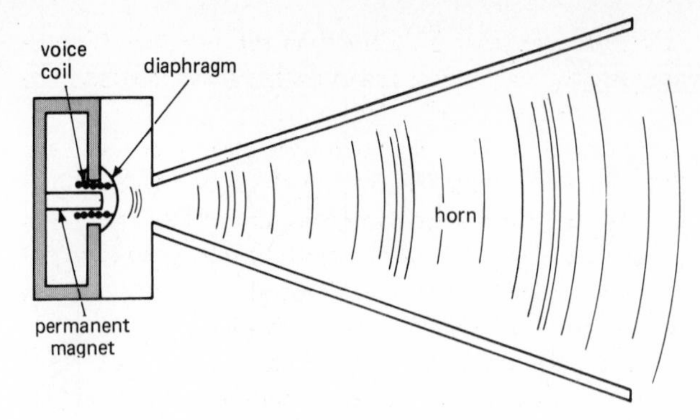

FIGURE 22-11
A horn-type speaker.

because of its high efficiency and rugged construction. Modern horn-type loudspeakers are also useful in high-fidelity systems due to their excellent performance in the middle and upper audio ranges.

As shown in Fig. 22-11, the voice coil of the horn-type speaker is coupled to a small diaphragm rather than to a large cone. The audio signal variations applied to the voice coil cause the diaphragm to move the air back and forth in the sound chamber. These acoustic waves are forced down the throat of the horn, fanning out as they progress through the horn and out the mouth.

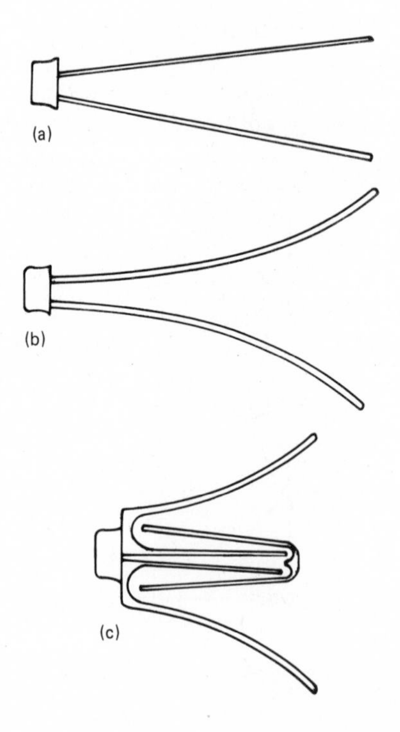

FIGURE 22-12
Various horn shapes: (a) a conical horn, (b) an exponential horn, and (c) a double-folded horn.

Since it is the small diaphragm that moves rather than the horn itself, good high-frequency response is produced, as well as generally good midrange response.

To provide a good impedance match to the listening room, the diameter of the horn mouth should be a significant fraction of the wavelength of the lowest frequency, and the flare should not be too rapid. Good low-frequency response would therefore require horns that are large and long. The problems related to the large size can be somewhat reduced by good design. A variable taper such as the exponential flare will generally allow an adequate bell opening in a shorter length than a simple conical flare. See Figure 22 – 12. By bending the horn, or using reentrant designs such as the folded horn, the required path length can be folded into a smaller volume. Still, the development of good low-frequency response for a horn-type speaker involves a large volume. And this brings us to an investigation of speaker enclosures.

22.6 Loudspeaker enclosures

When a speaker cone is moving forward, a compression of the air is produced in the forward direction and a rarefaction is produced in back. This rarefaction wave can then rush around the speaker and partially cancel the compression wave. When the speaker moves back, the two outgoing waves can

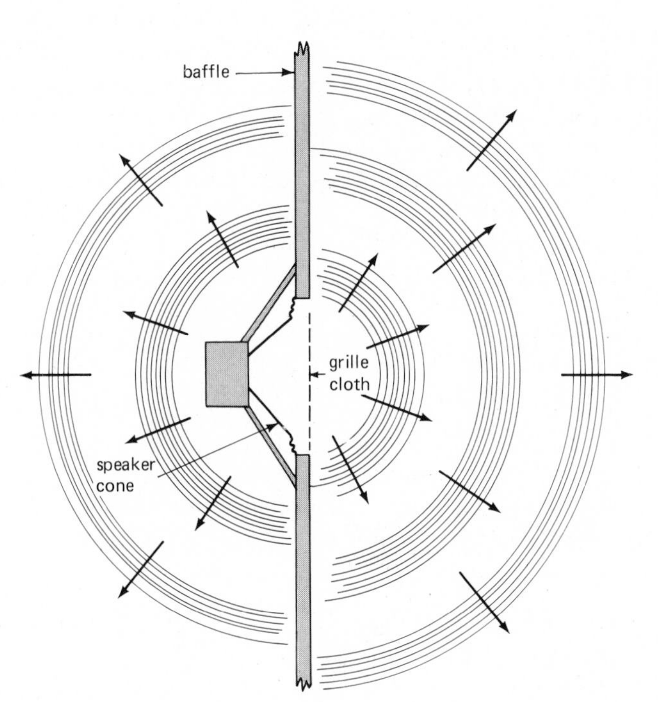

FIGURE 22 – 13
Baffle-mounted speaker permits isolation of back waves, preventing cancellation.

again produce cancellation. This results in partial suppression of the audible sound. This effect is particularly serious at frequencies below 500 Hz where the wavelength of the sound wave is considerably larger than the size of the speaker itself. However, if the loudspeaker is mounted on a rigid wall or on a **baffle,** as shown in Figure 22–13, the low-frequency back wave cannot get around the baffle at all. Such a baffle is the heart of an ideal speaker enclosure. This suggests that the best speaker mount would be in a wall separating the listening room from a large unused room.

22.7 Infinite baffle enclosures

Usually it is not convenient to use an entire room to "bury" the back wave. As a substitute, the speaker can be mounted on the wall of a closed box that is large enough so that the inside air does not restrict cone movement. If the box is sufficiently large and solid, and lined with absorbing material, a smooth and clear response can be produced. However, the required cabinet size is often too large for usage in normal rooms, particularly when several cabinets are needed for stereophonic or quadraphonic reproduction.

In order to accommodate the infinite baffle principle to small-volume enclosures, the **air suspension** principle has been developed. The speaker uses a cone that is loosely mounted at the rim to allow easy travel in response to the applied electric signal. Such a **high-compliance** speaker could normally be driven too far and cause distortion. However, when it is mounted in a small sealed enclosure, as shown in Figure 22–14, the back wave in the enclosure acts as a spring to restore the cone and prevents excessive excursions. For example, as the cone moves into the enclosure, the air inside is compressed, raising the pressure and driving it back out. Such small **bookshelf speakers** are capable of good bass response and yet they occupy little space. However, they are extremely inefficient and require high amplifier power to drive them enough to get a reasonable acoustic output.

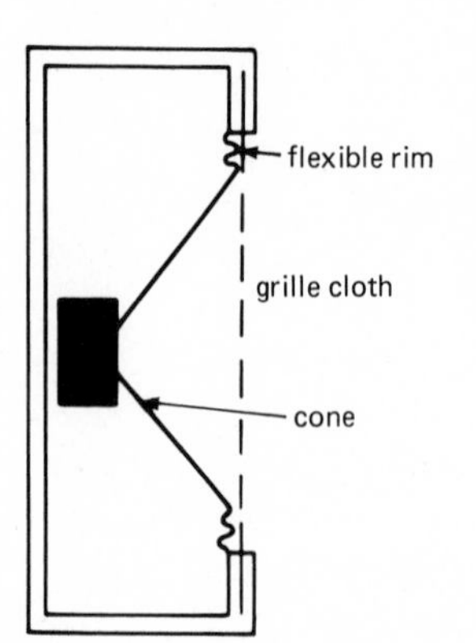

FIGURE 22–14
Air suspension, bookshelf speaker consists of a compliant speaker mounted in a small sealed enclosure.

In order to prevent the back wave from critically affecting the speaker response and efficiency, a variety of methods have been developed to allow the back wave to escape the enclosure. The simplest is the **bass reflex enclosure,** consisting of a moderately large cabinet with a hole or "port" in front of it in addition to the speaker hole. See Figure 22–15. Although it would seem that the problem of cancellation would again be present, such effects can be reduced by proper placement and shaping of the bass reflex port and by careful cabinet design.[6] The bass reflex design usually has a good solid "jukebox bass" response, but it is often too resonant at certain bass notes.

Several variations have been developed to require the back wave to travel longer paths before escaping the enclosure. The back wave can be further reduced by including absorbers such as glass wool, acoustic tiles, or sand. However, the increased quality of the sound produced may not be worth the extra cabinet space required.

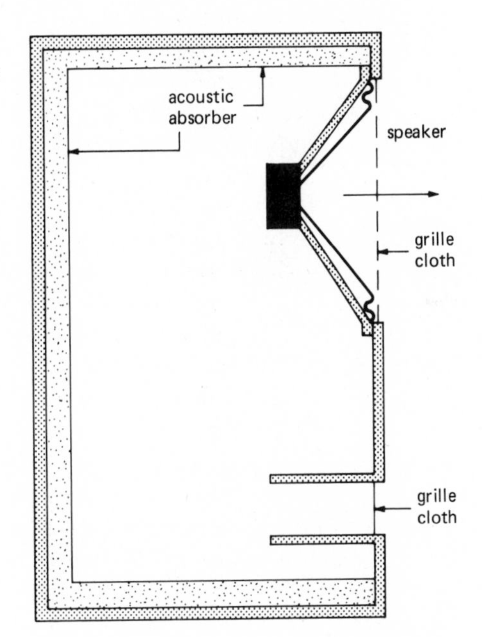

FIGURE 22–15
Bass reflex speaker enclosure.

22.9 Folded horns

Because of their high output efficiency and clarity, horns are widely used for high and midrange response. For the low-frequency range, large horn dimensions are required. For an exponential horn [see Figure 22–12(b)], a length of 3 m would be required to efficiently reproduce notes down to 30 Hz. Although this size is not unreasonable for a large theater, it would hardly fit well in an average listening room. However, by the use of reflecting surfaces, the sound can be caused to travel an equivalent distance with an appropriate flare, yet the horn will occupy a reasonable enclosure. See Figure 22–16. The enclosure of a low-frequency folded horn can be designed to fit in the corner of a listening room and to use the reflections from the floor and walls to improve the impedance match to the room. It is as if the walls and floor act as an extension of the horn itself. Such designs are generally efficient and of high quality, although they may still be large by some standards.

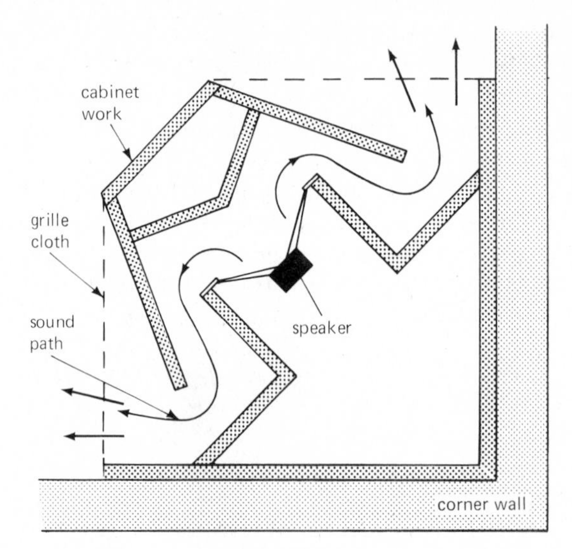

FIGURE 22–16
A low-frequency speaker in a folded horn enclosure, designed to be placed in a corner of a room.

22.10 Speaker system design

In order to match a speaker system to a given listening environment, the environment characteristics must be considered as well as the interests of the listener. Although the low-frequency speakers may occupy a large cabinet, the cabinet size is of minor importance to the midrange response, and the tweeter enclosure is of virtually no importance at all. However, the directional response of a given speaker may be of considerable importance, and so the arrangement of the speakers should be carefully considered.

It is often necessary to correct for speaker performance and balance the resulting room response with the speakers and listeners in place. For example, certain speakers are designed to project up to 90 percent of the sound energy back against the wall. The reflected sound will have good dispersion but will need correction for the wall absorption. The required equalization circuitry is often built into the amplifier, but it can also be placed between the amplifier and speakers. (See Section 21.7.) By appropriate test equipment, or by just listening, the response in the various frequency intervals can be electronically altered, octave by octave, to produce a uniform total room response.[7]

Ultimately, the individual tastes of the listener will determine the desired quality of musical sound. The most vociferous defender of "high-fidelity" sound may actually be uncomfortable with the crystal clarity of a system that turns his or her living room into a concert hall, preferring a more homey, "mellow" sound characterized by the suppression or absence of highs as well as perhaps lows.

QUESTIONS

1. List various situations in which an omnidirectional microphone may be useful.

2. What is the purpose of a crossover network?

3. An "acoustic labyrinth" speaker enclosure channels the back wave through a long path before allowing it to escape the cabinet. Explain the advantages and disadvantages of such an enclosure.

4. If a midrange and low-frequency speaker are both mounted in the same air suspension enclosure, intermodulation distortion may result. Explain how.

5. If a microphone picks up sound that is amplified and projected by nearby speakers, the sound can be repeatedly recycled through the microphone, causing loud howling. Explain how this runaway *acoustic feedback* can be reduced by proper choice and placement of microphone and speaker.

PROBLEMS

1. Using Figure 22–5 for a model, draw a directional response curve for (a) an omnidirectional microphone and (b) a beamed microphone.

2. What rules can you deduce from Figure 22–4 regarding the relation of the angle of reflection to the angle of incidence of the sound on the concave surface?

3. Using the beamed microphone system as shown in Figure 22–4, show that sound incident at 24° off the central axis will miss the microphone.

4. From the horn shapes in Figure 22–12, determine how much longer the conical horn must be to have the same opening as the exponential horn.

PROJECT

This is a striking demonstration and one that is well worth performing. Obtain an unmounted loudspeaker 20 to 25 cm in diameter, as well as a sheet of plywood about a meter square. Cut a hole in the center of the plywood about 4 cm smaller in diameter than the speaker. Connect the speaker to a good radio receiver and turn on any station that is broadcasting music. Now position yourself on one side of the baffle about 2 or 3 m away. Have someone else hold the speaker up to the hole in the center and then move it away from the hole. Do this several times and observe the marked change in the quality of the sound. What particular frequencies show the greatest difference? Explain these differences. This effect can also be demonstrated on a smaller scale using a pocket radio and a sheet of heavy cardboard.

NOTES

[1]D. L. Josephson, "Microphones, the Vital Link in the Recording Chain," *Audio* (December 1973, July 1974, and August 1974).

[2]*Understanding High Fidelity,* (Moonachie, N.J.: Pioneer Electronic Corporation, 1975).

[3]J. Turner, "Why Electrostatics?" *Audio* (May 1974):22.

[4]R. Timmins, "The Heil Driver," *HiFi Review* (August 1974):43.

[5]J. Moir, "Doppler Distortion in Loudspeakers," *Audio* (August 1974):42.

[6]D. B. Weems, "Taming the Bass Reflex," *Radio-Electronics* (February 1975):58.

[7]J. Eargle, "Equalization in the Home," *Audio* (November 1973):54.

Chapter Twenty-three

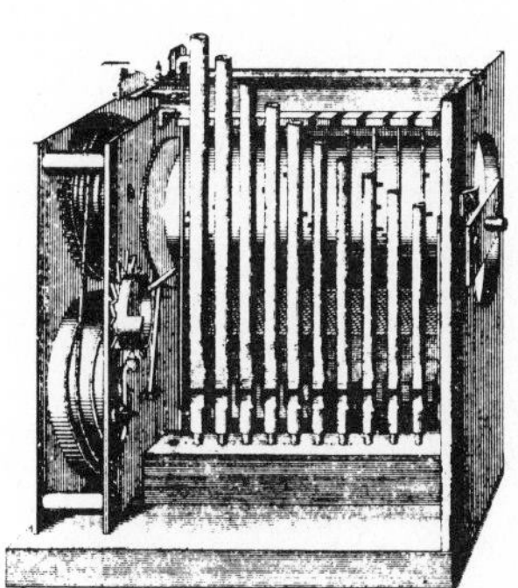

SOUND RECORDING AND TRANSMISSION

A variety of methods have been developed to record sound and transmit it from the studio to the home. Today, the systems capable of recording and transmitting high-fidelity sound include the phonograph record, magnetic tape, and FM (frequency modulation) radio.

23.1 Phonograph records

Sound recording really began with the gramophone, pioneered by Thomas Edison and others in the late 1800s. Over the years, the gramophone record evolved from a horizontally mounted rotating cylinder into a 78-revolution-per-minute (rpm) disc, or phonograph record. By 1948, the type of available discs had expanded to include 16⅔, 33⅓, and 45 rpm discs. Of these, the 33⅓ rpm long-playing (LP) record has survived as the major high-fidelity disc record.

The original basic technique of sound recording involved cutting a spiral groove into a disc such that the sound vibrations were transformed into horizontal undulations of the groove. See Figure 23 – 1. Playback of early discs was accomplished by allowing the record to turn at the proper speed while a needle attempted to follow the undulations of the groove. See Figure 23 – 2. The needle was attached to a lever arm that caused a flexible diaphragm to vibrate. This acted as the driver of a horn, allowing the vibrations to be transferred to the surrounding air as sound.

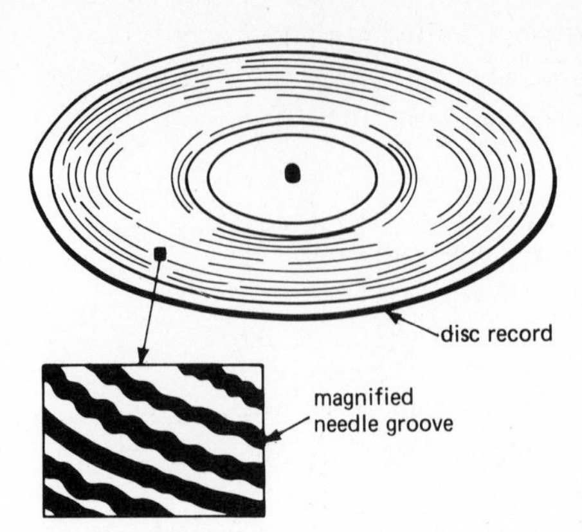

FIGURE 23–1
Disc record. Closeup view shows lateral groove undulations corresponding to the recorded audio sound.

This all-mechanical system has been modified over the years to the modern **turntable** and **cartridge pickup.**[1] See Figure 23–3. The record is placed on a heavy metal turntable that turns at a steady angular speed of $33\frac{1}{3}$ rpm. A diamond-tipped stylus rests in the record groove and follows the undulations.

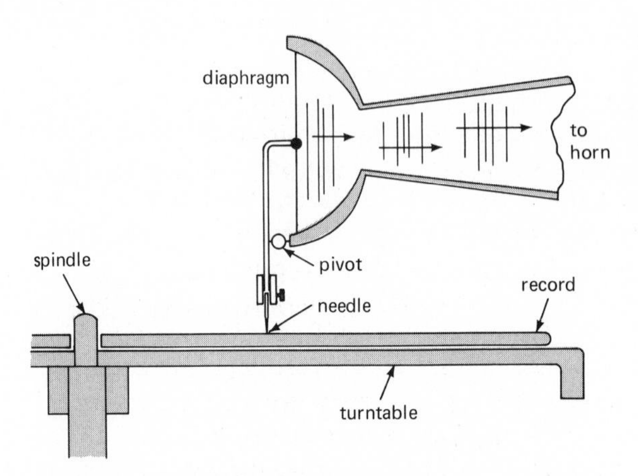

FIGURE 23–2
All-mechanical phonograph pickup. Vibration of the needle in the groove is transferred directly to the flexible diaphragm which produces sound waves in a horn.

The cartridge that holds the stylus is capable of turning the vibrations into electric signals. Of the many different ways in which this is done (moving magnet, moving coil, piezoelectric effect, etc.), the most common is the **moving magnet** method. The end of the stylus is attached to a small magnet that moves back and forth through a small coil in the cartridge, generating electric current in the coil. This low-level current is then routed to the preamplifier for amplification, using well-shielded wires to avoid noise pickup from the surroundings. The cartridge is mounted at the end of a tone arm, which is pivoted at the other end, allowing for both vertical and horizontal motion. Ideally, the axis of the cartridge should be parallel to the grooves across the entire playing track. Deviations from this alignment, called **tracking error,** can be minimized by proper tone arm length, placement of the pivot, angling of the cartridge, and other more elaborate additions and modifications.

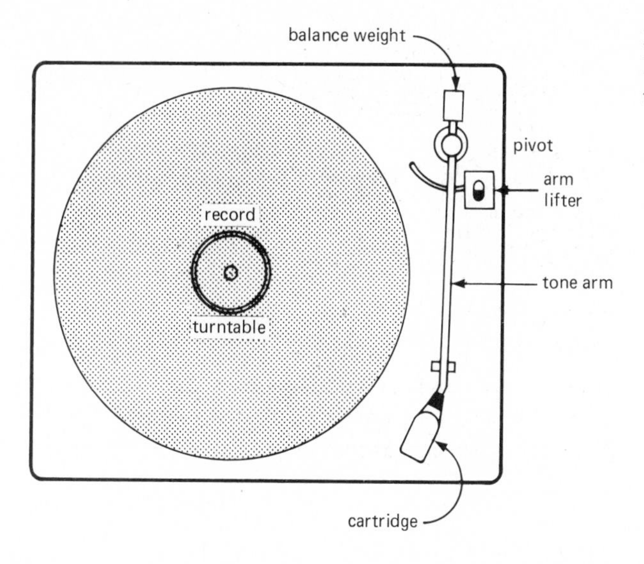

FIGURE 23-3 **Phonograph turntable.**

The stylus is generally tipped with diamond to ensure a long life, and it is shaped according to various formulas to follow the high-frequency track variations. The weight of the stylus on the record is usually adjustable and should not be too heavy if wear is to be minimized. However, if the weight is too light, the stylus will be thrown about, causing **shatter** in the reproduced sound and chipping of the sides of the groove, causing **scratch.**

Careful turntable design is essential in order to avoid any variations in frequency, mechanically generated noise, and loss of frequency response. The frequency can be caused to vary slightly if the grooves are not centered on the record properly (**wow**), if the record is warped (**warp-wow**), or if the turntable speed varies due to irregularities in the motor drive or pulley system (**flutter**). The latter can be minimized by using a massive turntable platter so that it has sufficient *rotational inertia.*

Unwanted noise is often created by low-frequency motor vibration (**rumble**), high-frequency wear on the record (scratch), and the effects of dust on the grooves and on the stylus (**pop**). These effects can be suppressed with the appropriate filters in the amplifier but at the expense of lowering the fidelity (see Section 21.6).

If the turntable is in the same acoustical environment as the speaker, there will usually be **acoustical feedback** of sound energy from the speaker to the tone arm, either through the floor and turntable structure or directly through the air. At certain frequencies, the acoustical wave will cause the tone arm to move in opposition to the direction of the groove, causing a suppression of the amplified signal and inducing distortion. At other frequencies, the acoustical wave can cause distortion by enhancing the motion of the stylus and can lead to **runaway acoustical feedback** or oscillation, causing piercing sounds and possible damage to the system. Acoustical feedback can usually be minimized by any of a variety of methods that isolate the turntable from its acoustical environment.

When a record is originally cut, the high frequencies are boosted in order that the groove variations be sufficient for the needle to follow. When the record is played, the signal is sent from the cartridge to a special preamplifier input that not only amplifies the signal but suppresses the highs to compensate for the original enhancement. This process of equalization (see Figure 23–4) has two advantages. First, it allows the sounds at almost all audible frequencies to be recorded without undue distortion. Second, since the highs in the preamplifier are reduced, any noise pickup is also reduced. Most noise comes from scratch and from pickup in the wires carrying the signal from the cartridge to the preamplifier.

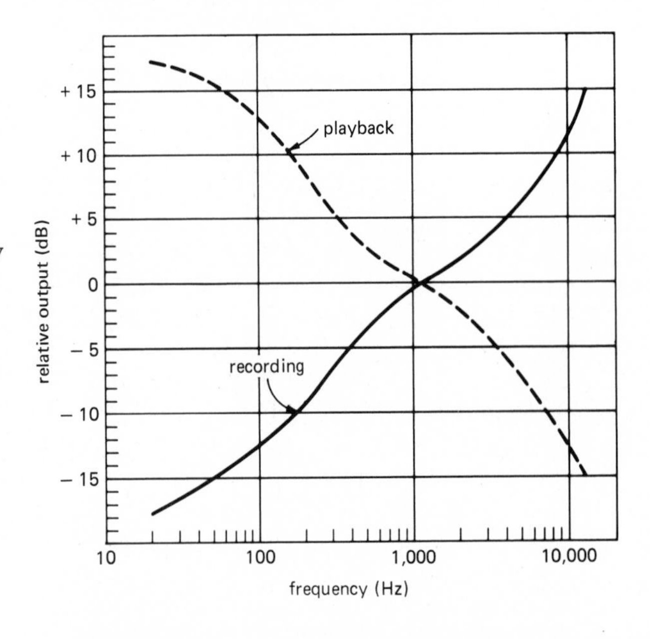

FIGURE 23–4 **Standard Record Industry Association of America (RIAA) recording and playback characteristics. Low frequencies are attenuated and high frequencies are boosted during recording. The reverse occurs in the preamplifier, resulting in a flat net response.**

Many modifications have been made to the basic turntable. Among them are the ability to vary the speed to accommodate different record types, the development of the high-quality automatic record changer, and, by 1958, the adaptation to two-channel (stereo) recording. The stereo system uses the two sides of the track for two different channels, recorded synchronously but distinct from each other. See Figure 23–5. The stereo cartridge then has two pickup coils, each designed to be sensitive to the variations of one of the sides of the groove but not the other. The two separated signals are then routed to the separate preamplifier inputs.

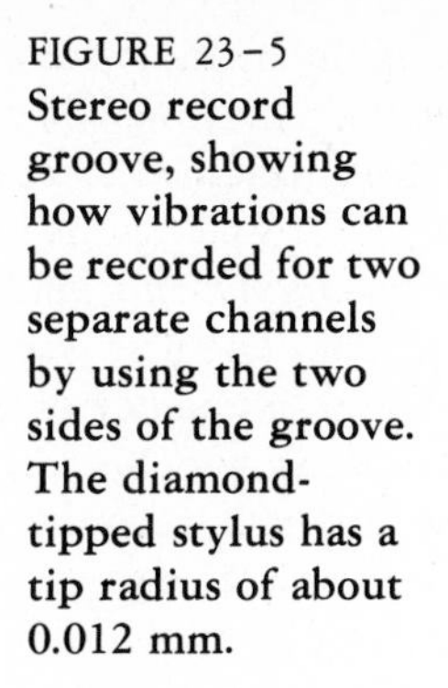
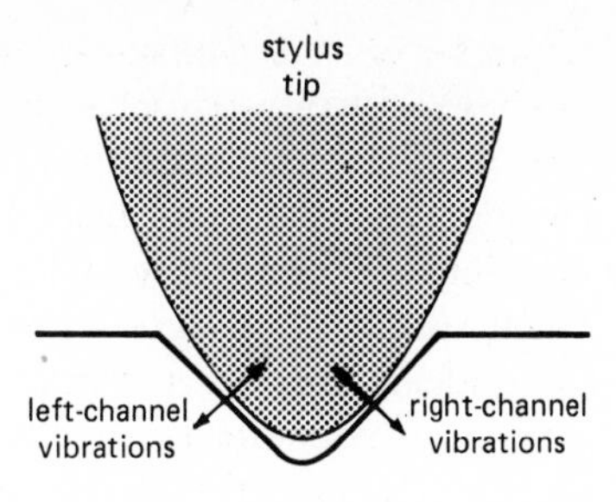

FIGURE 23–5
Stereo record groove, showing how vibrations can be recorded for two separate channels by using the two sides of the groove. The diamond-tipped stylus has a tip radius of about 0.012 mm.

Efforts to record four separate channels (quadraphonic recording) have raised additional demands on record technology, such as the development of a small line-contact stylus capable of following high-frequency vibrations.[2] (See Section 25.5.)

23.2 Magnetic tape

The most commonly used recording medium today is magnetic tape. This is an outgrowth of the wire recorders developed from 1920 to 1950. Relying on the magnetic properties of iron, it was found that sound can be recorded as varying states of magnetization of atoms or clusters of atoms (domains) of iron in the wire, each acting as a small magnet with north-south pole alignment. These microscopic magnets are oriented back and forth in larger regions in accordance with the sound wave. By the 1950s, the iron wire was replaced by plastic tape with a coating of iron oxide and, more recently, with a coating of chromium oxide. See Figure 23–6. Since magnetic tape can be made in any

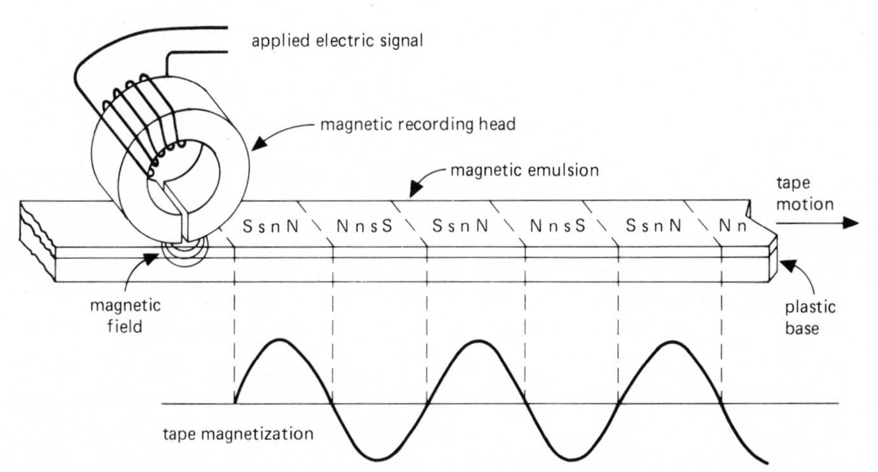

FIGURE 23–6
An audio signal is encoded on magnetic tape as a sequence of magnetized regions. Each region acts somewhat like a single bar magnet, with a north pole on one end and a south pole on the other. The pole strength varies smoothly from north to south: N–n–s–S.

width, several parallel tracks can be recorded at the same time, making it simple to use for stereo or quadraphonic recording.

A typical high-fidelity tape deck, designed to be used in conjunction with an amplifier, is shown in Figure 23–7. The tape is fed from one open reel to another at constant speed by the *capstan* and *pinch roller.* Three magnetic heads are positioned close to the tape. The first is an erase head E, consisting of an electromagnet driven by a high-frequency electric current. This causes ultrasonic magnetic variations to occur, which realign the magnetic domains on the tape, thereby erasing the previous recording. The recording head R consists of another electromagnet, which applies the desired signal to the tape.[3] The third head is for playback P, allowing the passing tape to induce electric signals in the small coil in the head, much as an electric generator generates electric current. This current is amplified and routed to the preamplifier for further amplification. The playback head also makes it possible to **monitor** a recording by listening to the recorded signal immediately after it is recorded.

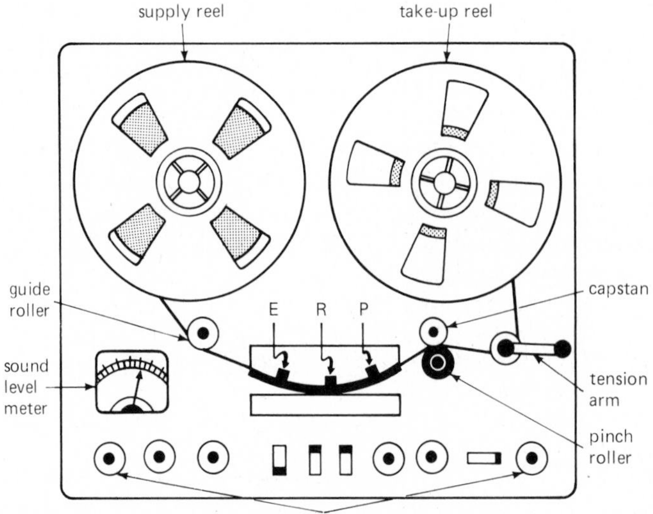

FIGURE 23–7 **Open-reel magnetic tape deck showing placement of the erase head (E), recording head (R), and playback head (P).**

Many tape decks have replaced the open reels with small packaged reels or cassettes. For use in the automobile, endless-loop cartridges are commonly used. Tapes generally consist of the magnetic oxide on a base of cellulose triacetate or polyester. Unfortunately, the cellulose triacetate tends to age if not properly cared for. The polyester tape does not age, but it stretches easily.

For best fidelity in the highs, the tape should be run at relatively high speed. This means $7\frac{1}{2}$ or 15 inches per second for music, although a lower speed of $3\frac{3}{4}$ inches per second is used for long-playing tapes. The higher speed causes the tape to be used more rapidly. Thinner tape will allow more tape to be wound on the reel, but thin tape is more susceptible to **stretch,** as well as

print-through—that is, the tendency for magnetization of one layer to induce magnetization of the adjacent layer. Another problem with magnetic tape is that unevenness in motor and roller speed, as well as tape stretch, can cause pitch variations called **drift** (long term), **wow** (moderate), and **flutter** (rapid).

23.3 Compression-expansion systems

Uneveness in magnetic tape coating can produce high-frequency tape noise called **hiss**. This noise is particularly noticeable in the presence of low-level signals. A second problem is that the magnetic coating may not respond adequately to high-level signals, resulting in reduced and distorted sound.

Both of these problems are reduced by a **compression-expansion** system.* In recording, compression circuits selectively reduce high-level passages and boost low-level passages. See Figure 23–8. The signals are there-

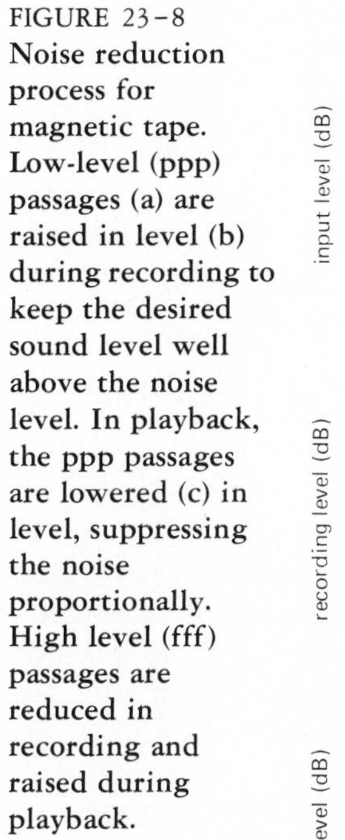
FIGURE 23–8 Noise reduction process for magnetic tape. Low-level (ppp) passages (a) are raised in level (b) during recording to keep the desired sound level well above the noise level. In playback, the ppp passages are lowered (c) in level, suppressing the noise proportionally. High level (fff) passages are reduced in recording and raised during playback.

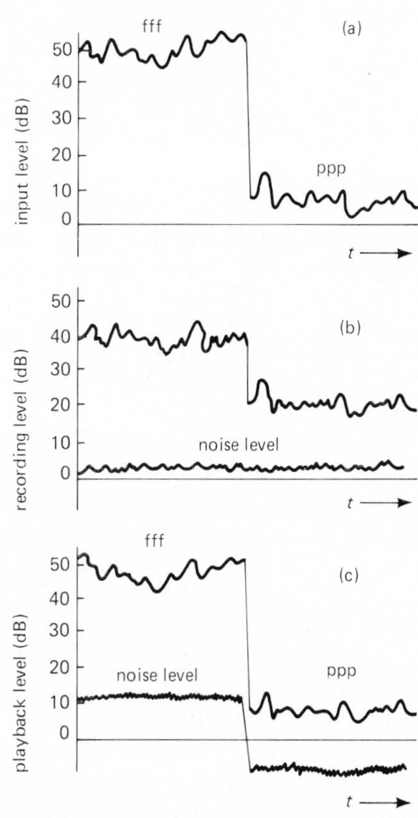

*The name *Dolby,* a trademark of Dolby Laboratories, Incorporated, is often identified with noise reduction methods, although other similar systems are also being used.

fore compressed into a narrower dynamic (decibel) range. Upon playback, the expansion circuits boost the high-level signals, allowing loud passages such as cymbal crashes to be properly heard. Low-level passages are appropriately reduced, reducing any tape noise proportionally. As a result, the signal-to-noise ratio is kept low. Such a system can also be applied to disc recordings.[4]

The compression and expansion system is often limited to high-intensity ranges or to high frequencies where tape hiss is most noticeable. It also gives the user more control over the sound of any system.

23.4 Radio transmission

Audio frequencies have been shown to span the range of about 20 Hz to 15 kHz. Electric signals of this frequency can be carried by wires for quite long distances. However, noise pickup and absorption of high-frequency signals make wires impractical for long-distance transmission, although it is adequate for telephone transmission.

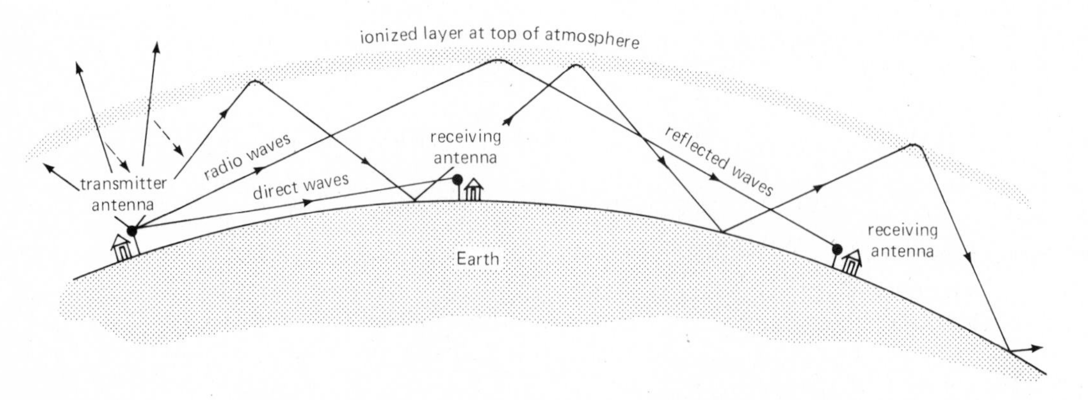

FIGURE 23-9
Radio waves from a transmitting station are picked up at receiving antennas in direct line of sight. Waves can also be propagated by reflecting from the upper atmosphere. Note the waves and skip distances.

Impressed variations of electric current from a sending station into its metal antenna can radiate signals outward through space in the form of electromagnetic waves. These waves travel with the speed of light, 300,000,000 m/s, and exhibit many of the properties of light waves, such as reflection, refraction, interference, and diffraction. These **radio waves** can also be picked up by a distant antenna through a tuning process at a receiving station. See Figure 23-9. However, the frequencies that are most easily transmitted are much higher than audio frequencies. These **radio frequencies** (RF) are divided into bands, with designations as shown in Table 23-1; see Figure 23-10.

All radio waves travel in straight lines in free space. If the transmitting antenna is in *line of sight* from the receiving antenna, radio waves can be picked up day or night. If the curvature of the earth prevents this, the waves are often reflected back and forth between an ionized layer of air high in the ionosphere and the ground. See Figure 23–9. Ionization of the upper atmosphere is attributed partly to ultraviolet radiation from the sun, and occurs in several layers of varying densities surrounding the earth. These layers change between night and day, time of year, and so on, and only certain frequency bands are reflected back toward the earth.

TABLE 23–1: Radio Frequency Bands for Radio Communications

SYMBOLS	DESIGNATION	BAND WIDTH (FREQUENCY)[a]
VLF	Very low frequency	3 kHz–30 kHz
LF	Low frequency	30 kHz–300 kHz
MF	Medium frequency	300 kHz–3 MHz
HF	High frequency	3 MHz–30 MHz
VHF	Very high frequency	30 MHz–300 MHz
UHF	Ultrahigh frequency	300 MHz–3 GHz
SHF	Superhigh frequency	3 GHz–30 GHz
EHF	Extremely high frequency	30 GHz–300 GHz

[a]kHz means kilohertz (10^3 Hz); MHz means megahertz (10^6 Hz); and GHz means gigahertz (10^9 Hz).

The majority of radio stations transmit at various radio frequencies in the MF band (300 kHz–3 MHz). The audio frequencies are carried by the process of **amplitude modulation** (AM). The audio frequency currents from a microphone, a tape deck, or a record are used to vary the amplitude of the continuous RF oscillations and waves from the transmitter. See Figure 23–11. These modulated waves propagate through space to the receiver antenna, where they are picked up and routed to a tuner. By turning the tuning dial on the tuner, the frequency of the desired station is selected. Subsequent circuitry converts the amplitude-modulated RF signal back to a pure audio signal.

An alternate method of radio transmission involves **frequency modulation** (FM) broadcast in the VHF band (88 MHz–108 MHz). In FM, the audio frequency currents are used to vary the frequency of the RF carrier wave yet

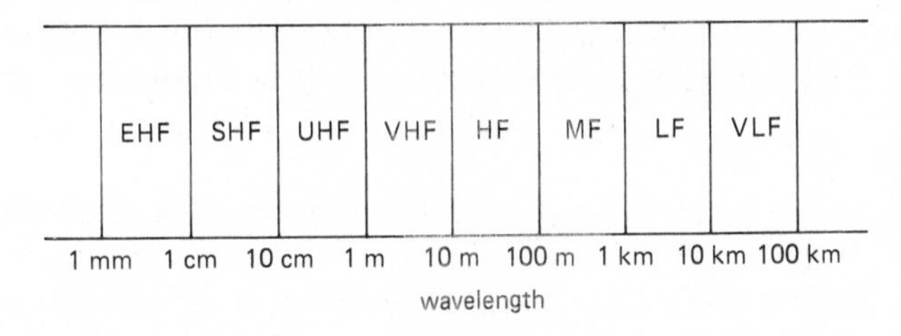

FIGURE 23–10 **Radio wavelength bands with their designations. See Table 23-1.**

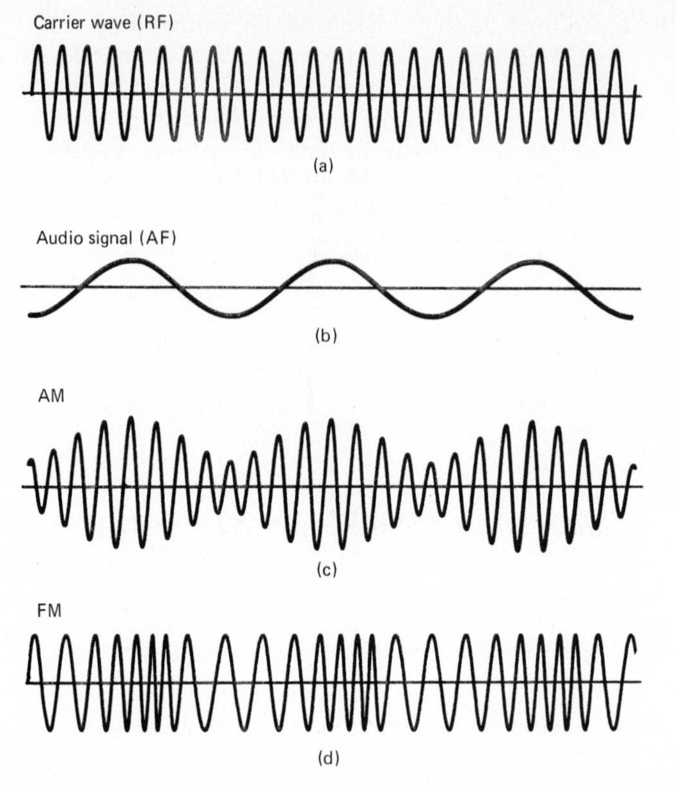

FIGURE 23–11 The audio frequency signal (b) can modulate the amplitude of a radio frequency carrier (a) and therefore produce an amplitude-modulated wave (c). Alternatively, the frequency can be modulated to produce a frequency-modulated wave (d).

keep its amplitude constant. See Figure 23–11(d). Louder sounds with AM mean greater variations in amplitude, while with FM they mean greater changes in frequency.

Good reproduction of voice and music can be accomplished with AM as well as with FM. However, FM has the advantage of eliminating noise pickup, **static**, due to local atmospheric disturbances. Frequent minor lightning discharges in the air induce undesirable current pulses in any radio-receiving antenna. When the tuner is designed to receive AM, these current pulses are amplified along with the broadcast and are heard as sharp crackling or frying sounds. These noises do not change the frequency, however, and thus will not affect FM reception. The resulting quietness of FM reception leads to its almost exclusive use for high-fidelity sound transmission.

23.5 FM multiplex stereo

For stereophonic sound, two separate audio signals must be transmitted. In order to transmit two audio waves on a single radio frequency carrier, the two audio channels are **multiplexed**, or mixed together in such a way that they can be separated later in the tuner.

One audio channel modulates the RF wave at the usual audio frequencies, 0–15 kHz. The other channel is combined with an ultrasonic (US) 38-kHz signal, resulting in modulation spanning the range 23–53 kHz. To this is added a **pilot signal** at 19 kHz. These audio and ultrasonic signals are combined as shown in Figure 23–12. The combination is then used to modulate the frequency of the RF carrier, which is then transmitted.

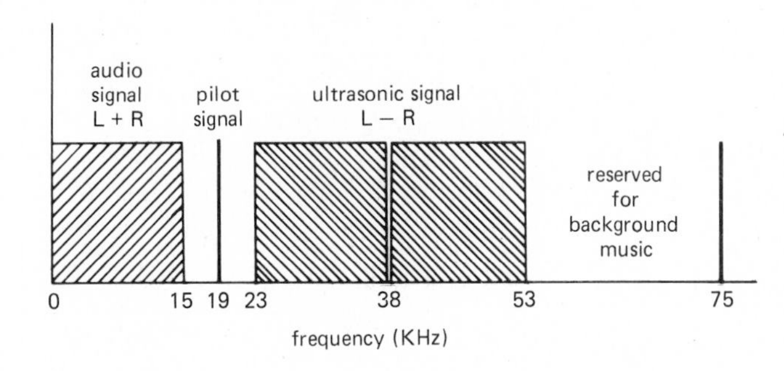

FIGURE 23–12 Signal spectrum for FM multiplex stereo transmission. The audio band carries the left-right sum signal, and the 23–53-KHz ultrasonic band carries the difference signal. The 53–75-KHz band is reserved for stations that transmit commercial-free background music services.

Following detection in the tuner, the audio signal is separated as one channel, the ultrasonic signal is reduced to the audio level as the second channel, and the 19-kHz signal turns on a light identifying the selection as a *stereo* station.

If the left channel is called L and the right channel R, it would seem reasonable to use either L or R for the audio channel and the other for the ultrasonic channel. In practice, L and R are added together for the audio channel, and R is subtracted from L (added with inverted amplitude) for the ultrasonic channel:

$$A = L + R$$
$$US = L - R$$

After these signals are detected in the receiver, the channels A and B are again separated out.

$$L = A + US$$
$$R = A - US$$

The reason for this extra complexity is that people with an FM tuner that is not designed for multiplex stereo reception can only pick up the audio channel. They will therefore get the sum of the two channels L and R, rather than just one of them.

QUESTIONS

1. Why should the weight of the stylus on a record be properly adjusted?
2. What is meant by tape monitoring?

3. List the advantages of disc records over magnetic tape, and vice versa.
4. Why is a turntable platter usually heavy?
5. Design a method for reducing acoustic feedback.
6. What happens if the pinch roller on a tape deck is not perfectly round?

PROBLEMS

1. How many turns will an LP record make while playing for 20 min? How far will the stylus be dragged along the plastic?

2. Sketch a RF signal (a) amplitude modulated by a square wave and (b) frequency modulated by a square wave.

3. Suppose the tone concert A (440 Hz) was broadcast via FM at a radio frequency of 110 MHz. How many cycles of RF occur during each cycle of the audio tone?

PROJECT

Locate several issues of recent hi-fi magazines and review the advertisements and articles on particular components: tape decks, tuners, turntables, and so forth. What characteristics are thought to be important for that component? What specifications are normally quoted? What quoted values are considered as good?

NOTES

[1]W. Strong and G. Plitnik, *Music, Speech, and High Fidelity* (Provo, Utah: Brigham Young University Press, 1977).

[2]R. F. Scott, "All About CD-4 Phono Cartridges," *Radio-Electronics* (June 1975):46.

[3]K. Johnson and W. Walker, *The Science of High Fidelity* (Dubuque, Iowa: Kendall/Hunt, 1977), p. 450.

[4]L. Feldman, "Noiseless Discs at Last," *Radio-Electronics* (February 1975):31.

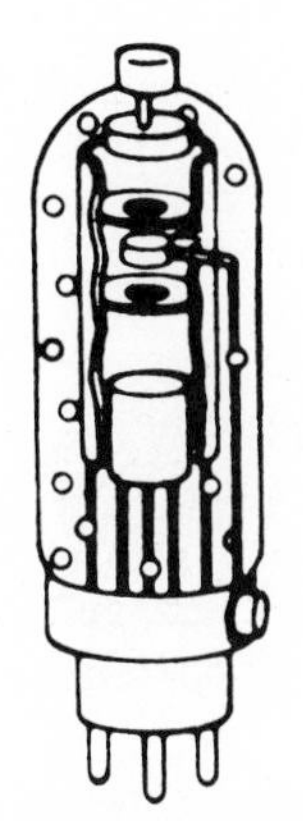

Chapter Twenty-four

ELECTRONIC MUSIC

Instruments can use electronics in a variety of ways. For example, one can simply add electronic amplification to a traditional acoustic instrument. At the other extreme are instruments designed to produce electronic signals directly, without first producing sound. Such instruments—the modular synthesizer, for example—can produce sound unlike any known traditional instrument. Use of electronics in musical composition not only has produced new instruments and sounds but has revolutionized the entire approach to composing music.

24.1 Electronically amplified instruments

Virtually every traditional instrument has been subject to electronic amplification. The more common of these are the electronically amplified accordion, guitar, piano, and violin.[1] Electronic instruments often use one or more microphones to pick up and amplify sound produced by standard acoustic instruments. In many cases, such as an accordion, microphones are placed directly on specific portions of the instrument. By proper amplification, the audibility of the instrument is enhanced and usually changed in timbre as well.

24.2 Electronic stringed instruments

An acoustic (nonelectric) guitar is designed so that the vibration produced by the strings is carried through the bridge to the top plate, where the motion of the wooden surfaces creates the audible acoustic waves. Although the sound waves can be picked up by microphones, other methods have been developed to pick up the string vibrations directly. One such method, used on violins and guitars, involves placing a strain gauge between the bridge and top plate. See Figure 24–1. This transducer intercepts the vibrations passing from the strings to the body and converts them to electric signals that can be amplified.

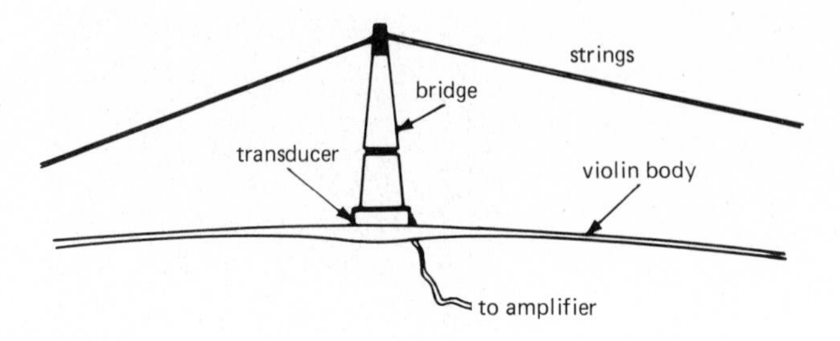

FIGURE 24–1
A small pressure-sensitive transducer under a violin bridge serves as an electronic pickup.

The more common form of guitar pickup uses magnetic heads placed near the bottom of the fingerboard. Each head consists of a small magnet surrounded by a small wire coil. See Figure 24–2. The motion of each steel string in the presence of the corresponding magnet causes changes in the magnetic field at each head. These changes generate electric currents in the coils surrounding each magnet, which are routed to a preamplifier.

Often two or more sets of pickups are used: one near the bridge and the other further away. The pickup closer to the bridge tends to pick up more overtones and will produce a brighter timbre. The player can then select the

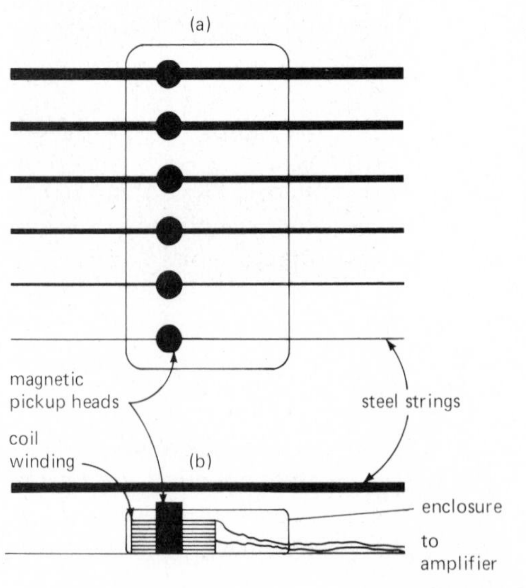

FIGURE 24–2
Electromagnetic pickup for each steel string of a guitar: (a) view from above and (b) view from the side.

preferred output or mix them as desired. See Figure 24–3. The response of body and air modes characteristic of the acoustic instrument are largely bypassed by such pickups and will be virtually inaudible to a large audience. Such acoustic response can be picked up by microphones, usually exterior to the instrument, and amplified along with the string pickups.

FIGURE 24–3
Typical electroacoustic guitar with two pickups for each string and built-in preamplifier. (Courtesy of Norlin Music, Inc.)

Since an **electroacoustic guitar** usually depends on electronic amplification for its sound projection, the acoustic projection function of the guitar box is of secondary importance, allowing it to be considerably smaller. Many designs of the electric guitar or electric bass involve mounting the strings directly on a rigid backing and shortening the dimensions radically. Such **electric guitars** have virtually no direct acoustic output and often bear little resemblance to their traditional counterpart.

24.3 Electronic modifiers

The signal generated by the pickup on an instrument is routed through amplifiers to the recorder or speakers. Part of the amplifier on an electric guitar is often built into the instrument, allowing the player to conveniently adjust the amplification and tone. See Figure 24–3. The player can electronically modify the sound by a variety of methods. These include adding a **fuzz box**, an electronic module that causes distortion. A **wa-wa pedal** allows the player to rapidly vary the timbre. A throw of a switch can add tremolo (amplitude modulation). *Reverberation* is produced by a series of very closely spaced echoes and can be created by another simple circuit. See Figure 24–4(a). The phenomenon enriches the sound considerably and gives the illusion of hearing the instrument played inside a cave. The use of a continuous magnetic tape loop, as shown in Figure 24–4(b), allows the generation of *echoes,* which further enhance the large-enclosure effect.

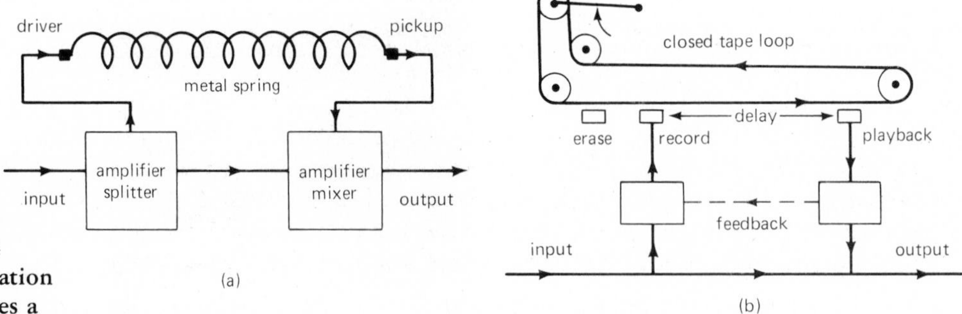

FIGURE 24–4
(a) Reverberation unit generates a reverberant standing wave in a metal spring, which is mixed with the normal sound. (b) Echo unit records sounds, which are picked up a fraction of a second later and mixed with the normal sound. Multiple echos are generated by returning part of each echo to be recycled.

Although electronic modification is most often applied to plectrum and Hawaiian steel guitars, it can be applied to other instruments as well, and it generally gives the player a wider range of playing options than are available with traditional acoustic instruments.

24.4 Electronic organs

Since the early 1900s, a variety of devices have used a piano or organ keyboard to directly produce electronic notes, entirely bypassing the initial acoustic production. These instruments led to the development of the **Hammond organ** in 1929. This was a mechanical system in which rotating toothed steel wheels induced currents in adjacent magnetic heads. See Figure 24–5(a). Each pitch was determined by the speed of rotation of the tone wheel and the number of teeth.

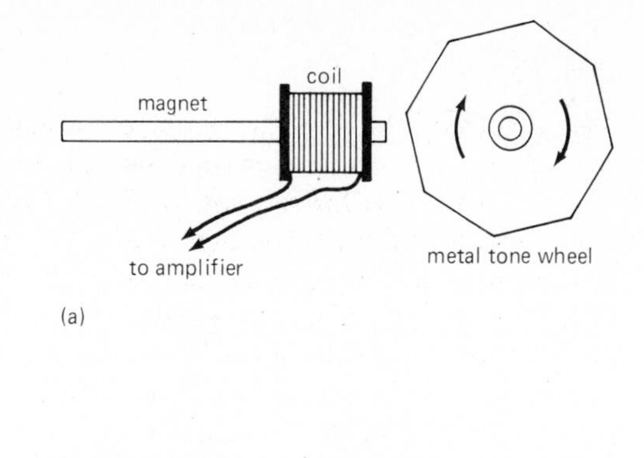

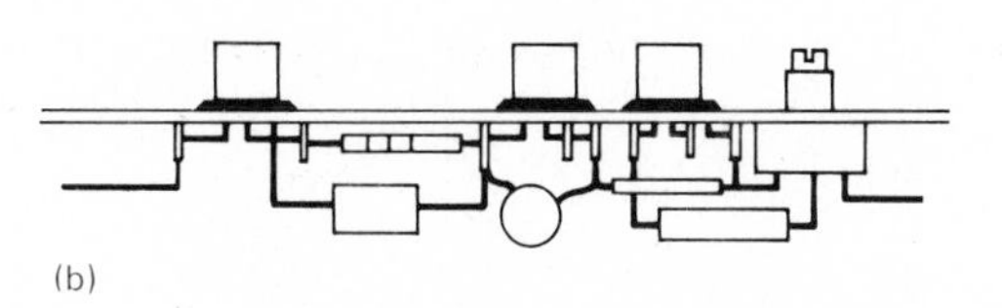

FIGURE 24–5
The tone wheel and magnetic pickup of the mechanical-electronic organ (a) have been largely superceded by all-electronic components using transister circuit cards (b) and integrated circuits.

Other electromechanical organs were later developed using rotating discs and sensors of various sorts. By the 1960s, the electromechanical organs were superceded by all-electronic systems, in which the notes are produced by electronic oscillators.[2] See Figure 24–5(b). Tones can be generated with a variety of waveforms, chosen mainly to simulate specific types of organ pipes. By activating electronic circuits with switches and pedals, the player can change the amplitude and timbre, as well as the attack and decay, of the note. It is a simple matter to add vibrato, tremolo, and reverberation. See Figure 24–6.

24.5 Electronic instruments

While the modern electronic organ is quite different in design from a pipe organ, it is still designed basically to simulate a pipe organ. Electronic circuits have also been designed to closely simulate the rhythmic tap of a snare drum, the click of castanets, the crash of cymbals, and many other familiar sounds, and some electronic organs have incorporated special features such as these.

The advent of electronics has also opened the opportunity to create musical instruments capable of producing sound unlike anything produced acoustically. Figure 24–7 shows one such instrument, the *Theremin.* This device, first built in 1919, consists of two high-frequency oscillators, typically 100,000 Hz each. One of the oscillators is coupled to an antenna, so that when a hand is placed near the antenna, the frequency is shifted by a small amount, dependent on the closeness of the hand. The difference between the frequencies of the two oscillators lies in the audio frequency range and is caused to glide over a wide range of audio frequencies by moving a hand or other object

FIGURE 24–6
Self-contained organ console that combines both wind-blown pipes and electronic tone generation as a source of musical sounds. Note the two keyboards, foot pedals, and many stops. (Courtesy of Rodgers Organ Company, CBS Musical Instrument Division)

in the proximity of the antenna. Dynamic control (loudness) is achieved by moving the other hand in the proximity of a loop. Although there are traditional instruments, such as a sitar, whose pitch can be varied continuously over limited ranges, the wide range of the continuous pitch of a Theremin is

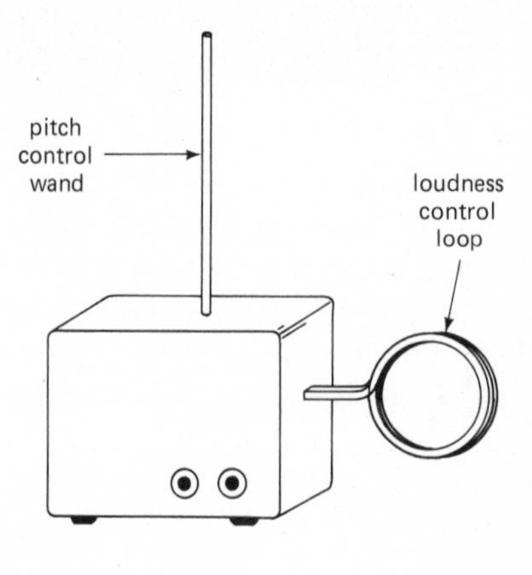

FIGURE 24–7
The Theremin electronic instrument generates a continuous tone whose pitch and loudness depend upon the proximity of the performers hands to a short wand and a loop, respectively.

new to music. It has an ethereal sound that lends itself nicely to background music for science fiction and space scenarios.

24.6 Modular synthesizers

Traditional musical instruments have specific forms that, in turn, determine the playing characteristics of the instrument. With electronics, a great deal more flexibility is possible. The composer can then concentrate more on what he or she wants to create, with less concern as to the availability of the desired sound sources. This flexibility is possible through the use of a **modular synthesizer.**[3] See Figure 24–8. This device consists of special function modules

FIGURE 24–8 Sound wave synthesizer with integrated keyboard. (Courtesy of Electronic Music Studios of Amherst)

that can be interconnected in any desired arrangement by using switches, wire patches, or a computer. See Figure 24–9. Oscillator or wave generator modules are available to produce a variety of basic waveforms such as sine, square, rectangular, triangular, and sawtooth. See Figure 24–10. The sine wave produces a pure tone, whereas the others include different mixes of harmonics. The symmetric waveforms—square, rectangular, and triangular— have only odd harmonics and produce a hollow sound similar to a closed-end organ pipe. The sawtooth wave, being asymmetric, produces all harmonics. Similar waveforms are generated in electronic organs.

Noise has no specific frequency but is a random mixture of all audio frequencies. **White noise** has equal amounts of available intensity at all audio frequencies. Other frequency distributions called **colored noise** can be generated, or they can be produced by passing white noise through appropriate filters. When broken into pulses, the noise can be made to sound similar to wire brushes on a drum head or the crash of cymbals, depending on the frequency range selected by filters.

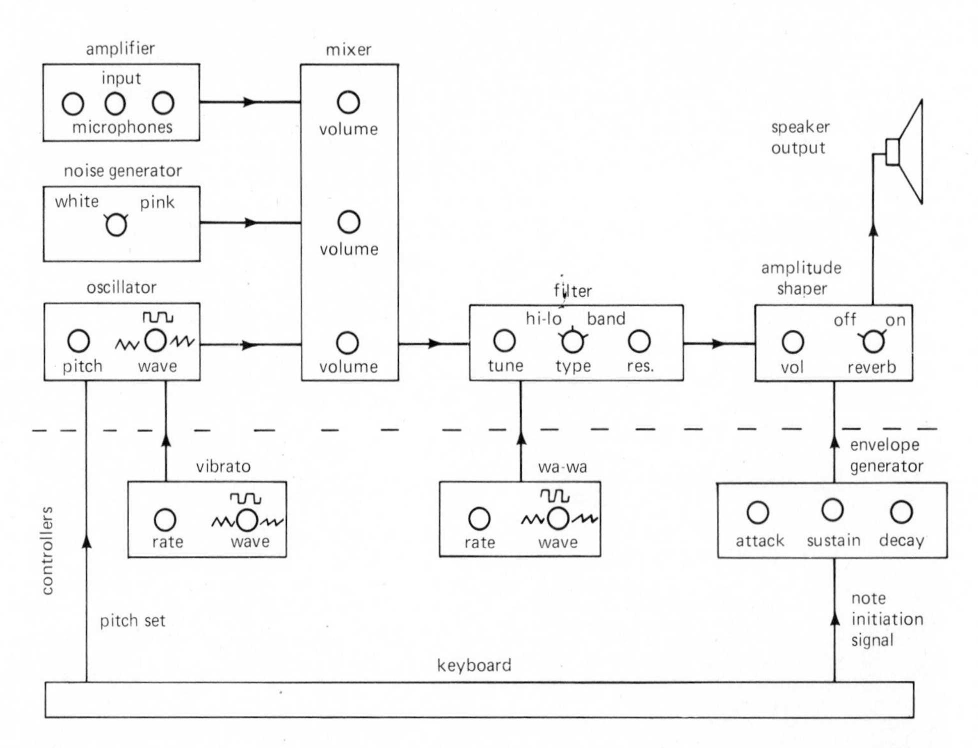

FIGURE 24–9
Block diagram for the operation of a typical electronic sound wave synthesizer.

In the synthesizer, the generation modules can be controlled in a variety of ways. For example, a sine wave or white noise can be controlled by an envelope generator that breaks up the input into an individual note and determines its transient properties—that is, its attack, sustain, and decay. See Figure 24–11. Each note can be keyed from push buttons, switches, or a piano-type keyboard. The resulting output can be routed through a filter, whose frequency response can be made to change according to some other generator module. To this, one can add tremolo or reverberation. Two or more sine waves can be added in a mixer module. Alternatively, if the two signals are routed to a ring modulator, the output will consist of two signals whose frequencies are the sum and the difference of the input frequencies.

A synthesizer studio generally includes a modular synthesizer, stereo amplifier, speakers, and tape recorders with equipment for mixing, editing,

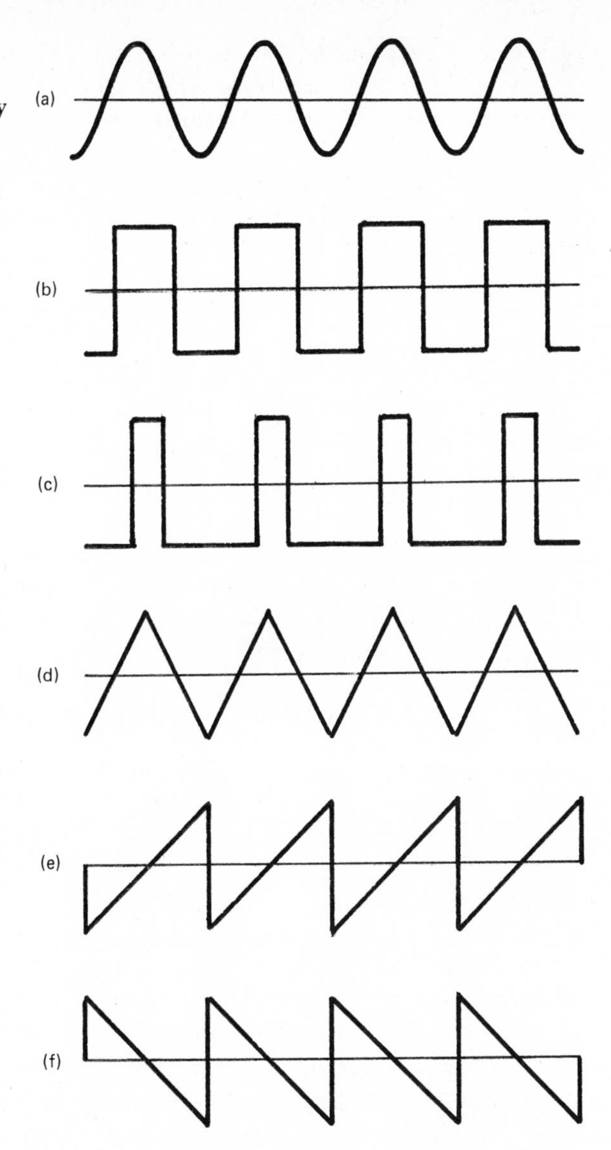

FIGURE 24–10
Various wave forms generated by a synthesizer, (a) sine, (b) square, (c) rectangular, (d) triangular, (e) sawtooth (rising), (f) sawtooth (falling).

and synchronization. Composition of synthesizer music is a time-consuming process, and the results are limited only by the composer's ingenuity, time, and patience. "Switched-on Bach," composed by Walter Carlos using the Moog synthesizer, is an example of the adaptation of the synthesizer to a traditional score. Orchestral style is exemplified by "Snowflakes Are Dancing" by Tomita; the jazz style is exemplified by "Headhunters" by Herbie Hancock. "Touch" (1970) and "Silver Apples of the Moon" (1967) by Morton Sobotnick, using the Buchla synthesizer, are examples of the new musical style made possible by the synthesizer.

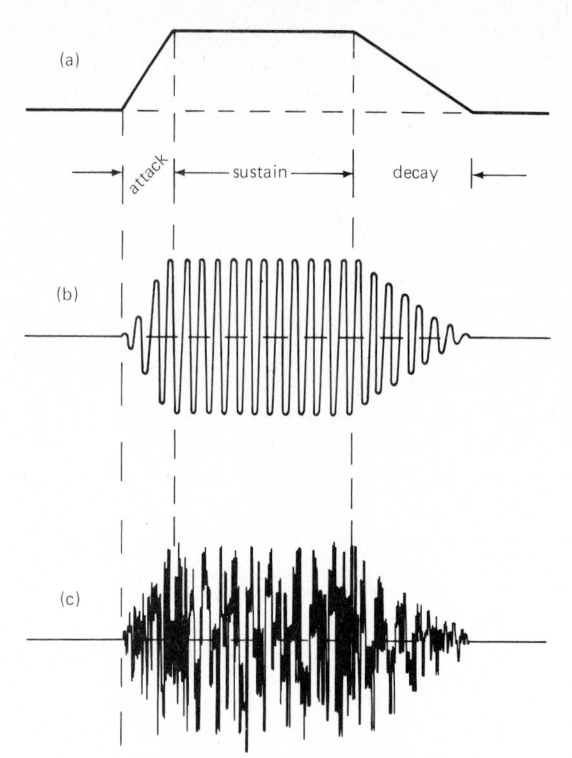

FIGURE 24–11
The trapezoid-shaped wave form of an envelope generator (a) can be used to produce a pure percussive note by modulating a sine wave (b) or the crash of cymbals by modulating white noise (c).

24.7 Computer music

Digital computers have been used in various aspects of music since the late 1950s. A digital computer is an electronic device that carries out simple arithmetic operations and simple logical decisions. Its usefulness is due to the speed with which it calculates.

In a computer, a sequence of calculations that repeats itself is called a loop. Suppose the computer is programmed so that one loop takes $1/440$ s, and the loop is repeated 440 times. Now if a radio receiver is placed close to the computer, it will pick up a 440-Hz note (concert A) lasting for 1 s.

By programming the computer properly, one can produce a simple tune, although it will be primitive by most musical standards. The computer composer has also been known to program the computer to operate the card reader, line printer, magnetic tape transport, and other peripheral units in rhythm. Arrangements of appropriate pieces such as "Stars and Stripes Forever" or the "Anvil Chorus" create a cacophony that must be seen and heard to be appreciated.

More serious electronic music is created by computer programs that calculate the desired composition, note by note.[4, 5] Using the computer, the composer can structure all the components of his composition—including timbre, tempo, waveform, envelope (attack, sustain, and decay)—with great flexibility and accuracy. Moreover, since compositions can be coded onto computer cards, the composer can vary the performance by the change of a few cards. Unlike most other musical media, the computer composer has almost complete control over the performance, leaving little room for interpretation by others.

The composer of computer music must provide waveforms representing the characteristic sounds, which are stored in the computer memory. The composer then designs "instruments" that control the generation of notes based on the waveforms. He must provide for the desired time structuring of his generated notes or sounds. Since the computer is digital, the output is a series of specific numbers (digital signals). However, these numerical steps can be converted to proportional voltages (analog signals), which are then passed through a low-pass filter to smooth out the steps. See Figure 24–12. These signals are finally sent through an amplifier to be recorded on disc or tape. By programming the computer properly, one can create a sequence of notes that simulate almost any known instrument, or one can create new sounds. To get an idea of the capabilities, listen to a computer piece, such as J. K. Randall's "Lyric Variations for Violin and Computer" (1968) or L. Hiller's "Computer Cantata" (1963).

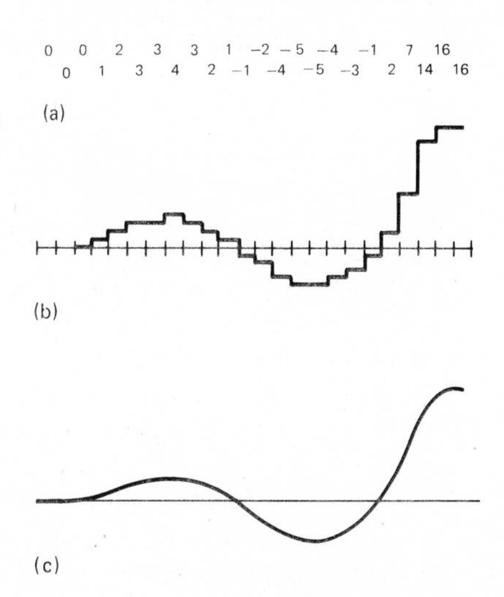

FIGURE 24–12 The attack portion of a musical note is synthesized, starting with a series of numbers stored in a digital computer (a), converted to electric voltages (b), and smoothed with a low-pass-filter circuit (c).

The computer can also be used in conjunction with a synthesizer. It can control the oscillators, amplitudes, mixing, and other synthesizer functions. Such computer control increases the versatility of the synthesizer dramatically. Also, a keyboard can be used in conjunction with a computer to provide a wider range of sounds than is available with a "hard-wired" synthesizer.

Electronic music is still in its infancy, and it will undoubtedly continue to expand and develop. One may worry that persons attending a future concert may find the orchestra replaced by a rack of electronic circuitry and the conductor replaced by an electronics technician—or perhaps a computer. However, musical performance is a personal experience, and it is likely that the traditional acoustic instruments will be with us for some time to come.

QUESTIONS

1. What is meant by an acoustic instrument?
2. What are the transient properties of a percussive note?
3. Which waveforms in Figure 24–10 include even harmonics? Why?
4. In what way would the sound of a triangular wave be different from the sound of a sawtooth wave?
5. In what way would the sound of a triangular wave be different from the sound of a square wave?
6. "Pink noise" contains equal sound intensity in each octave. How does this compare with white noise? How would it sound compared to white noise?
7. How can a closed loop of magnetic tape be used to add echoes to an electronic guitar output? Why is the feed back loop only weakly active?

PROBLEMS

1. If the two frequencies of 330 and 440 Hz are routed to a ring modulator, what frequencies are produced? What notes are these?
2. What must be the frequency range of the variable oscillator of a Theremin to span the entire audio range?
3. Express pink noise and white noise in units of decibels per octave (see Question 6).

PROJECT

Locate a synthesizer and experiment with its capabilities.

1. Compare the timbre of a square wave, a triangle wave, and a sine wave. How do these sounds compare with what you would expect (see Section 8.8)?
2. Experiment with the envelope generator. Compare a note with a sharp attack with a note with a slow attack. What sort of instruments do each sound like?
3. Compare vibrato with tremolo. Do they sound similar? Which do you like best?
4. Try the white noise generator. Can you simulate a cymbal clash? Wire brushes on a drum head? What other instruments can you simulate?

[1]N. H. Crowhurst, *Electronic Musical Instruments* (Blue Ridge Summit, Pa.: Tab Books, 1971).

[2]R. H. Dorf, *Electronic Musical Instruments* (New York: Radiofile, 1968).

[3]A. Strange, *Electronic Music* (Dubuque, Iowa: Brown, 1972).

[4]M. V. Matthews et al., "Computers and Future Music," *Science* (25 January 1977):263.

[5]H. V. Foerster and J. W. Beauchamp, ed., *Music by Computers* (New York: Wiley, 1969).

Chapter Twenty-five

STEREOPHONIC SOUND

If you are attending a live concert in a concert hall, you are experiencing the performance in a particularly intimate manner. You identify the various instruments with their locations on stage. Moreover, the echoes and reverberation of the hall help to engulf you with the sound and give the hall its characteristic **ambience.** Each concert hall has its own ambience. A trained musician can often identify the unique characteristics of a hall by listening to music played in it.

One of the major purposes of a sound reproduction system is to recreate the sense of presence of the live performance. To accomplish this, it is necessary that the quality of the reproduction be sufficiently high so that the sound sources are distributed in the room as they would be in a concert hall. All the characteristic reflections and reverberations should also be reproduced. In this chapter, we will examine this problem.

25.1 Directional perception of sound

When a nearby sound is produced, acoustic waves are channeled down each ear canal and through the inner ear and sensory system. The resulting sound is perceived in terms of its pitch, timbre, and loudness. Yet one can usually also determine the direction from which the sound arrives and its distance of production, even with eyes closed and often with one ear blocked. The ability to discern direction is related to a variety of subtle cues,[1] not all of which are fully understood. The two most important will be discussed first.[2]

If a listener receives a sound from 45° to the right in a horizontal plane, as shown in Figure 25–1, the sound is easily channeled into the right ear. However, the left ear is in the sound shadow of the head, and loudness will be severely depressed, especially at high frequencies where the sound will not readily diffract around the lee side of the head. There will therefore be a difference in loudness response of the two ears. This is called the interaural amplitude difference cue.

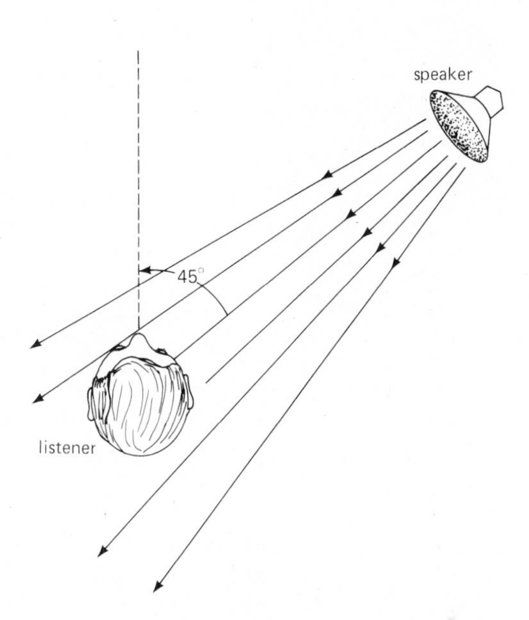

FIGURE 25–1 Sound from a source at a horizontal angle arrives at the two ears at slightly different times and with different intensities.

A second sound cue—the interaural time delay cue—involves the difference in time of arrival of the sound at the two ears. In the example here, the sound will arrive at the left ear by less than a thousandth of a second later than at the right. Experimental studies show that below about 1500 Hz, this difference in time of arrival will result in a perception of the direction of incidence. At higher frequencies, the sense of direction is discerned from the difference in relative loudness.

Transient sounds, such as clicks, contain many frequencies, bringing both audio cues into play. As a result, transient sounds are much more easily localized than continuous tones.

25.2 Other audio cues

A person is generally able to identify the direction in elevation of a sound source along the medial plane equidistant from the two ears. See Figure 25–2. For example, a sound produced directly overhead is seldom confused with a sound directly forward. Moreover, some directional sensitivity occurs even when one ear is blocked. These abilities are not explained by the interaural amplitude difference cue nor the interaural time delay cue.

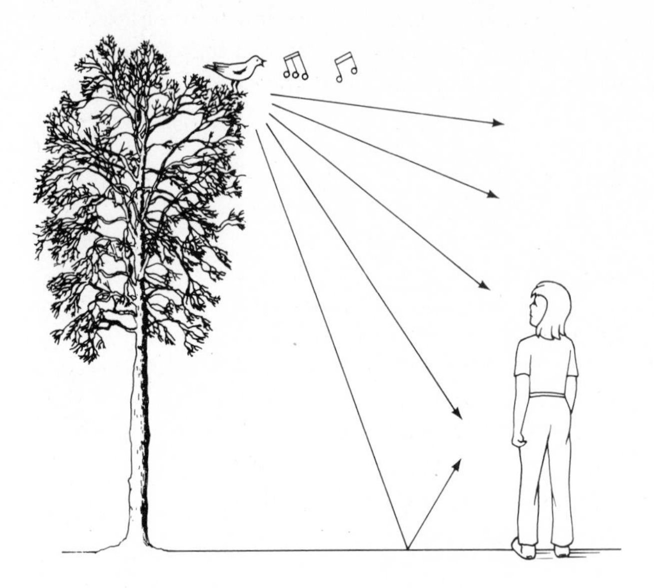

FIGURE 25-2
Reflections from environmental and body surfaces enable the listener to discern the vertical angle of sound arrival.

When microphones are placed in the ear canal, it is found that the frequency response to a sound source varies with the direction in the vertical as well as the horizontal plane and may be different for the two ears.[3] Other studies show that sound sources with frequencies above 5000 Hz will allow some sense of direction for the monaural (single-ear) listener.[4] Those frequencies above 5000 Hz have wavelengths less than 15 cm, which is dimensionally similar to the outer ear (pinna) structure. Studies have shown that when the structure of the outer ear is modified, the ability to sense direction is impaired.[5] The conclusion must be drawn that the outer surface of the head, particularly the labyrinth of cartilage and cavities comprising the outer ear, modifies the sound in the process of channeling it down the ear canal. Since the outer ear is very asymmetrical, the modification will be different for different source directions, allowing the interpretation of the modification to be used to identify the source direction.

25.3 Distance perception

The perception of distance is also determined by a variety of audio cues.[1] Since the perceived loudness of a sound increases as the source approaches the listener, apparent loudness is one such cue, and it is particularly important for continuous tones. However, apparent loudness affects perceived distance only in the absence of more definitive physical cues, such as are provided by pulsating or transient sounds. These include the relation between the air pressure and the air motion in the arriving pulse, the relation of the direct sound with its reflections, and possibly the change in timbre, which results from the high frequencies being more readily absorbed in the intervening air. There are undoubtedly other cues that aid in sound source ranging, and much remains to be learned about the ability to utilize these cues.

25.4 Stereophonic (two-channel) reproduction

Monophonic or single-channel sound can be presented with excellent fidelity, but it cannot give a natural sense of direction to the sound. All the sound seems to emanate from the speaker system. In order to give a sense of *presence* it would be necessary in principle to recreate all the audio cues: phasing, shielding, echoes, and so on. The ideal solution, then, would be to place a microphone at each sound source during recording, as well as at various other locations in the concert hall. It would then be necessary to reproduce the sound through an equal number of speakers, appropriately arranged.

Actually, multichannel pickups are commonly used for studio recording.

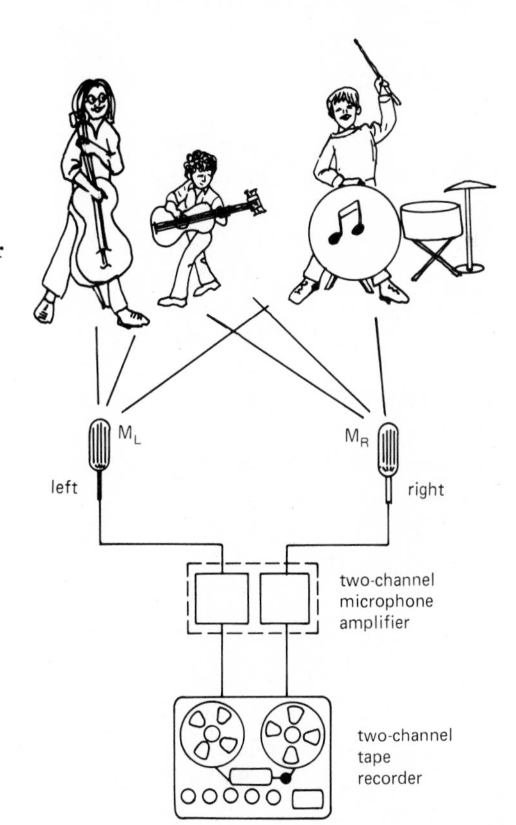

FIGURE 25–3 **Multichannel recording involves placing two or more microphones at different locations in the recording studio or concert hall. Each microphone picks up primarily the closest sound sources.**

With up to 24 recording channels, a sound technician can later construct a recording to give the specific results desired. However, to separately reproduce all recorded channels in a normal listening environment is generally considered impractical.

The increase to two-channel or **stereophonic reproduction** goes surprisingly far in recreating the sense of presence. In this procedure, two well-separated microphones are placed in the recording environment, as shown in Figure 25–3. (If more than two microphones are used, the audio technicians can still mix the sound to produce two separate channels.) These two channels are synchronously recorded on records and tapes. The two channels are then reproduced, separately but synchronously, from a stereo turntable pickup or two-channel tape.[6] Alternatively, they can be combined in a somewhat complex manner, transmitted over FM multiplex radio, and separated again into two audio channels by a FM stereo tuner. See Section 23.5. The two signals are separately amplified, response-shaped, and fed into two separate speakers or speaker systems.

Consider two speakers and a listener forming a triangular arrangement, as shown in Figure 25–4. If the recording instrument had been located near the right microphone, the sound would have been picked up primarily in the right channel and will be reproduced primarily by the right speaker. This will give the listener an illusion of hearing music coming from the proximity of the right speaker. The same applies to the left speaker. If the source had been

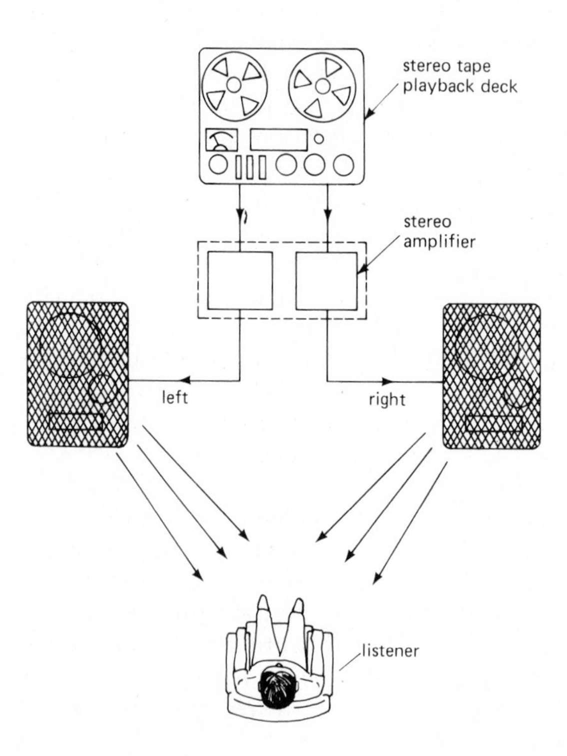

FIGURE 25–4
Stereo playback involves separately routing each of two recorded channels to the appropriate loudspeaker.

centered between the microphones, the sound level of each channel will be about equal, and the speakers will reproduce the sound equally. Although the audio cues are not all consistent here, the equality of response of the two speakers usually leads to a sensory interpretation of the sound emanating from a location midway between the speakers.

In a similar manner, an instrument produced at any point along a line between speakers will produce a response in the two channels related to the closeness to the corresponding microphone, resulting in a proportional loudness ratio in the two speakers, finally giving the impression of its presence at about the appropriate position. As a result, the listener will perceive the sound sources distributed along a line between the two speakers in a pattern similar to the arrangement in the recording environment. Combined with the sense of distance that comes through stereo as well as monophonic reproduction, a sense of two-dimensional sound is reproduced.

The stereo system is particularly effective when the sound source is distributed along a horizontal plane in front of the listener. This arrangement works well to simulate actual presence in an audience listening to a performance on stage. However, it fails completely in reproducing sound distribution in the vertical plane, as well as sound arriving from the sides and rear. The natural reverberations that approach from all directions in a concert hall will be routed through the speakers and thus will approach the stereo listener from the front. This gives one the sensation of being seated at the rear of the concert hall or slightly outside it. This problem can be partially remedied by having the listener placed midway between the speakers. However, in this case other audio cues become modified, leading to an unnatural sense of spacial distribution. Listening with stereo headphones is one example of midpoint listening. The listener does have the experience of being enveloped in music, but its original distribution is all but lost, leaving the sensation that the sound is originating at the center of the listener's head.

25.5 Quadraphonic (four-channel) reproduction

Two-channel stereo reproduction can effectively reproduce the spacial distribution of the major musical instruments in a concert hall performance. Careful design of the home acoustical environment is also quite important in producing a comfortable, reverberant listening atmosphere and can partially simulate certain recording environments.

However, to properly recreate the specific ambience of each original environment, especially a large concert hall, more sophisticated techniques are called for. Of various methods that have been tried, the one system that has withstood the test of time is the rectangular or quadraphonic arrangement. See Figure 25 – 5. Recording can be carried out by placing four microphones at four roughly symmetric points in the concert hall. The signals are then synchronously routed through four separate or discrete systems. Finally,

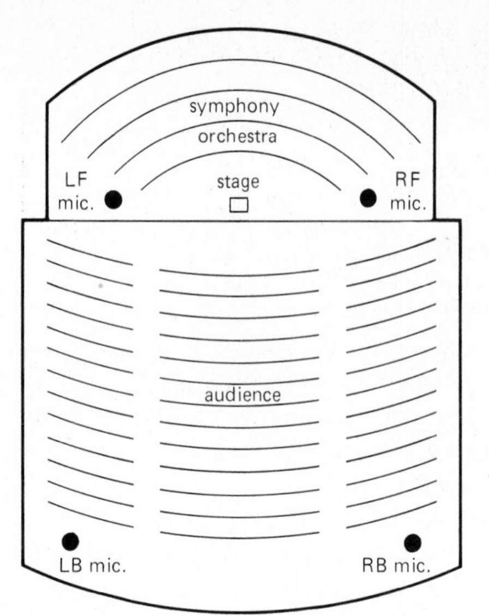

FIGURE 25-5
Arrangement of four microphones in a concert hall.

the sound is reproduced through four separate speakers or speaker groups, appropriately arranged. See Figure 25-6. With the listener seated approximately at the center, the listening environment is turned into the audio equivalent of a concert hall: orchestra, reverberation, and all.

In four-channel reproduction of sound, there may be a few missing cues, such as reflections from the ceiling, but these are of secondary importance. Of more importance is the added complexity of the four-channel sound. The apparatus is quadrupled, requiring four parallel amplifiers, four speaker sets, and so on. Although the size and cost of the electronics can be held down by housing the parallel components in a common cabinet, the four speakers present some serious space problems. A single base reflex cabinet may be tolerated in front of the room; two such speakers for stereo reproduction may be slightly intrusive; but four such speakers could well be too much. And if corner folded horns are used, the listening room must have four corners that can be utilized. Actually, the rear two speakers need not necessarily be of the same size or quality of the front two, leaving the possible compromise of using smaller speakers for the rear—and possibly for all four.

Four-channel sound reproduction also places severe demands on the transmission and recording systems, which already have to strain somewhat to produce stereo. Four-channel FM radio broadcasting requires additional bandwidth on the already crowded FM broadcast band. Four-channel magnetic tape is no serious problem, but the phonograph record is not readily amenable to recording four separate channels. However, methods similar to FM multiplexing have been developed to record four sound channels onto the two sides of the groove.[7] The cartridge and stylus must be designed to respond to the ultrasonic vibration frequencies of up to 45 kHz. So although discrete four-channel sound is desirable, it does place severe demands upon the present technology, as well as on the pocketbook.

To overcome some of these problems, a variety of **matrix-multiplexing** techniques have been developed to combine the four channels into two prior to recording and to then separate them back into four at the preamplifier. It is not surprising that the resulting separation is not complete. It may be surprising, however, that the separation may be adequate for practical purposes.

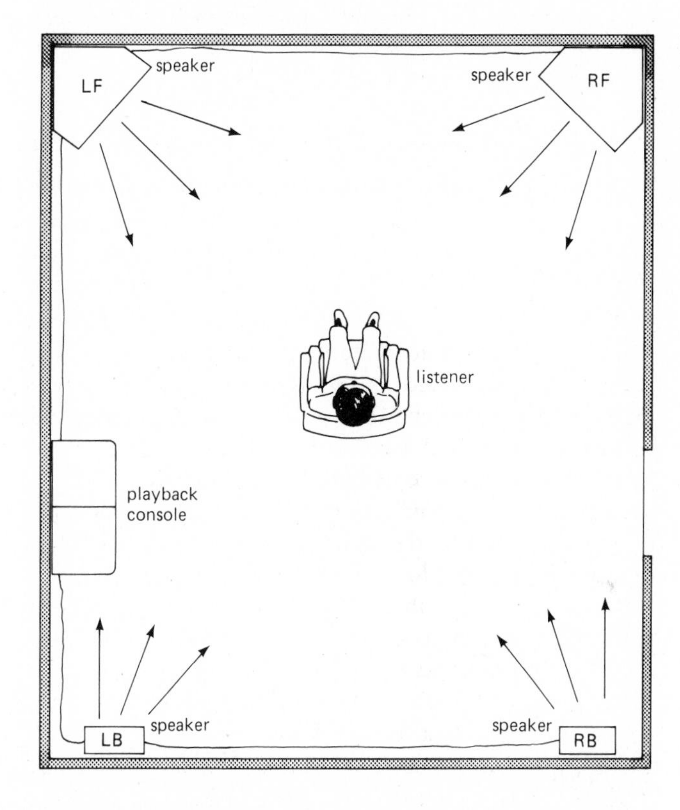

FIGURE 25–6
Top view of a listening room wired for quadraphonic playback.

There are a variety of mixing and decoding recipes used in matrix four-channel sound reproduction.[8] Although each has its own advantages and disadvantages, most of them give good sound in practice. Even encoding with one recipe and decoding with another usually works well, although the apparent placement of the sound sources in the listening environment may not correspond well with the original arrangement.

Most listeners feel that quadraphonic sound is more or less better than stereo. However, there is plenty of debate as to whether such improvement in ambience is worth the extra effort or expense.

Ambience can also be *simulated* by a variety of methods. For example, a reverberation circuit can be built into the stereo system and the reverberant part of the output can be played through the rear speaker pair of a quadraphonic speaker arrangement.[9] Other methods of simulating ambience are limited only by the ingenuity of the listener.

The advent of high-fidelity multichannel sound reproduction has made it possible to bring the performances of the world's masters into the living room at the convenience of the listener. Ultimately, however, the most elaborate sound reproduction system cannot replace the visual and participative experience of an evening at the theater, and it is likely that the live performance will continue to be a meaningful musical experience for the foreseeable future.

QUESTIONS

1. What audio cues aid in discerning the direction of a sound source?

2. What audio cues aid in discerning the distance of a sound source?

3. Suppose headphones were placed on a student, and a 500-Hz sound was fed to the left ear slightly later than to the right ear. The student will discern the source of sound to be at about 45° to the right of center. Why?

4. Discuss the audio cues that are not correctly reproduced by a two-speaker stereophonic sound system.

5. Discuss the relative advantages and disadvantages of monophonic versus stereophonic sound systems.

6. Discuss the relative advantages and disadvantages of stereophonic versus quadraphonic sound systems.

PROBLEM

If the ears of the student in Question 3 are 20 cm apart, what is the approximate time delay in the response of the left ear?

PROJECT

In an open field, have a blindfolded subject try to discern the direction of a sound source. Make measurements at a variety of different angles. For the source, try a low-frequency tone (a clarinet, for example), a high-frequency note (a whistle), and transient sounds (clapping). Have the subject listen with both ears and then with only one ear. For monaural listening, try modifying the ear structure with adhesive tape or putty.

NOTES

[1]G. von Békésy, *Experiments in Hearing* (New York: McGraw-Hill, 1960).

[2]J. H. Appleton and R. C. Perera, *The Development of Electronic Music* (Englewood Cliffs, N.J.: Prentice-Hall, 1972), p. 66.

[3]C. L. Searle, L. D. Breida, D. R. Cuddy, and M. F. Davis, "Binaural Pinna Disparity: Another Auditory Localization Cue," *Journal of the Acoustical Society of America* 57 (1975):448.

[4]K. Belendink and R. A. Butler, "Monaural Localization of Low-Pass Noise Bands in the Horizontal Plane," *Journal of the Acoustical Society of America* 58 (1975):701.

[5]M. Gardner, "Binaural Localization," *Journal of the Acoustical Society of America* 54 (1973):1489.

[6]*Understanding High Fidelity* (Moonachie, N.J.: Pioneer Electronic Corp., 1975).

[7]J. A. Giovenco, "How CD-4 Works," *Audio* (November 1976):38.

[8]H. Kitahra, "QS Matrix Simplified," *Radio-Electronics* (October 1975):16.

[9]"Time Delay for Ambience," *Audio* (December 1976):40.

PART FIVE

Acoustical Architecture

Chapter Twenty-six

REFLECTION, ABSORPTION, AND REVERBERATION

Musical sounds from a single musical instrument or an orchestra are modified considerably by the presence of large obstacles, walls, partitions, or complete enclosures. By suitable design, these architectural features can enhance the listening enjoyment of an audience and prevent unwanted sounds from interfering. In fact, a great deal of music has been composed with a specific listening environment in mind. It is therefore worthwhile to consider in detail how architectural systems affect musical sound in order to determine how best to design and utilize them. In this chapter, we will study the reflection, absorption, and reverberation of sound.

26.1 Noise

We have already discussed noise as a random combination of various frequencies. See Section 24.6. White noise, generated in an electronic synthesizer, is a combination in which all the audible frequencies present are sounded with equal intensity. Noise composed mainly of low frequencies is called **rumble**, whereas high-frequency noise is usually called **hiss**. See Figure 26–1. To most people, however, any unwanted sound is called noise,[1] just as any unwanted plant in a garden is called a weed. Such noise can be sounds of a busy city heard in a concert hall or the leakage of music from one practice room into another practice room. Most of these noises can be minimized by appropriate barriers.[2]

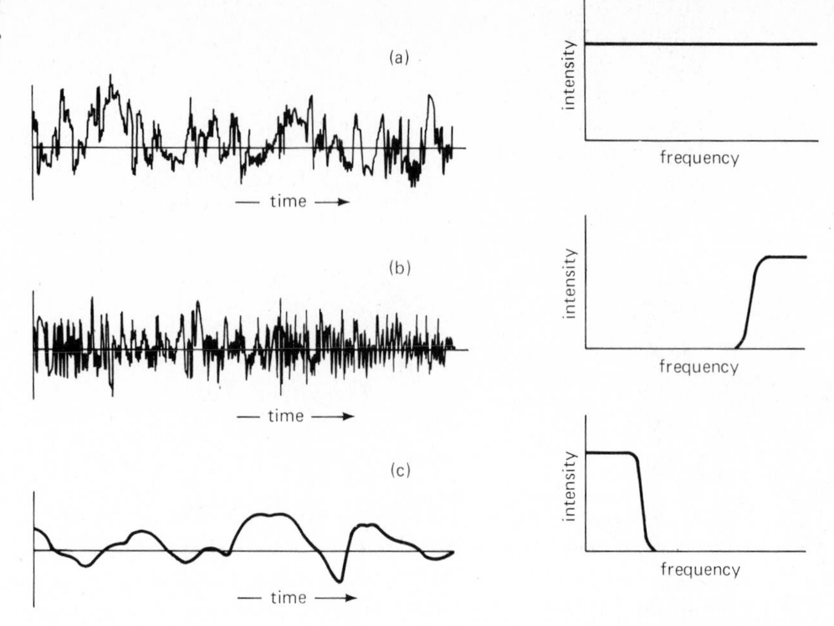

FIGURE 26–1
Noise is generally classified by its frequency spectrum. Typical noise types include (a) white noise, (b) hiss, and (c) rumble.

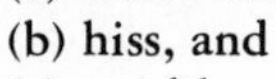

26.2 Surfaces

When a sound is produced in the open, the sound wave spreads outward in all directions. At increasing distances from the source, the sound energy is spread over an increasingly large area. The intensity therefore drops in inverse proportion to the square of the distance from the source. This $1/R^2$ (inverse square law) effect simply means that noise can be reduced by being sufficiently far from its source. (See Section 9.5.) And this is why open-air theaters are usually built at locations far from city noises.

Noise can also be reduced by placing a barrier between the source and the listener. Since a sound wave is a form of energy, it cannot simply disappear, but, as energy, it can change its direction of propagation or its form. When the sound wave strikes the barrier, there are three processes that modify the sound energy: *reflection, transmission,* and *absorption.* See Figure 26–2.

A smooth granite slab will reflect about 99 percent of the sound energy striking it. In fact, almost any solid surface that is sufficiently stiff and massive will reflect at least 95 percent. Even light smooth walls will reflect 90 percent. Of the sound energy that is not reflected, part will be transmitted out the far side. The remaining sound energy will be absorbed in the material itself. This absorption is actually a conversion of sound energy into thermal energy (heat). The process that converts energy of motion to thermal energy is called friction—something must rub against something else. There is very little such rubbing within a solid wall, and, as a result, very little sound energy is absorbed.

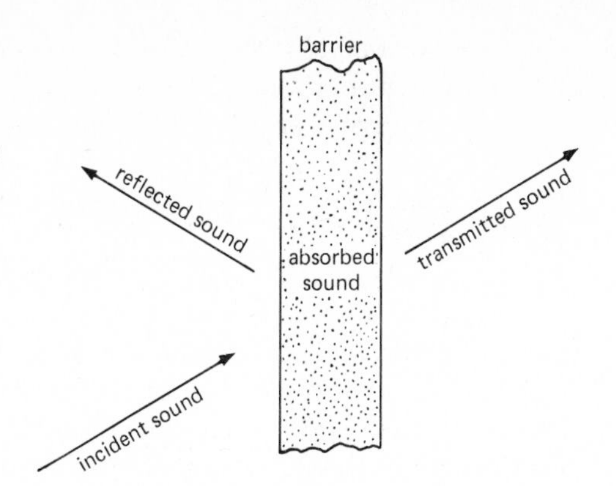

FIGURE 26–2
When sound energy is incident on a surface, some is reflected, some is absorbed, and some is transmitted.

26.3 Sound insulation

A material that minimizes the sound transmission is called a **sound insulator.** This minimization must be accomplished by maximizing the reflection or absorption or both. Many different materials and techniques have been devised to do this.[3] Most studies indicate, however, that good insulation is determined primarily by a single parameter, the mass of the wall. According to this **mass law,** a thick concrete wall or a brick wall makes a fine sound insulator, particularly at high frequencies. The mass, or inertia, of the wall simply makes it resist being pushed back and forth by the sound wave.

Although the mass of the wall is the main feature of importance in insulation, it is not the whole story. The stiffness of a wall is important as well. Stiffness is related to the wall's ability to resume its original shape after being displaced. A pile of sand, for example, has plenty of mass but is not stiff. Stiffness is helpful in suppressing the transmission of low-frequency sound. On the other hand, a stiff wall can resonate at certain frequencies (**resonance**

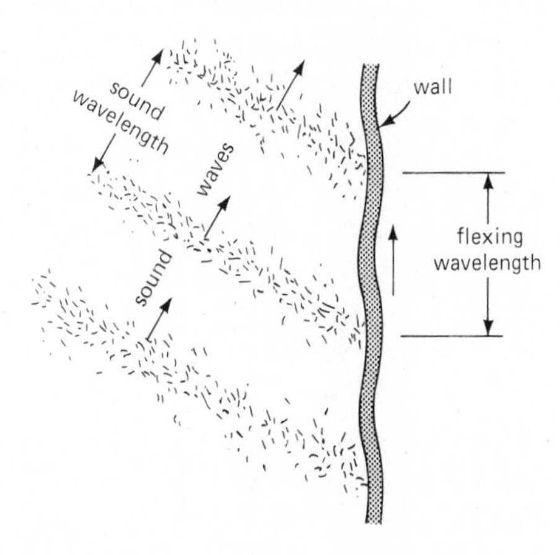

FIGURE 26–3
The coincidence effect (flexing is shown greatly exaggerated).

effect). Moreover, at sufficiently high frequency, sound waves striking the stiff wall obliquely may cause waves to run along the wall coincident with the wave (**coincidence effect**). See Figure 26–3. Both these flexure modes of vibration can increase the transmission of sound. Unfortunately, most walls must be somewhat stiff to support themselves as well as other parts of the room structure.

For good sound insulation, it is best, in many cases, to reduce the stiffness as much as possible and use materials that have some sound-absorption ability. For example, attaching sheets of damping material to the wall will help reduce transmission. In fact, such composite sandwich panels can usually be designed to have better insulation than the mass law predicts. Most remaining transmission through sandwich panels are related to resonance and coincidence effects.

Since most buildings will not permit the expense of constructing and supporting brick walls between rooms, it is necessary to use less mass than desired. One way to do this is to use a double wall with light, absorbing material between. See Figure 26–4. Each surface serves to reflect sound, thereby reducing transmission. The absorber material between the walls decreases air resonance, which would otherwise be set up at certain frequencies in the space between the walls. The double-wall concept can also be applied to acoustic windows, which are composed of at least two thick panes of glass in cushioned mounts and separated by at least 10 cm.

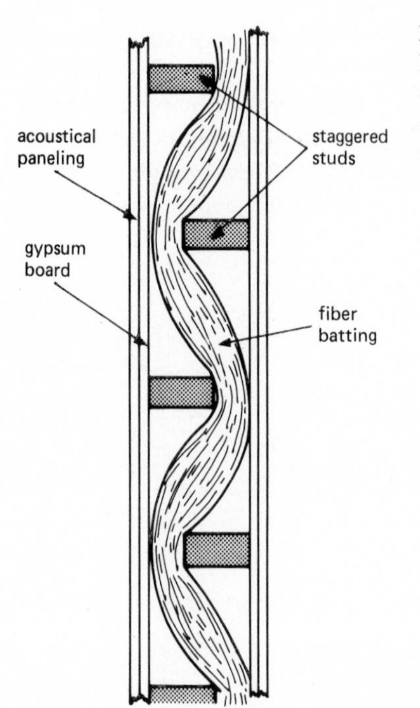

FIGURE 26–4
Double-wall construction for acoustical insulation.

Even with care in wall design, acoustic insulation efforts will fail if sound is carried along wall supports or through floor, ceiling, air ducts, or small gaps in the construction. Moreover, sound reflected back into the source enclosure will continue to rattle around, thereby having repeated opportunities to strike the walls, unless efforts are made to absorb the sound from within.

26.4 Reflection and absorption

In a noisy room, it is usually desirable to minimize reflection from the walls. An openining is the ideal nonreflector since sound striking the opening does not return. This is usually an unacceptable solution since the sound is transmitted and will likely cause problems beyond the opening.

Absorption of sound is the most practical way of minimizing reflection. Since absorption involves converting sound energy to heat, a mechanism of friction must be provided. One of the best friction devices is a material made of many small fibers. The back-and-forth rush of air among the fibers will cause heating of the fibers and the air and thus damp the sound wave. Although the thermal energy is not enough to produce a measurable rise in temperature, it is easily possible to absorb 80–90 percent of the reflected sound wave. Since the motion of air is greater slightly away from the reflecting surface, absorption is best if the absorbing panels are spaced somewhat away from the reflecting surface. See Figure 26–5.

A second method of sound absorption involves covering the surface with resonating cavities. Just as blowing across a bottle opening can set up resonance, the presence of a sound wave over a series of cavities will enhance the rush of air in and out of the cavities. (See Section 26.2.) Friction against the cavity walls and opening will then absorb the sound.

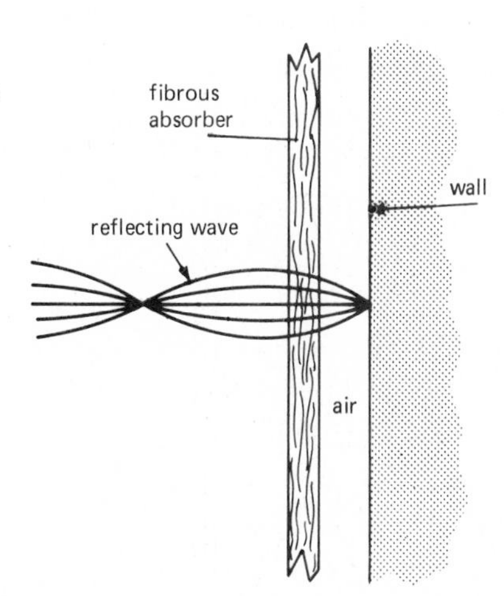

FIGURE 26–5
Fibrous absorbers, preferably spaced away from reflecting walls, will absorb sound by frictional damping.

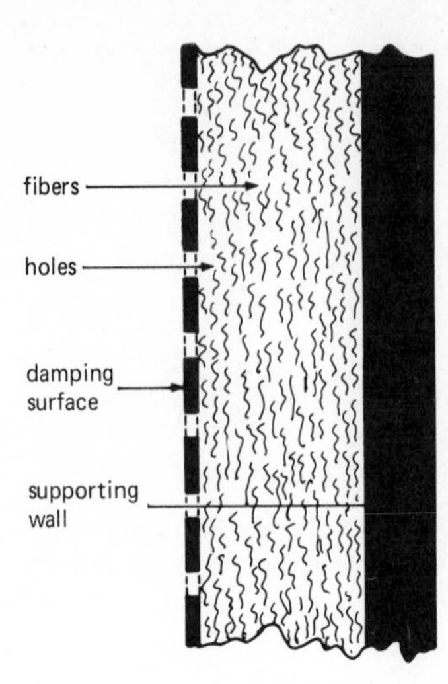

FIGURE 26–6
A perforated surface above a fibrous panel absorbs sound through friction in the fibers; friction at the hole walls, and friction of the vibrating surface.

These two methods of sound absorption can be used and combined in a variety of ways. The resonating cavities can be created by using a perforated sheet to create the cavity openings. When this sheet is placed some distance away from the reflecting surface, each hole acts as a cavity opening, but without cavity walls. Air friction against the holes does the absorption. By placing fibrous material between the holes and the wall, as shown in Figure 26–6, the absorption is enhanced. These principles are used in acoustic tiles, as shown in Figure 26–7; the tiles consist of fibrous panels with flat surfaces in which holes are formed. These holes need not be round or of uniform size. Acoustic tiles have the advantage of being pleasing to look at yet being effective absorbers. Even the flat surface between holes acts as a good low-frequency absorber if it is not too stiff.

The **absorption coefficient (absorptivity)** of a surface, the fraction of the incident sound intensity that is not reflected, usually varies with frequency. Table 26–1 lists the absorption coefficients of a few common materials rated at 500 Hz. More extensive tables are available.[4] (Also see Appendix III.)

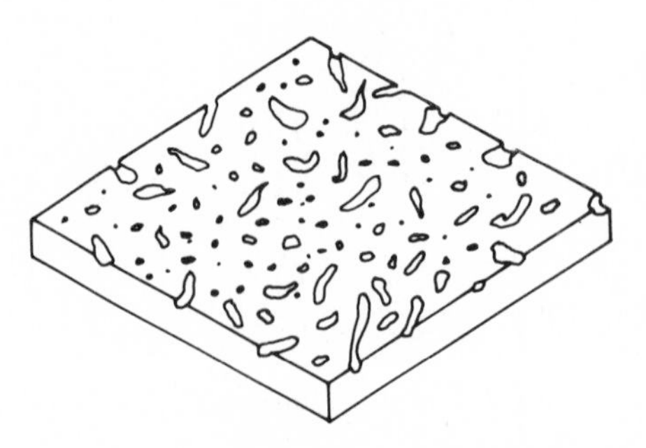

FIGURE 26–7
Typical decorative acoustical tile with holes of random sizes and shapes; it is used largely on ceilings.

TABLE 26–1: Absorption Coefficients for a Few Common Building Materials, at 500 Hz[a]

MATERIAL	ABSORPTION COEFFICIENT
Acoustical tile (Fibretex)	0.50
Brick, unglazed	0.03
Carpet, heavy, on concrete	0.11
Fabric, medium velour, 0.034 kg/m^2	0.49
Floors, parquet, on concrete	0.07
Glass, ordinary window	0.18
Plaster, gypsum or lime	0.05
Plywood paneling, 6 mm thick	0.17

[a]See Appendix III for a more complete listing of materials and absorption coefficients.

26.5 Standing waves

The shower-bath virtuoso is aware that a series of notes sounds much richer when sung in the shower than in the open air. In part, this is because the walls reflect most of the sound energy, allowing a large build up of sound energy. Beyond this, certain notes sound better than others. This can be understood by considering the enclosure as a three-dimensional resonating pipe. See Figure 26–8. The shower, for example, can first be viewed as a vertical pipe closed at top and bottom. By singing the proper note, the air will vibrate vertically in its fundamental mode with a node at top and bottom and an antinode in the middle. The higher harmonics can also be excited. Furthermore, standing waves can be set up between the two side walls and between the front and

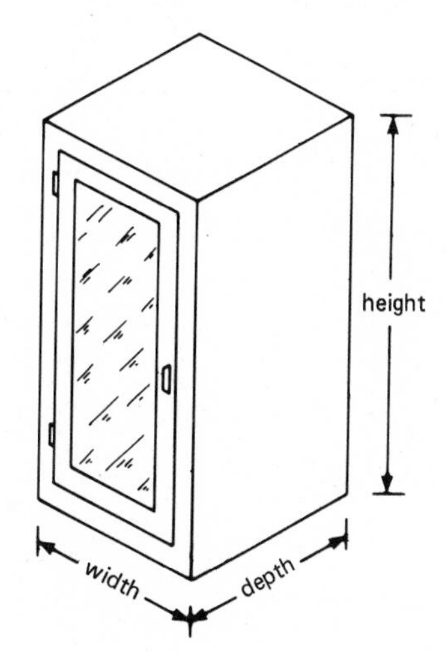

FIGURE 26–8
A shower enclosure has three dimensions determined by parallel walls, and it can resonate in all directions.

back. These latter two dimensions are about half the top-to-bottom distance in a typical shower. The pitch of the corresponding fundamental will therefore be about an octave higher. Combination modes can also be excited.

Such standing waves can also be set up in larger rooms and auditoriums. However, the number of resonant modes in such a large enclosure becomes enormous. The room need not have perfectly rectangular shape to have resonant vibration modes, but their calculation becomes quite complex in that case. Measurements show that large enclosures have many regions of high and low responses, dependent more or less on the sound source, position, and frequency.[5]

In any case, with the sound source distributed in space, and distributed in frequency, most of the small-scale spottiness washes out, particularly at the higher frequencies. This results in a **diffuse sound field.** Under such conditions, the approach of **wave acoustics** may be unnecessarily complex in practice. Alternatively, one can consider the sound waves to travel along lines called **rays.** Such an approach, called **geometrical acoustics**, ignores certain wave effects such as diffraction.[6] However, it provides a simpler approach to the description of many features of the acoustical environment.[7]

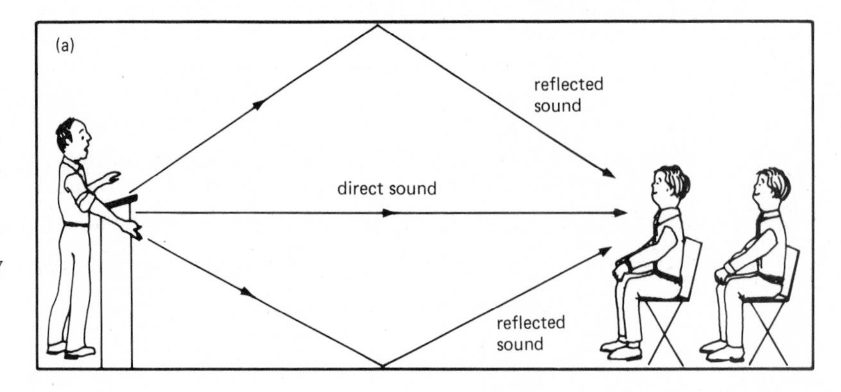

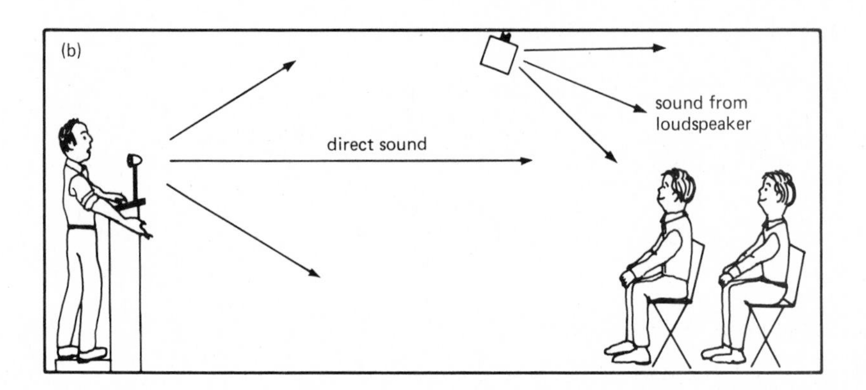

FIGURE 26–9
The listener identifies the source of sound as that which arrives first. This will usually be the true source (a) but may be a secondary source, such as a loudspeaker (b).

26.6 Echoes

If a sudden sound is created, a listener will first hear the directly transmitted sound. Sound heard after reflection from nearby surfaces will be heard slightly later due to the increased distance of travel. See Figure 26–9. If the time between these two sounds, the **initial time delay** gap, is less than 35 milliseconds* (ms), they will combine to produce a single acoustic response to the listener that is of greater loudness and quality than the direct sound alone. In fact, any number of similar sounds originating within an interval of 35 ms will be heard as a single sound. Even so, studies show that the listener usually identifies the source of the earliest sound as the source of the combined sound. This is called the **precedence effect.** If the initial time delay gap exceeds 50 ms, the two sounds are heard separately. The second sound is called an **echo.** In the presence of a series of reflecting surfaces, a series of echoes may be heard.

Suppose a sudden sound is produced between two well-separated, hard, parallel walls. See Figure 26–10. A series of regularly spaced echoes will be heard, usually as a buzz, as a result of successive reflections from the walls. This **flutter echo** will have a characteristic pitch if the echo rate is rapid enough.

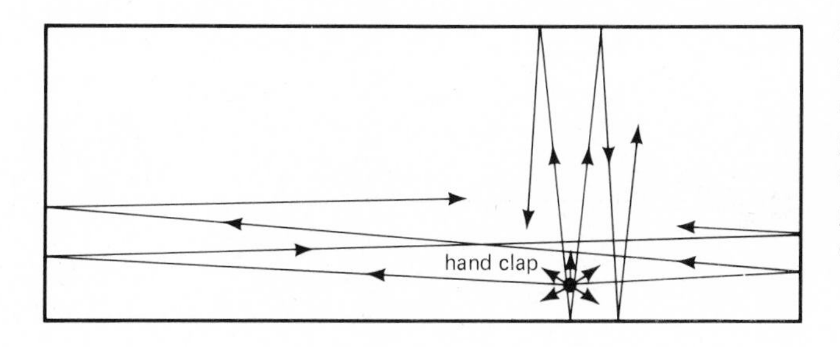

FIGURE 26–10 **Sudden sounds created between pairs of parallel surfaces can produce flutter echos.**

Example

What is the pitch of the flutter echo heard halfway between a tile floor and the standard 8-ft (2.44-m) ceiling?

Solution

Take the time between successive echoes to be the time sound travels to either wall and returns. Equation (3a) gives

$$f = \frac{V}{x} = \frac{348 \text{ m/s}}{2.44 \text{ m}}$$

$$f = 142 \text{ Hz}$$

This is about a D_3 on the musical scale.

*A millisecond equals one-thousandth of a second.

26.7 Transient reverberation

Good auditorium design avoids unbroken parallel walls. Nevertheless, the many walls of an enclosure will produce a series of reflections. The arrival of successive sounds will usually be so close in time that they cannot be distinguished. They therefore will not be heard as a series of echoes but as a continuous **reverberation.**

Consider a musician on stage striking a single percussive note, as shown in Figure 26–11. The first sound heard by a listener in the audience is the direct sound, which travels in a straight line from the musician to the listener. After an initial time gap, the first-reflection sound arrives. This reflection usually occurs from a side wall, the ceiling, or other overhead panel. This will be followed by successive sounds from increasingly distant surfaces and from multiple reflections. As the reflected sounds become more numerous, they also become fainter. As a result of absorption at each reflection, the loudness will soon decrease to an inaudible level. See Figure 26–12(a).

By general agreement, the **reverberation time** of an enclosure is the time it takes for the intensity to decrease to 10^{-6}, or one-millionth, of its initial value (60 dB). After this time, typically 1–3 s in most situations, the sound is considered to be inaudible.

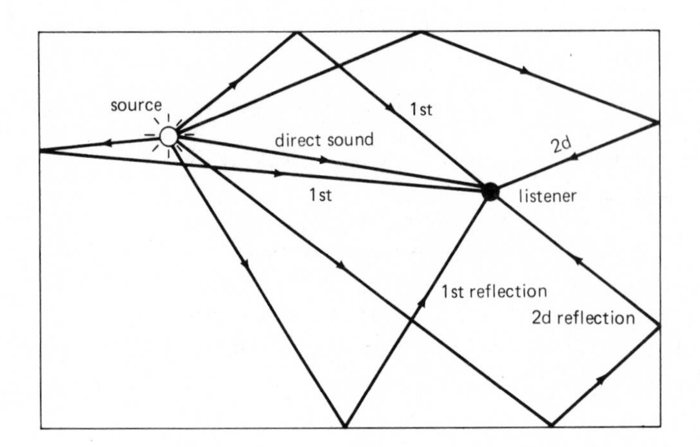

FIGURE 26–11
The listener hears the direct sound first, followed by a series of reflected sounds.

26.8 Sustained reverberation

Consider a tone that is sustained for a time comparable to the reverberation time. At first, the input of the sustained sound will cause a uniformly increasing reverberation through a room. See Figure 26–12(b). After the reverberant energy has built up for a while, the rate of absorption will approach the rate of production. The sound level then approaches an equilibrium level in the room. When the note stops being sounded, the reverberation will fade away at

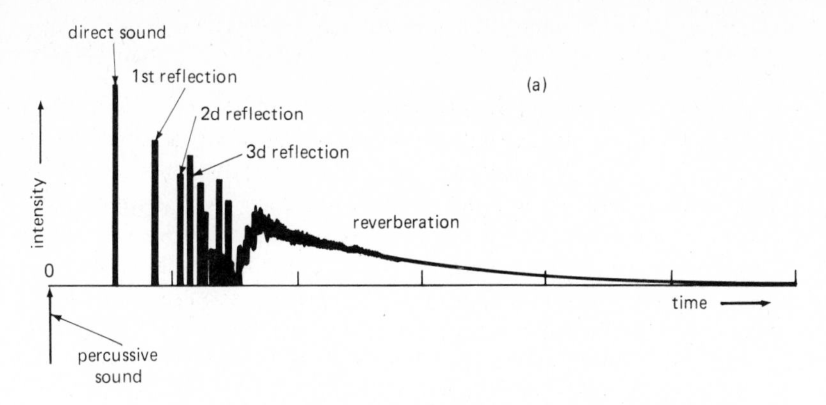

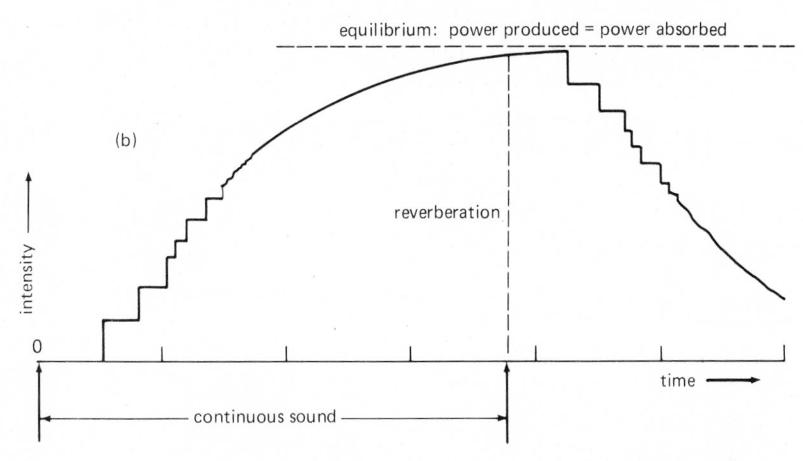

FIGURE 26–12 **Reverberation intensity in a room resulting from (a) a single percussive sound and (b) a continuous sound for a fixed duration.**

a rate given by the reverberation time. Moreover, the rise to equilibrium is also characterized by the same reverberation time, as shown in diagram (b). If the initial note is louder, the ultimate equilibrium loudness level is higher. However, the times associated with the rise and fall of the reverberant sound are the same. In the next chapter we will discuss what the times of rise and fall should be.

QUESTIONS

1. What is the mass law?
2. How does reverberation differ from echoing?
3. Suppose the voice of a vocalist is amplified and projected from a loudspeaker so that the listener can hear the vocalist directly as well as from the loudspeaker. What problem may arise if the loudspeaker is considerably closer to the listener than the vocalist? Explain.
4. If drapes are used to absorb room noise, why are they hung somewhat away from the wall?

5. Explain the poor transmission through solid walls in terms of impedance mismatching. (See Section 5.5.) Why are two walls better than a single wall of the same total thickness?

PROJECTS

1. Measure the dimensions of a shower enclosure, and calculate the resonant frequencies for all three dimensions. Test your calculations with tuning forks or by singing.

2. An interesting echo effect is produced by the reflection of a sound pulse from a series of steps. The reflection from the vertical surface of each successive step occurs later than the last, resulting in a series of closely spaced echoes. If each step has a width W, the time separation between successive echoes is given by Equation (5b):

$$T = \frac{2W}{V}$$

where V is the speed of sound. One should therefore hear a musical tone of frequency

$$f = \frac{1}{T}$$

$$f = \frac{V}{2W}$$

Try this experiment. Clapping your hands at the center of a circular outdoor amphitheater may work the best.

NOTES

[1]W. Burns, *Noise and Man* (London: John Murray, 1968). Discusses effects of noise, including normal and defective hearing, auditory health, and noise sources.

[2]R. Taylor, *Noise* (Baltimore: Penguin Books, 1970). A good survey for all aspects of noise.

[3]W. G. Hyzer, "Acoustic Materials," *Research/Development* (February 1977):74.

[4]M. Rettinger, *Acoustics: Volume I* (New York: Chemical Publishing, 1976).

[5]A. H. Benade, *Fundamentals of Musical Acoustics* (New York: Oxford University Press, 1976), chap. 11.

[6]V. O. Knudsen, "Architectural Acoustics," *Scientific American* (November 1963): 78.

[7]M. R. Schroeder, "Computer Models for Concert Hall Acoustics," *American Journal of Physics* 41 (1963): 461.

Chapter Twenty-seven

ACOUSTICAL PARAMETERS

Since musical enjoyment is a sensory experience, evaluation of a musical performance is usually expressed in subjective terms. With a knowledge of the modern science of sound, this experience can be translated into an objective description. Finally, an understanding of acoustics allows the objective description to be related to acoustical design:

sensory experience → objective description → acoustical design

For example, a good auditorium for symphonic music should have **fullness of tone**. This subjective expression is related mainly to a relatively long reverberation time, typically 1.5 to 2.0 s. This reverberation time is in turn related to the amount of absorbing surface present in the auditorium. A good acoustician can therefore determine how much absorber to install in order to achieve the desired fullness of tone.

It has been argued, however, that the middle step is not necessary. In fact, most of the early concert halls, as well as the instruments used in them, were designed mostly on intuitional grounds. The art of design was passed down from master to apprentice, improving slowly as a result of experience. The era of modern acoustics, however, has added a new dimension to environmental design. Together with past experience, this makes it more likely that a design will have the behavior that is desired.

27.1 Reverberation in a concert hall

Reverberation is generally considered desirable for a musical performance. In fact, it is probably the most important parameter in listening enjoyment. It increases the general loudness in an auditorium. It also causes the listener to feel enveloped in the music. If the reverbation time is comparable with the time between adjacent notes, the notes seem to blend smoothly into each other. If the reverberation time is too short, the music lacks fullness as well as loudness. If the reverberation time is too long, the individual notes blend together, causing a loss in clarity or definition. The reverberation time of the concert hall should therefore ideally be matched to the type and tempo of the music, or vice versa. Since sound takes longer to travel in a large room than in a small room, reverberation times tend to be longer in auditoriums. This in turn means that slower music will sound better in larger halls.

Many composers have undoubtedly considered the acoustics of the performance environment in composing their works. It is no accident that the string quartet is usually expected to be performed in the confines of an intimate chamber, or that the slow strains of a Gregorian chant should be heard in the vast expanses of a cathedral.

These considerations are summarized in Figure 27 – 1, where generally agreed upon reverberation times are seen to depend upon concert hall size.

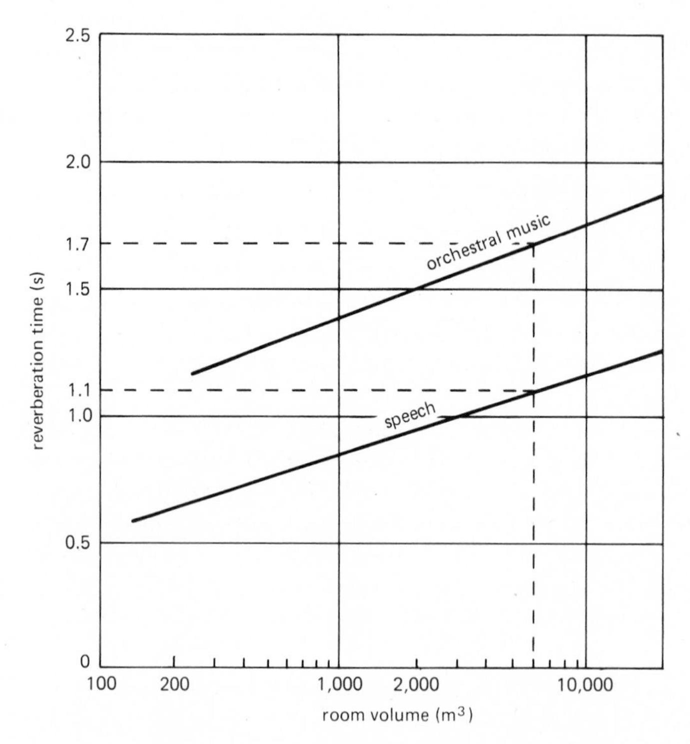

FIGURE 27 – 1
Graphs for preferable reverberation times versus room volume at 500 Hz.

Shorter reverberation time is generally desired for speech—and for opera to some extent, since better clarity is needed for the vocalization. These curves represent gross averages over a variety of types of music and over the opinions of various listeners. They do not include a variety of other effects that will be treated later in the chapter.

27.2 Determination of reverberation time

The reverberation time of a room is determined by a complex combination of its features. Early studies by W. C. Sabine, the father of modern architectural acoustics, showed that there were but two essential parameters, the total volume of the room and the total absorbing area. Other features such as room shape were of secondary importance and could be ignored in the initial evaluation.

Sabine summarized his findings in a single equation which can be written as

$$T = 0.16 \frac{V}{S} \qquad [27a]$$

where T is the reverberation time, in seconds, V is the room volume, in cubic meters, and S is the *total absorption area,* in square meters.* This area includes walls, ceiling, and floor. A significant part of the floor absorption is accounted for by the audience (or lack thereof).

27.3 Total absorbing area

If the room surfaces consisted of various sections that were either perfect absorbers or perfect reflectors, the total absorption area S would just be the sum of all areas of the perfect absorber. However, only an opening acts as a perfect absorber, and it is therefore necessary to consider the absorption coefficients of each surface individually.

If a surface of 6 m^2 has an absorption coefficient $\alpha = 0.5$, only half the sound striking it will be absorbed. The rest will be reflected back into the room. The absorption is therefore equivalent to a perfect absorber surface of 3 m^2. In general, then, the equivalent area of perfect absorption is obtained by multiplying the true area S by α. If the room is composed of surfaces S_1, S_2, S_3, . . . , each with absorption coefficients α_1, α_2, α_3, . . . , then the total equivalent absorption area is given by $S = S_1\alpha_1 + S_2\alpha_2 + S_3\alpha_3 + \ldots$. See Figure 27–2.

The absorption by the audience presents another problem. According to

*The acoustics industry in the United States, which is still using the English system of units, uses the "sabin" as the unit of absorption. One sabin = 1 ft^2 (0.0929 m^2) of perfect absorber.

Sabine, each person acts as an absorption area of about 0.4 m^2. Where seats are not occupied, the unoccupied seat accounts for some absorption. In order that the room acoustics not change with audience occupancy, empty seats should have the same absorption as a person who would otherwise be sitting in it. Designers seldom achieve this much absorption. As a result, the reverberation time usually lengthens when the concert hall is only partially filled.

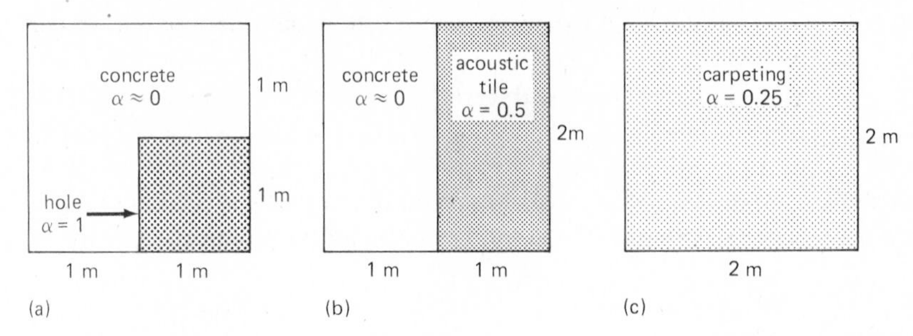

FIGURE 27-2 Of the incident sound waves on a surface, arbitrarily taken to be 100%, (a) a hole in the surface has $\alpha = 1.0$ and transmits 100%, (b) acoustical tile with $\alpha = 0.50$ absorbs 50%, and (c) carpeting with $\alpha = 0.25$ absorbs 25%, while concrete with $\alpha = 0$ absorbs practically none. The total absorption area is approximately 1 m^2 for each of these three examples.

Sabine's prescription assumes audience absorption depends only upon the total number of people. Studies by Beranek[1] indicate, however, that the absorption depends upon the total area covered by the audience. Audience absorption would therefore be increased if they were more spread apart, as is the modern trend. Other studies have shown more complex relations between absorption and audience makeup. Since the audience accounts for most of the absorption in a concert hall, serious attention should be given to correctly determining its absorption. For the purposes of this book, we will use Sabine's prescription, keeping in mind that if the seats are large and well spaced, the true reverberation time may be smaller than calculated.

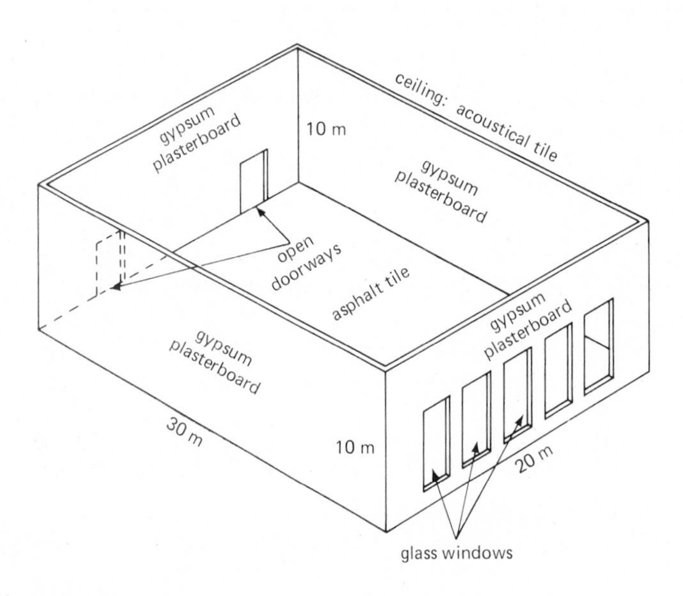

FIGURE 27-3 View into a rectangular music hall.

27.4 A sample calculation

Consider a rectangular music hall 20 m wide, 10 m high, and 30 m long, as shown in Figure 27–3. The concrete floor is covered with asphalt tile. The ceiling is covered with acoustical tile. See Table 26–1 and Appendix III. Long side walls and both end walls are gypsum plasterboard. One end wall has five plate glass windows 2 m × 6 m, and the other end wall has two open doorways 2 m × 3 m. Normally 350 people occupy the room during a group performance. Determine the reverberation time at 500 Hz.

Solution

Tabulate the total area of the surfaces and list the absorption coefficients for 500 Hz (from Table 26–1) followed by their respective products, as follows:

	AREA	ABSORPTION
Floor	20 m × 30 m = 600 m²	600 m² × 0.03 = 18 m²
Ceiling	20 m × 30 m = 600 m²	600 m² × 0.50 = 300 m²
Two long walls	2 × (10 m × 30 m) = 600 m²	600 m² × 0.05 = 30 m²
End wall 1	10 m × 20 m − 5 × (2 m × 6 m) = 140 m²	140 m² × 0.05 = 7 m²
End wall 2	10 m × 20 m − 2 × (2 m × 3 m) = 188 m²	188 m² × 0.05 = 9 m²
Windows	5 × (2 m × 6 m) = 60 m²	60 m² × 0.04 = 2 m²
Doorways	2 × (2 m × 3 m) = 12 m²	12 m² × 1.00 = 12 m²
Persons		350 × 0.40 m² = 140 m²
		Total 518 m²

The volume of the room is 10 m × 20 m × 30 m = 6000 m³. The reverberation time is, by Equation (27a),

$$T = 0.16\left(\frac{6000}{518}\right) = 1.85 \text{ s}$$

A music hall of this size should have a reverberation time more like 1.70s. See Figure 27–1. From Equation (27a), we can calculate the desired absorption:

$$S = 0.16\,\frac{V}{T} = 0.16\left(\frac{6000}{1.70}\right) = 565 \text{ m}^2$$

The absorption should therefore be increased by 47 m² to bring the total up to this value. By placing an additional area A of acoustical tile over gypsum plasterboard, we have the additional absorption:

$$0.5A - 0.05A = 47 \text{ m}^2$$

Solving for A:

$$A = \frac{47}{0.45} = 104 \text{ m}^2$$

Therefore 104 m² of acoustical tile, placed over gypsum plasterboard, would reduce the reverberation time to 1.70 s.

Although Sabine's formula is quite crude, and sometimes tends to overestimate the reverberation time, it is usually adequate for rough estimates. More exacting calculations require consideration of a variety of other factors that affect the reverberation time.[2] For instance, the room shape has some significance, and the air itself will absorb significantly above 1000 Hz, depending upon the temperature and humidity. More inclusive formulas have been developed to take these and other features into account,[3] and they should be used for an accurate acoustical evaluation.

27.5 Characteristics of reverberation

Characteristics of reverberant sound are often described in perceptual terms, such as dryness, warmth, and brilliance. To the trained ear, these terms take on generally agreed-upon meanings.

However, acousticians generally tend to use objective descriptions, such as *reverberation time, initial time delay gap,* and so on. Unfortunately, many acousticians tend to criticize the use of perceptual terms as being vague or, in some cases, meaningless. Such objections most likely stem from the difficulty that the acoustician finds in trying to translate the perceptual terms into measurable quantities.

Since enjoyment of music is a perceptual experience, we take the position that perceptual terms are meaningful to those who are adequately experienced. Such characteristics of reverberant sound affect the listeners' enjoyment, as well as the musicians' ability to perform. The acoustician cannot dictate what they should be but must try to translate the subjective sensory expressions into objective descriptions, which in turn relate to acoustic design.

A room with a short reverberation time is called **dead** or **dry**, whereas one with a long reverberation time is called **live**. In addition to the total reverberation time, reverberation can be characterized by its *frequency dependence, time dependence,* and *spacial dependence.* Reverberation time is usually defined for midfrequency, 500–1000 Hz. Above 2000 Hz, the reverberation time usually falls, due in part to absorption in the air. If the reverberation is sufficiently long at low frequencies—an increase of say 20 percent at 250 Hz and below—the room will have a sensation of warmth. This is a desirable characteristic for music (but not for speech) and it requires the surfaces in the room to be sufficiently solid. Unfortunately, many modern auditoriums use light panels that tend to absorb low-frequency sound, reducing the warmth of the room, producing the sensation of **brittleness**.

Time dependence of the reverberation pattern is also important. If the ratio of **direct-to-reverberant loudness** is sufficiently high, clarity is enhanced. Alternatively, low direct-to-reverberant loudness gives better fullness of tone. The sensation of **intimacy** that is associated with a small room is due mainly to a short **initial time delay gap,** the time between the direct sound and the first reflected sound. A room with an initial delay gap of less than 20 ms and a high direct-to-reverberant ratio has a quality of **presence** that is generally considered desirable, especially for chamber music. If the gap is greater than 50 ms, the presence is poor.

If the auditorium is broken up with surface irregularities such as niches and columns, the reverberation will be composed of a large number of closely spaced sounds. The reverberation will have a smooth **texture.** If the walls are smooth, there will be fewer and more widely spaced reflections, resulting in a harsh texture.

27.6 Spacial distribution of sound

Reflection from many surface irregularities will cause the reverberant sound to approach the listener from a multitude of directions, giving a sense of being bathed in sound. This sound **diffusion** is usually related to live halls designed for good texture.

An auditorium has **uniformity** if the quality of the sound is independent of the location of the listener. This characteristic is difficult to achieve, particularly in locations under balconies or at the rear of the hall. Careful attention to the room shape is essential to produce good uniformity.

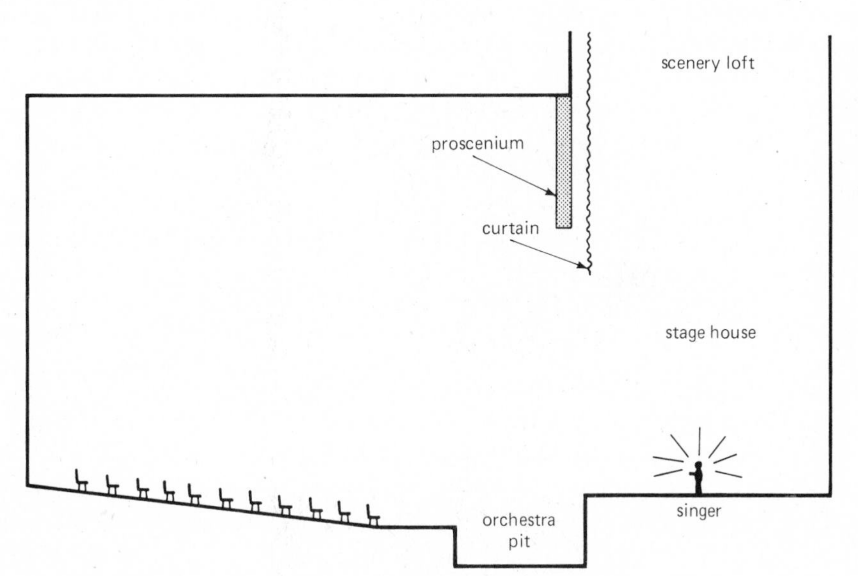

FIGURE 27–4
A multipurpose theater may have a stage house that is not well matched acoustically to the listening environment.

27.7 The stage house environment

Modern concert halls usually have a stage house, separated from the audience by the proscenium and often topped by a scenery loft. See Figure 27–4. Although this arrangement adds to the versatility of the concert hall, it may detract from the acoustics.

If the sound from all parts of the orchestra are projected to the audience in equal proportion, the stage house has good balance. This usually requires that the walls and ceiling near the performers be appropriately reflecting. If the sound from the orchestra mixes in a pleasing manner, the listener experiences a good blend. This depends to some extent upon the musicians' performing skills and their arrangement. But the closeness of the stage house walls is important, too, in providing that the time of travel from each area of the orchestra to the listener not be too different.

If the musicians are able to play in unison, they are said to have good ensemble. This depends in part on the quality of the musicians and their conductor. However, good ensemble is aided by a good acoustical environment. There must be an immediacy of response between musicians in the various parts of the orchestra. This can be achieved by close reflecting surfaces on the sides or the ceiling of the stage house or both. The musicians must also be able to hear some of the concert hall reverberation in order to sense the music as the audience hears it.

27.8 Extraneous sound

An auditorium should be designed to respond well to very loud passages as well as very quiet passages. Such an auditorium is said to have a wide dynamic range. Too often, a pianissimo passage is inaudible, lost in the sound of a noisy air conditioner or the rattle of a passing train. The lower limit of audibility is set by the normal audience noise, the coughs, cleared throats, clothing rubbing, and so on. Other extraneous noise should be kept below this level. This usually requires that careful attention be paid to the ventilation system, as well as to good sound insulation in the walls and ceiling. Needless to say, the auditorium should not be built next to a busy airport.

27.9 Optimization of design

Ideally, one might hope to design an auditorium that meets all the desired standards mentioned. Unfortunately, several of the desired attributes have mutually exclusive acoustical characteristics. For example, a live hall may have the fullness of tone desired for nineteenth-century romantic music but will therefore fail to provide the clarity desired for a baroque quartet. Design of a concert hall will therefore require some compromises, or else it must be tailored to a specific type of music.

QUESTIONS

1. Why is an objective description of acoustical characteristics generally desirable?

2. What is meant by the term *warmth?*

3. What are the major parameters that determine the reverberation time of a room?

4. What is meant by equivalent absorption area?

5. Why are theater seats usually designed to be good absorbers?

6. Why does an opening act as a perfect absorber?

7. Why might a room be designed with an acoustic tile ceiling, such as that in the example?

PROBLEMS

1. Suppose that in the example problem, 200 m^2 of the acoustic tile were removed from the ceiling and replaced with plaster. What would be the new reverberation time?

2. A rectangular room has dimensions 4 m × 12 m × 20 m and a total absorption area of 110 m^2. (a) What is its reverberation time? (b) Is this most appropriate for music or speech? Explain.

3. A rectangular room has dimensions 6 m × 16 m × 25 m. What total absorption area is necessary to produce the proper reverberation time for average music?

4. A square room has walls 25 m long and 6 m high. The floor is asphalt tile. The walls are concrete with a plate glass window of 2 m × 3 m in one wall and an opening of 1 m × 2 m in another. The ceiling is gypsum board. The room normally holds 200 people. Determine the amount of acoustic tile that must be added to produce a reverberation time of 1.6 s.

PROJECT

Evaluate the reverberation time of a local music hall, and discuss the improvements desirable to improve its acoustical characteristics.

NOTES

[1]L. Baranek, *Music, Acoustics, and Architecture* (New York: Wiley, 1962).

[2]For more on acoustical calculations, see M. D. Egan, *Concepts in Architectural Acoustics* (New York: McGraw-Hill, 1972).

[3]See, for example, J. Fitzroy, "Reverberation Formula Which Seems to be More Accurate with Nonuniform Distribution Absorption," *Journal of the Acoustical Society of America* 31(1959): 893.

Chapter Twenty-eight

ARCHITECTURAL DESIGN

A concert hall is a setting in which musicians and listeners can get together. The proper design can enhance the enjoyment of all. With an understanding of acoustics, the proper design is best accomplished by valid engineering. In addition to providing the acoustical aspects already discussed, the concert hall must often be designed to suit multiple purposes, to comfortably accommodate a sufficiently large audience, to provide convenient access and a clear view, and to be visually pleasing and structurally sound. Architects prefer an original and creative design, and yet they must observe local building codes. In addition, economics must not be forgotten. The resulting design will usually be a compromise.

28.1 Concert halls of the past

A study of architectural design must begin with a study of the structures already in existence.[1] Many of the best date back many hundreds of years. In fact, many people believe that old concert halls are characteristically better than new ones. This may be related to the feeling that a concert hall, like a fine wine, mellows with age. Or that "they just don't build buildings like they used to." A more likely explanation is that poor halls were torn down long ago, whereas good halls have escaped the bulldozer as a result of the support of the local music lovers and the musicians who play in them. Carnegie Hall in New York is an example of a good acoustical structure that is still standing as a result of the determination of the citizens of New York City.

A study of acoustics, then, begins with a study of concert halls built over the span of history. The structural designs of these halls are intimately related to the type of music that was in vogue at the time of construction. The music, in turn, was related to the type of instruments that were available. As to which determined which, the issue is not clear. It seems most likely that the three developed together.

28.2 The open-air theater

Some two thousand years ago, the Greeks and Romans used open-air theaters for speeches, drama, songs and choruses, and a few instrumental recitals. These theaters were usually built on hillsides away from major population centers—and that was in an era when environmental noise was much less than it is today. The seats were built in semicircles centered on the stage to keep the audience as close to the stage as possible. The tiers were built upward, allowing each listener to have a clear line of sight to the stage. See Figure 28–1. The direct sound from the performers was augmented by reflections from the stage surface and the intervening heads of the audience and empty seats.[2] The arrival times of these small-angle reflections were well below the 35 ms necessary to ensure the clarity needed for drama and speeches. (See Section 26.6.) These early reflections, and the virtual absence of reverberation, resulted in surprisingly good hearing conditions at the rear of many of the large classical outdoor theaters.

FIGURE 28–1
Outdoor theater at Orange, France. It is considered to be the most perfectly preserved Roman amphitheater remaining today. The wall behind the stage area is more extensive than in most other such structures.

In these open-air theaters, a reflecting surface was often placed behind the raised stage, which aided in reflecting sound back to the audience and reducing unwanted background noise and distracting sights. However, unless this surface was quite close to the performers (say with 3 m), the time delay of the reflected sound would have detracted from the performance clarity. Such reflecting structures, however, would have helped to increase intensity and simulate a sense of reverberation, making the theater much more desirable for orchestral music.

Outdoor amphitheaters are still used today for many presentations, giving us the pleasant experience of a Sunday afternoon band concert in the park or a symphony under the stars. See Figure 28 – 2. However, the lack of control of the acoustics and loudness in these structures tends to put special pressures on the performers to play loudly, as it must have done to the ancient Greeks and Romans. Enjoyable as an outdoor presentation can be, the acoustical quality suffers from a lack of fullness of tone, which only comes from controlled reverberation.

FIGURE 28 – 2
The Hollywood Bowl. (Courtesy, Los Angeles Philharmonic Association)

28.3 Sizes of concert halls

As we have mentioned, the design of concert halls is related to the type of music that is to be performed in them. During the baroque period (1600 – 1750), music was generally written for and played in relatively small rooms for small audiences. Reverberation times were typically less than 1.5 s, giving the rooms the needed clarity and brilliance. This type of setting was particularly appropriate for the prevalent instruments such as the harpsichord and recorder, which had low sound output and required good definition. Of course, chamber music is still written today, although it is often recorded in the studio in order to allow the listener to hear it in the proper acoustical environment.

The classical period (1750–1810) was characterized by the development of the symphony orchestra, with larger audiences to match. The concert halls were necessarily larger, and the reverberation times lengthened to 1.5 s and longer. The music tempo written during this time was typically somewhat slower.

The music composed during the romantic period (1810–1900) was char-

TABLE 28–1: **Reverberation Times in Opera Houses and Concert Halls Known for Good Acoustical Characteristics**[a]

	ROOM VOLUME (m^3)	REVERBERATION TIME (s)	YEAR OF DEDICATION
OPERA HOUSES			
Theatre National de la Opera, Paris, France	9,960	1.1	1875
Festspeilhaus, Bayreuth, Germany	10,300	1.55	1876
La Scala, Milan, Italy	11,245	1.2	1778
Royal Opera House, London, England	12,240	1.1	1858
War Memorial Opera House, San Francisco, California	20,900	1.6	1932
CONCERT HALLS			
Musikhochschule Konzertsaal, Berlin, Germany	9,600	1.65	1954
Tivoli Koncertsal, Copenhagen, Denmark	11,890	1.3	1956
Palais des Beaux-Arts, Brussels, Belgium	12,500	1.42	1929
Benjamin Franklin Kongresshalle, Berlin, Germany	12,950	1.2	1957
Herkulessaal, Munich, Germany	13,600	1.85	1953
Neues Festspeilhaus, Salzburg, Austria	15,500	1.5	1960
Academy of Music, Philadelphia, Pennsylvania	15,700	1.4	1857
St. Andrew's Hall, Glascow, Scotland	16,100	1.9	1877
Symphony Hall, Boston, Massachusetts	18,740	1.8	1900
Carnegie Hall, New York, New York	24,250	1.7	1891
Tanglewood Music Shed, Lennox, Massachusetts	42,450	2.05	1938
Royal Albert Hall, London, England	86,600	2.5	1871

[a]See Reference 1 for a more extensive table.

acterized by a lack of definition and required fullness of tone. These requirements were met by large concert halls with low direct-to-reverberant sound and reverberation times of 1.8 – 2.2 s.

In earlier eras, music was performed mostly in the churches. Since the medieval cathedral was usually a voluminous structure with low absorption, reverberation times were as long as 5 – 10 s. Although a symphonic presentation would sound quite muddy in a cathedral, slow choral selections and Gregorian chants match the acoustics nicely. Again, the music undoubtedly developed in the context of the environment in which it was presented.

Modern halls have been built in many cases for the music of these past eras and hence will have reverberation times of 1.8 – 2.0 s at 500 – 1000 Hz. However, there is also a trend toward "hi-fi" halls with a relatively high definition made possible by reverberation times less than 1.5 s. See Table 28 – 1.

28.4 Shapes of concert halls

Although reverberation time depends mainly on size, shape will determine many other important characteristics, such as intimacy, texture, diffusion, and uniformity.

The **Morman Tabernacle**, built in Salt Lake City in 1869, has a semicircular transverse cross section and a longitudinal cross section approximating an ellipse. Such a structure having a perfectly elliptical cross section has two points called *foci*. See Figure 28 – 3(a). Sound radiating in all directions from one focus will, after one reflection, converge on the other focus. A whisper at one focus will be plainly heard at the other. Such an enclosure is called a **whispering gallery**. See Figure 28 – 3(b) for another type of whispering gallery.

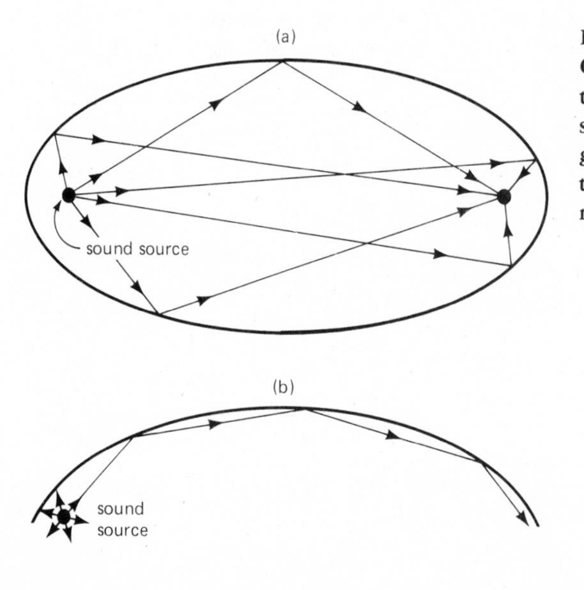

FIGURE 28 – 3
Concave surfaces tend to (a) focus sound and (b) guide sound around the room walls by many reflections.

Approximately elliptical forms such as the Morman Tabernacle, and even spherical **geodesic dome** shapes such as the Pacific Cinerama Theater in Hollywood, exhibit diffuse focusing effects. Such shapes generally have good fullness of tone but can have poor uniformity. The celebrated acoustical excellence of the Tabernacle is undoubtedly a result of taking advantage of the projection capabilities of the convex walls while taking care of the placement of the choir and organ in order to minimize focusing effects.

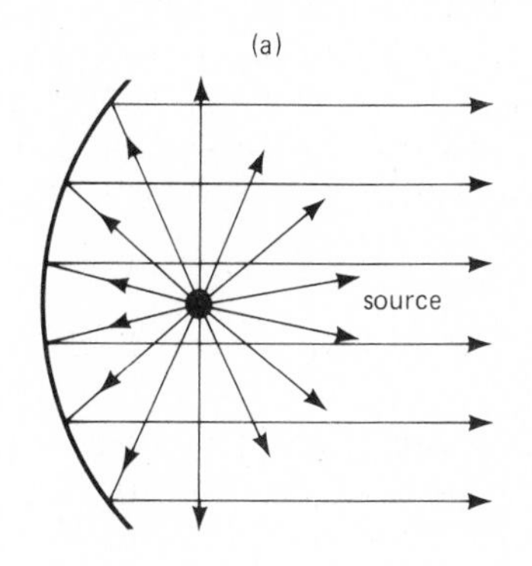

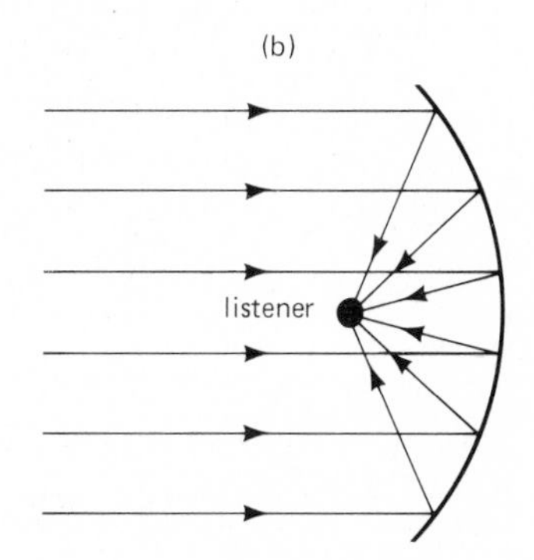

FIGURE 28–4
A concave surface (a) channels sound from a divergent source or (b) focuses sound into specific regions.

A concave surface behind a church pulpit, as shown in Figure 28–4(a), may aid the dynamic impact of a sermon. However, the concave surfaces in auditoriums, as shown in Figure 28–4(b), will only aid the audibility in one section of the audience and may create echoes. Convex surfaces, on the other hand, will spread the sound, enhancing diffusion. See Figure 28–5.

Many concert halls built during the eighteenth and nineteenth centuries are rectangular and relatively narrow. See Figure 28–6. As a result, the reflection from the walls arrives relatively soon after the direct sound. This short, initial, time delay gap results in the desired sense of intimacy. To make the

FIGURE 28–5
Sound waves reflected from a large convex surface will diverge upon reflection.

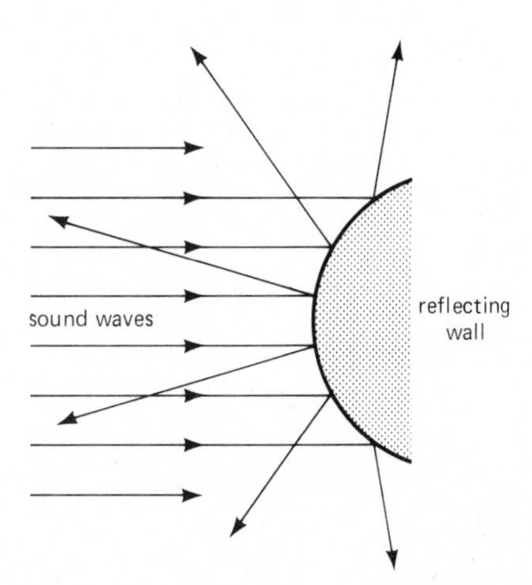

reverberation time sufficiently long, the room volume is made large by high ceilings.

Opera houses, on the other hand, are traditionally built in a horseshoe shape. The La Scala Theater in Milan, Italy, Figure 28–7, is a typical example. In such a building, everyone has a clear view of the stage, and vision is an important part of opera enjoyment. Reverberation time is relatively low (1.4–1.5 s), allowing good vocal definition. Opera music is usually composed to be consistent with this. The direct-to-reverberant sound ratio is high, giving poor diffusion, but this is of little consequence in opera houses since attention is usually focused on stage. Wagnerian opera is somewhat of an exception, requiring longer reverberation times. The Festspeilhaus in Bayreuth was built, under Wagner's direction, to have an appropriate reverberation time of 1.6–1.8 s. American opera houses also tend to have longer reverberation times. The reason is not clear, but it may be because most Americans cannot follow European languages anyway, and pick up their cues from the action, libretto, or memory.

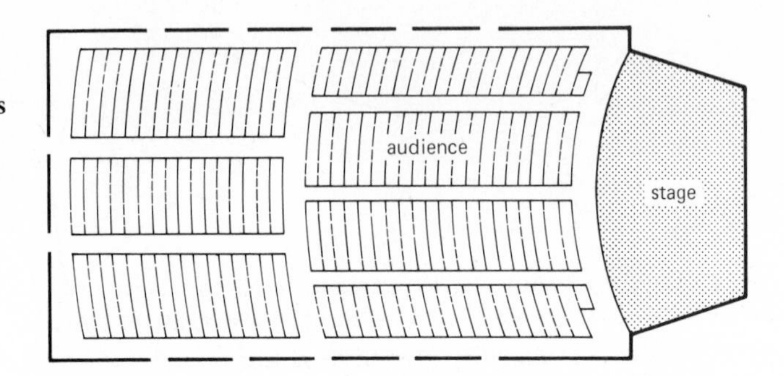

FIGURE 28–6
Many traditional music halls, such as Symphony Hall in Boston, were built in a rectangular shape.

The modern trend in concert halls has been toward the seating of large

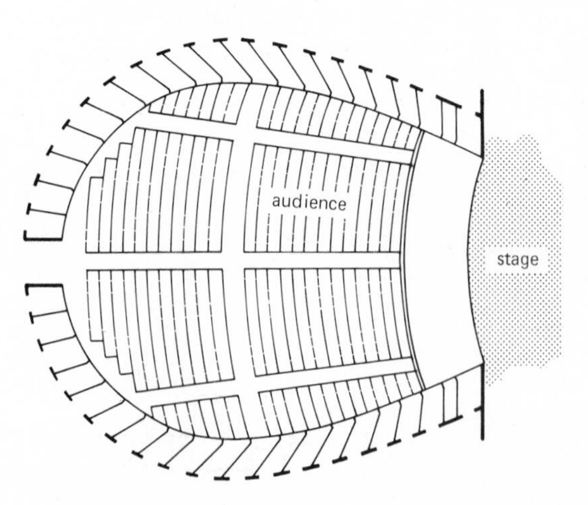

FIGURE 28–7
A traditional opera house, such as the La Scala in Milan, Italy, is built in the shape of a horseshoe.

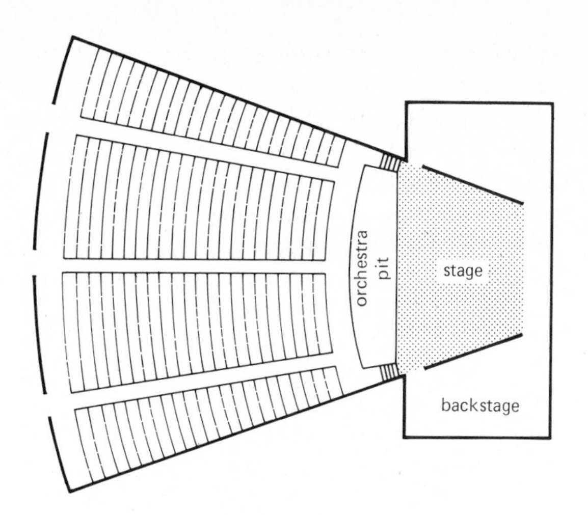

FIGURE 28–8
Modern fan-shaped auditoriums are usually designed for large audiences and multiple purposes.

audiences, resulting in fan-shaped auditoriums. See Figure 28–8. An understanding of acoustics makes it possible to devise methods for optimizing the desired characteristics.[3, 4, 5, 6] However, fan-shaped auditoriums have not always been found to be acoustically satisfactory, and there seems to be a trend back to the rectangular shapes of the nineteenth century.

28.5 Reflecting surfaces

In most concert halls, the audience is the dominant absorber. To provide enough reverberation time, designs should avoid adding extraneous absorbers and should use solid wall and ceiling materials that reflect well. Thick wood, plaster, and concrete are all good reflectors. Thin wooden slats, which are visually pleasing, are poor reflectors at low frequency and detract from the warmth of the hall. Good diffusion and texture require that the side walls, balcony front, ceiling, and stage walls be irregular. The stage house should be

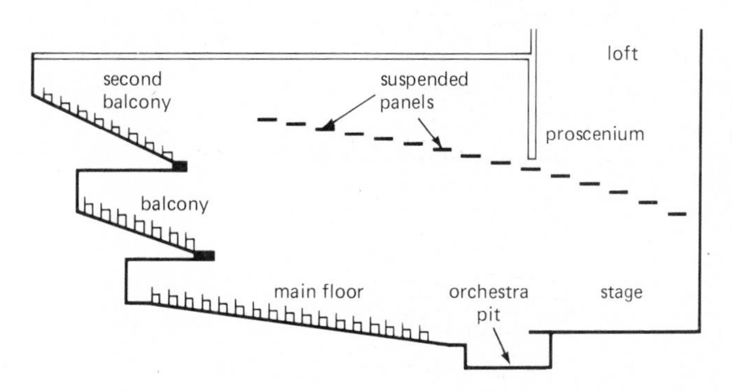

FIGURE 28–9
Suspended panels serve to reflect stage sounds to the audience and to decrease the initial time delay gap.

well coupled to the main hall via a wide and high proscenium and should have close side and ceiling walls. Too often, a concert is given on a stage capable of storing scenery above it. Such a stage house will act somewhat as a separate acoustic chamber. (See Figure 27–4.)

Many of these problems are solved by reflecting panels. See Figure 28–9. A canopy partially covering the stage house will reflect sound back to the musicians and out to the audience. Panels placed above the proscenium and the audience will reflect early sound to the back of the auditorium, where the direct-to-reverberant ratio is usually low. Most important, "clouds" (overhead panels) will decrease the initial time delay gap, giving a sense of intimacy yet maintaining the high ceiling necessary for a sufficiently long reverberation time. Panels should be of varied size and orientation for good diffusion and texture. Panel placement can, in principle, be done easily and, perhaps, automatically. Only recently has it been realized that a good concert hall can be designed in which the acoustics can be changed within a matter of minutes to match the performance.

All the available acoustical knowledge does not guarantee an architectural success. The construction and reconstruction of Avery Fisher (New York Philharmonic) Hall is a classic case in point.[7] The reader will find this story an excellent example of the problems found in applying the art and science of acoustics to the design and construction of a high-quality music hall.

28.6 Electronic amplification

The addition of an electronic sound amplification system to a concert hall can be a real asset, especially when used to aid a relatively weak source such as a vocalist or an instrumental soloist. Moreover, the direct-to-reverberant sound ratio can be kept high, resulting in good clarity.

Careful design of equipment and its placement is needed to ensure good high-fidelity performance. Sufficiently high amplification is needed, but acoustic feedback (see Section 22.2) must be avoided. The sound from the speakers must not differ from the input sound by more than 50 ms if they are to sound in phase. Observe in Figure 28–2 that the twenty-speaker array over the front of the stage is beyond the orchestra to avoid feedback and minimize time delay. Finally, the amplification should be suppressed at frequencies that have high reverberation times or enhanced at frequencies at which a room is dead. This process of equalization (see Section 21.7) illustrates a major advantage of electronics in correcting for acoustical defects.[8]

A cathedral presents special problems for electronic sound amplification systems because of its great volume, high ceilings, hard surfaces, and long reverberation time. Figure 28–10 shows side and plan views of the relatively new St. Mary's Cathedral in San Francisco; the placement of a cluster of seven loudspeakers in a single enclosure provides good sound reproduction over the entire congregation. This speaker system, designed with great care and

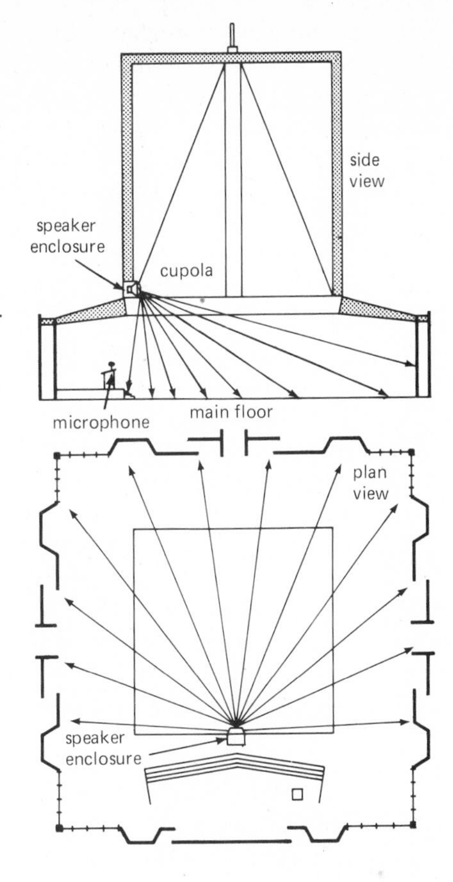

FIGURE 28–10
Two views of the unconventional St. Mary's Cathedral in San Francisco, showing the loud-speaker cluster placed at the lower edge and center of the cupola, over and ahead of one of ten microphones. (Courtesy of Charles Catania)

expertise by an audio engineer,[9] consists of three large sectoral horns H and three large bass speakers B, 15 in. in diameter. See Figure 28–11. Through electronic mixer circuits, and impedance-matching transformers T, the power to the three speaker groups is balanced. The speaker enclosure is centrally located high above the altar, at the lip of the cupola. The sounds from the three speaker groups are directed toward the two side naves and toward the center, where the loudness and the direct-to-reverberant sound ratio would normally be poor. The distance from the average listener to the speakers is similar to the distance to the altar, so the sound from each is in phase. By the use of band-pass filters (see Section 21.6), the amplification is separately adjustable for each one-third octave range. Equalization of frequency response is attained by tuning each separate range. Sounds within the cathedral are picked up by from one to ten microphones: three in the choir gallery at the altar, three at the organ console (see Figure 12–11), one at the altar, one at the pulpit, and so on. The altar position carries a single voice, as does the pulpit, and therefore requires good amplification and clarity. In each of the three speaker groups, the sectoral horns reproduce the high frequencies above 800

Hz, while the large bass speakers reproduce the low frequencies from 800 Hz down.

With careful design, the installation of an electronic sound system can improve the acoustical quality of an auditorium, as well as provide a flexibility for tuning the acoustical environment to meet the desired needs.

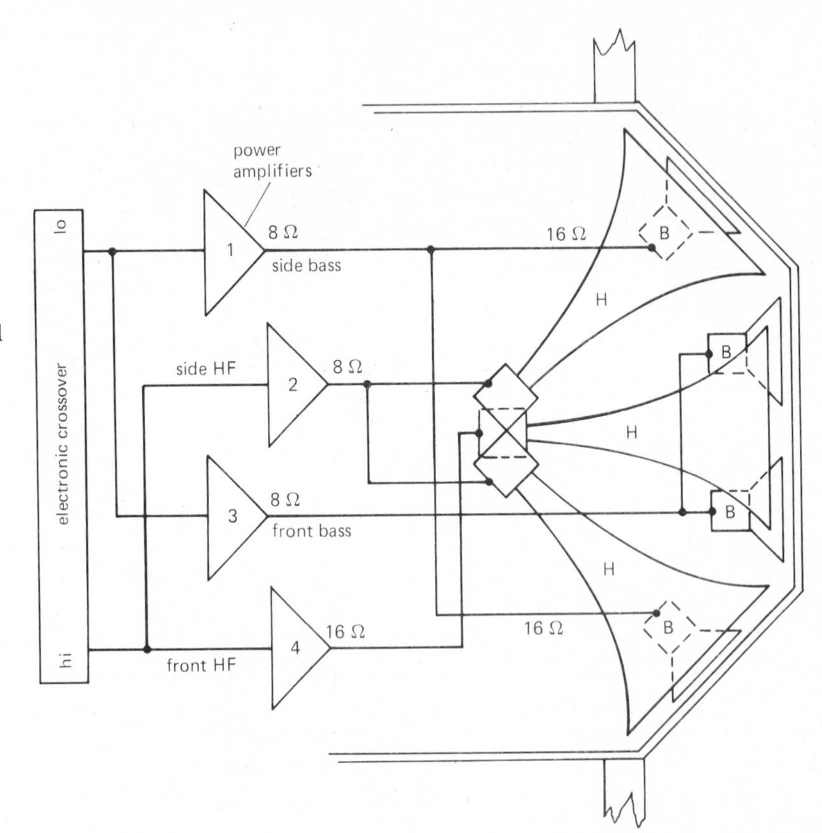

FIGURE 28–11 **Bi-amplified system, showing the placement of three multicellular horns H operating above 500 Hz and four 15 inch diameter bass speakers B, in St. Mary's Cathedral in San Francisco. Each amplifier contains an impedence matching transformer with its output at 8 ohms. (Courtesy of Charles J. Catania)**

28.7 The Sydney opera house

This now famous edifice for the performing arts, completed at a cost of over 100 million dollars, stands in Sydney Harbor, Australia. The clamlike structure contains four main performing halls, as well as hundreds of other areas, and took over 20 years to build. See Figure 28–12. Opera, orchestral, and chamber music concerts, ballet, drama, choral works, jazz, folk concerts, recitals, films, and variety shows are among the performances presented in this unique building. The excellent acoustics in the large halls is attributed to the structural irregularities in the walls and ceilings and the surface materials used. The two woods used extensively to decorate the interiors are brush box and white birch plywood.

As closing remarks to this book, we emphasize that in order to present a wide range of topics in music and acoustics, many of the topics were treated briefly. However, much can be learned by reading the articles and selected topics from the texts listed in the references found at the ends of the chapters. The tremendous increase in research in this field in the last few years has brought about many new discoveries and serves as an indication that much is still to be learned from future experimentation.

FIGURE 28-12
The $100 million opera house on the Bennelong Peninsula at the entrance to Sydney Harbor, Australia. This building contains four major auditoriums: a concert hall (seating 2690 people), a theatre (1547), a drama theatre (544), and a music hall (419). (Courtesy of Sydney Opera House Trust)

QUESTIONS

1. What is a whispering gallery?

2. What are the characteristics of an acoustically warm room? What structural considerations are needed to achieve warmth?

3. Why are suspended panels often used in a large auditorium?

4. Why does a live room usually have good uniformity, diffusion, and fullness of tone?

5. What are the disadvantages of a live room?

6. If an architect believes an auditorium design will result in a dead room, what changes could he make in the design to correct this?

PROBLEMS

1. Two loudspeakers, 10 m apart, are driven by a common amplifier. A listener close to one of them will hear the sounds from the two speakers arrive at different times. (a) What is the time difference? (b) Will the distant speaker sound like an echo of the near speaker? Explain.

2. Calculate the frequency and pitch of the fundamental vertical resonance in a standard 2-m-high shower.

3. How far will sound travel in 50 ms, the time for a clearly discerned echo?

PROJECT

Evaluate a local concert hall in terms of the design features described in this chapter. Determine whether or not the acoustic behavior is consistent with your expectations from the architectural design.

NOTES

[1]L. L. Beranek, *Music, Acoustics, and Architecture* (New York: Wiley, 1962).

[2]R. S. Shankland, "Acoustics of Greek Theatres," *Physics Today* 26 (October 1973):30.

[3]L. L. Beranek, "Acoustics in the Concert Hall," *Journal of the Acoustical Society of America* 57(1975):1258.

[4]G. Plenge et al., "New Methods in Architectural Investigations to Evaluate the Acoustic Qualities of Concert Halls," *Journal of the Acoustical Society of America* 57 (1975):1292.

[5]M. Rettinger, *Acoustics, Volume 1—Acoustic Design* (New York: Chemical Publishing, 1976).

[6]V. O. Knudsen and C. M. Harris, *Acoustical Designing in Architecture* (New York: Wiley, 1950).

[7]B. Bliven, Jr., "Annals of Architecture: Avery Fisher Hall," *New Yorker* (8 November 1976):51.

[8]D. Davis and C. Davis, *Sound System Engineering* (Indianapolis: H. W. Sams, 1975).

[9]C. J. Catania, "Sound System Design for St. Mary's Cathedral, San Francisco," *Journal of the Audio Engineering Society* (October 1975).

Appendix I

BASIC PHYSICAL PRINCIPLES

A.1.1 Fundamental units

The subject of acoustics involves many of the simplest principles of mechanics. Like all other branches of science, mechanics is based upon objective measurements of different kinds, using measuring instruments like a meter stick, a set of scales, or a stopwatch. From recorded measurements, one tries to set up simple rules that, when applied to the data, not only account for all the observations that have been accurately made but hopefully for all others of a similar nature yet to be made. If this extension is successful, one assumes one has discovered a **law of nature.**

Every measurement, whether it is a *distance,* a *speed,* an *acceleration,* a *velocity,* a *quantity of heat,* the *energy of a sound wave,* the *frequency of a radio wave,* or the *electric current in a wire,* requires two entities: a **number** and a **unit.** One might, for example, measure the length of a room and record 12.85 meters (m), determine the speed of an airliner and record 920 kilometers per hour (km/h), or find the intensity of a sound wave to be 0.0250 watts per square meter (W/m^2). In each of these measurements, the units following each number are just as important as the number itself, for without them the numbers would have little or no meaning.

In all of mechanics there are *three fundamental units of measurement.* They are

length mass time

All other units, of which there are many, can be expressed in terms of these. For example, if we are driving along the highway and we look at the speedometer and the pointer indicates 75, we realize we are going 75 km/h. The units for speed then are *distance divided by time.*

All units, other than the three fundamental units, are called *derived units.* Speed is therefore expressed in derived units. The different units of length, mass, and time recommended for use today are as follows:

length	1 kilometer = 1000 meters	1 km = 1000 m
	1 meter = 100 centimeters	1 m = 100 cm
	1 centimeter = 10 millimeters	1 cm = 10 mm
mass	1 kilogram = 1000 grams	1 kg = 1000 g
	1 gram = 1000 milligrams	1 g = 1000 mg
time	1 hour = 60 minutes	1 hr = 60 min
	1 minute = 60 seconds	1 min = 60 s
	1 second = 1000 milliseconds	1 s = 1000 ms

A1.2 Speed and velocity

Speed and velocity are both measured in the same units, *distance traveled divided by the elapsed time.*

$$\text{speed} = \frac{\text{distance traveled}}{\text{elapsed time}} \qquad [1]$$

Distance traveled is illustrated in Figure A1 – 1. If we specify the direction in which a body is moving, we refer to its motion as a *velocity,* but if its direction is of little or no consequence, we refer to its *speed.* Both are measured in the same units, and for both we write

$$V = \frac{x}{t} \qquad [2]$$

For example, if we speak of a car as traveling 200 km in 2.5 h, its speed V is

$$V = \frac{200\ \text{km}}{2.5\ \text{h}} = 80\ \frac{\text{km}}{\text{h}}$$

If however, we say the car travels 200 km north in 2.5 h, its velocity V is

$$V = \frac{200\ \text{km}}{2.5\ \text{h}} = 80\ \frac{\text{km}}{\text{h}} \quad \text{to the north}$$

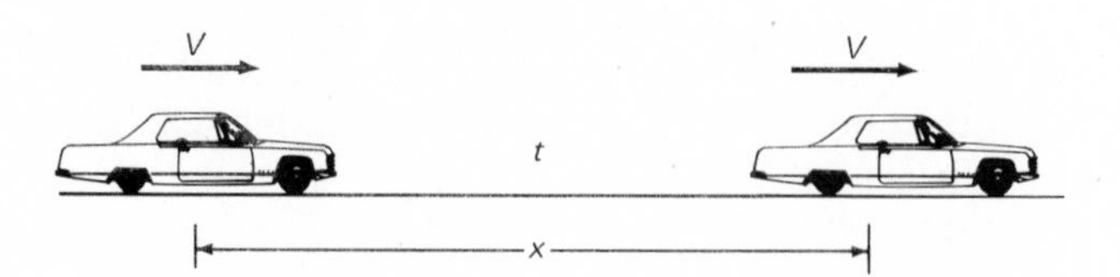

FIGURE A1 – 1 Diagram of a car moving with a constant speed and constant velocity.

A1.3 Distance traveled

If the speed or velocity of a body is constant, the distance traveled is given by Equation (2). Solving for x as the unknown in the equation, we obtain

$$x = Vt \qquad [3]$$

Example

If a body moves with a speed of 6.2 m/s, how far will it travel in 15 s?

Solution

Direct substitution of the given quantities in Equation (3) gives

$$x = Vt = 6.2\,\frac{\text{m}}{\text{s}} \times 15\ \text{s} = 93\ \text{m}$$

Observe that the units of seconds cancel, giving the distance in meters.

A1.4 Acceleration

Acceleration is defined as *the time rate of change of velocity.* A car picking up speed has a positive acceleration in the forward direction, while another slowing down has a negative acceleration in the forward direction. By this definition we may write as an equation

$$\text{acceleration} = \frac{\text{change in velocity}}{\text{elapsed time}} \qquad [4]$$

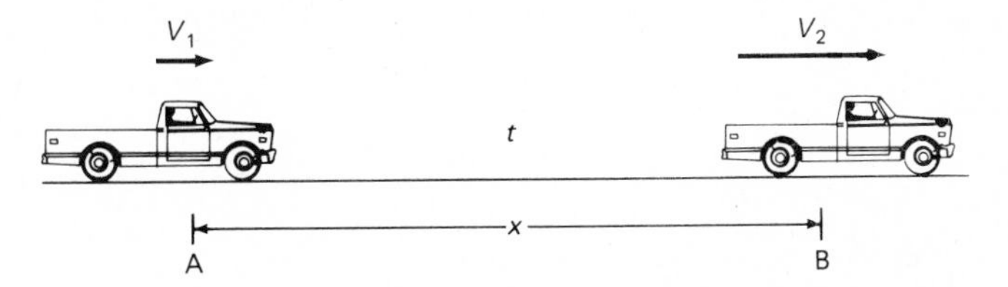

FIGURE A1 - 2
A small truck accelerates along a straight and level road.

Consider as an illustration of accelerated motion the truck shown in Figure A1 - 2. Due to a constantly acting force exerted through the drive wheels, the truck is continually accelerated as it goes faster and faster along a line from A to B. As it passes A, it has a relatively low velocity V_1, while farther along its path at the point B, it is moving faster and has a velocity V_2. With these symbols, V_1 is called the *initial velocity* and V_2 is called the *final velocity.* If the time it takes the truck to go from A to B is called the elapsed time t, Equation (4) can be written as

$$a = \frac{V_2 - V_1}{t} \qquad [5]$$

where a is the acceleration and $V_2 - V_1$ is the *change in velocity.*

Example

Suppose at A in Figure A1 - 3 the velocity of a bus is 10 m/s and that it takes 4 s to go from A to E. At E its velocity has increased to 30 m/s. What is the acceleration?

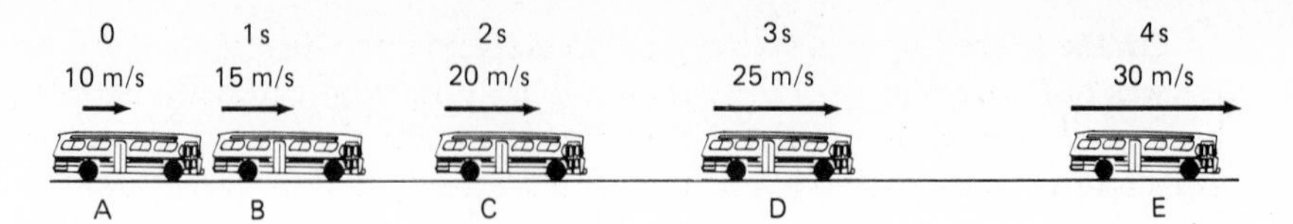

FIGURE A1-3
The instantaneous velocity at the end of each second for a uniformly accelerated bus.

Solution

By direct substitution of the known quantities in Equation (5), we obtain

$$a = \frac{V_2 - V_1}{t} = \frac{30\ \text{m/s} - 10\ \text{m/s}}{4\ \text{s}}$$

$$a = \frac{20\ \text{m/s}}{4\ \text{s}} = 5\ \frac{\text{m}}{\text{s}^2}$$

We say the bus has a constant acceleration of five meters per second squared, or five meters per second per second. To better understand what this means, consider the diagram in which the position of the bus is shown for each second of time. Initially the bus is at A, and the time $t=0$, so the velocity is 10 m/s. In 1 s time, the bus is at B and the velocity has increased by 5 m/s and is therefore traveling at 15 m/s. In another second of time, the bus is at C, and the velocity has increased another 5 m/s, so the bus is traveling at 20 m/s. At the end of 3 s, the bus is at D and the velocity is 25 m/s. At 4 s, it is at E, and the velocity is 30 m/s.

A1.5 Negative acceleration

When a body is slowing down, the initial velocity is greater than the final velocity, and the acceleration given by Equation (4) is negative.

Example

In going up a long hill, a car slows down from 20 m/s to 8 m/s in 4 s. Find the acceleration.

Solution

Using Equation (5), we make direct substitution of known quantities and obtain

$$a = \frac{8\ \text{m/s} - 20\ \text{m/s}}{4\ \text{s}}$$

$$a = \frac{-12\ \text{m/s}}{4\ \text{s}} = -3\ \frac{\text{m}}{\text{s}^2}$$

This answer may be read as minus three meters per second per second.

In the preceding two sections on acceleration, the direction of the velocity was not specified in any examples. The term *speed* could therefore have been used throughout with no other changes. In Section A1.11, the direction aspects of velocity will be taken up under the heading of *vector addition.*

A1.6 Free fall

If a mass m is dropped, it falls to the floor or ground. It does so because the earth is pulling upon it with a downward force, the force due to gravity. Neglecting air friction, the *acceleration* of a body in free fall is constant and independent of the mass. This may be demonstrated by the illustration shown in Figure A1 – 4. A glass tube about 1 m long, containing a coin and a feather, is connected by a hose to a vacuum pump. If after evacuation the tube is turned end to end, the coin and feather will be observed to fall side by side with the same downward acceleration. If the air is now admitted by opening a valve, and the tube again turned end to end, the coin will fall quickly, while the feather will float slowly downward. In the absence of air friction, all bodies fall with the same acceleration.

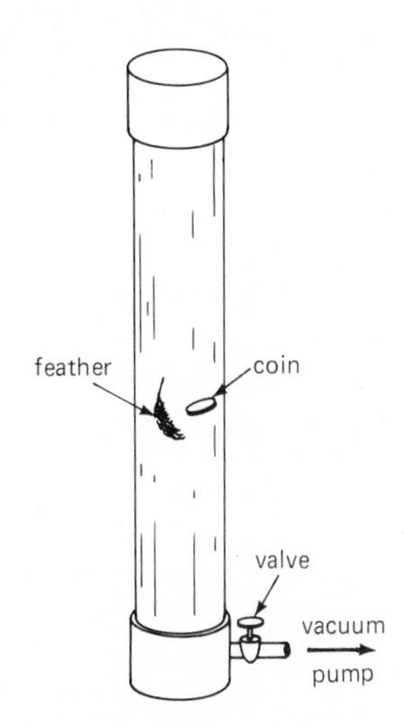

FIGURE A1 – 4
In a vacuum, a feather and a coin fall with the same acceleration and strike the bottom at the same time.

Thousands of experiments have been performed to accurately measure the acceleration of freely falling bodies. Over the earth's surface the value is found to be very nearly constant and is specified by the letter g.

$$g = 9.80 \frac{\text{m}}{\text{s}^2} \qquad [6]$$

In general, the value of g at sea level lies between a minimum of 9.7804 m/s² at the earth's equator and a maximum of 9.8321 m/s² at the North and South poles. The International Committee on Weights and Measures has adopted a standard, the value 9.80665 m/s^2. For general purposes, the value 9.80 m/s^2 is commonly used.

A1.7 Mass and inertia

To Sir Isaac Newton we owe the credit for presenting to the world the three *basic laws of mechanics.* These laws are called *Newton's laws of motion.* The first law introduces the concepts of mass and inertia and states that a body at rest or in uniform motion will remain at rest or in uniform motion unless some external force is applied to it.

Newton's first law can be demonstrated by many experiments. Three simple ones are shown in Figure A1 – 5. In diagram (a) is a small car with ball-bearing wheels—much like a roller skate—resting on a smooth horizontal glass plate. If the car is given a slight push, it will coast across the glass plate with very little friction. In the absence of all friction, the car would continue to move with constant velocity. If the car is at rest and the glass plate is jerked quickly to the right or the left, the wheels of the car will turn, but the car itself will tend to remain at rest.

In the second experiment, a 500-g mass is suspended by a piece of thread

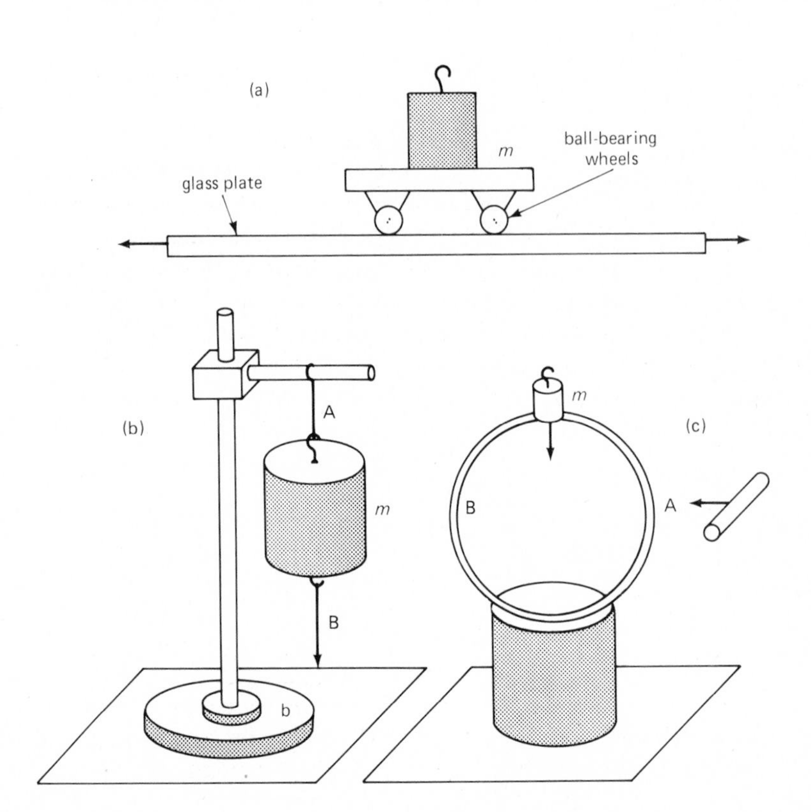

FIGURE A1 – 5 Three demonstrations illustrating the concepts of inertia and mass that illustrate Newton's three laws of motion. (a) Moving the glass plate rapidly to right or left will find the car remaining stationary. (b) An appropriate force can break the thread support at A or at B at will. (c) Striking the cardboard hoop properly can remove the hoop, causing the mass m to fall in the container below.

A, then pulled downward by another piece of the thread B from the same spool. If the force F is a slow and steady pull, the thread will always break at A. If thread B is given a sudden downward jerk, it will always break at B. In the first instance, the tension in the upper thread A is always greater than that in B and is equivalent to the applied force F plus the weight of the mass m. In the second case, the force F can be made momentarily very large, causing the thread to break before the mass m has had time to move far enough to break the upper thread.

In the third experiment, one places a cardboard hoop about 20 cm in diameter on top of an open tin can. On top of the hoop is placed a small 5- or 10-g mass. With the forefinger of one hand extended, one counts to three, each time going through the motions of striking the hoop at A, hoping the hoop will be pushed from under the mass allowing m to fall into the open can below. It never does. Upon repeating the experiment, but striking the hoop at B (too fast to be noticed by the audience) the mass will fall straight down and into the can. (This makes an excellent party trick.)

In all three of the experiments above, the mass at rest tends to remain at rest, and in the first experiment, the moving car tends to remain in motion, in agreement with the first law. Mass is the quantitative measure of inertia. In the metric system of units, inertia is measured in grams or kilograms.

A1.8 Force equation

The force equation is undoubtedly the most important relation in all of mechanics and is often referred to as *Newton's second law of motion.* It states that when a body is acted upon by a constant force, its resulting acceleration is proportional to the force and inversely proportional to the mass. To push or pull on anything is to exert a force upon it. To accelerate an automobile along a level road requires a greater force than the force required to give a boy's small wagon the same acceleration. The acceleration of an automobile or a small wagon depends upon the mass as well as the applied force.

Newton's second law of motion as stated above is often written as

$$a = \frac{F}{m}$$

or

$$F = ma \qquad [7]$$

force = mass × acceleration

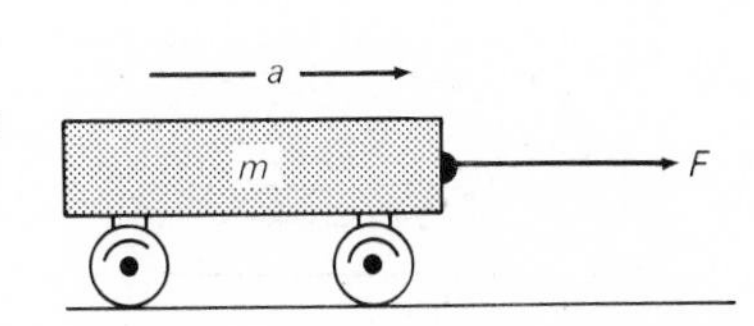

FIGURE A1-6
A mass *m* acted upon by a constant force *F* is given a constant acceleration *a*.

where a is in meters per second squared, m is in kilograms, and F is in newtons (N). Each of these factors is shown in Figure A1 – 6.

Example

A force F applied to a mass of 5 kg gives it an acceleration of 3.5 m/s². What is the magnitude of the force?

Solution

By substituting the given quantities into Equation (7), we obtain

$$F = ma = 5 \text{ kg} \times 3.5 \frac{\text{m}}{\text{s}^2}$$

$$F = 17.5 \frac{\text{kg} \cdot \text{m}}{\text{s}^2} = 17.5 \text{ N}$$

As a force, the newton has the dimensions in absolute units of kilogram meters per second squared.

A1.9 Weight and mass

In Section A1.6, we have seen that if a mass m is allowed to fall freely, it is the downward force due to gravity that gives rise to its constant downward acceleration g. When we apply Newton's second law of motion, there is a downward force that we call its **weight**, and we assign to it the symbol F_g. In the case of freely falling bodies, the force equation $F = ma$ is written in gravitational symbols as

$$F_g = mg \qquad [8]$$

weight = mass × acceleration due to gravity

See Figure A1 – 7.

In the metric system, the weight of a body is in newtons. Note carefully that the weight of a body of mass m is given by mg, whether it is falling freely or at rest on the ground.

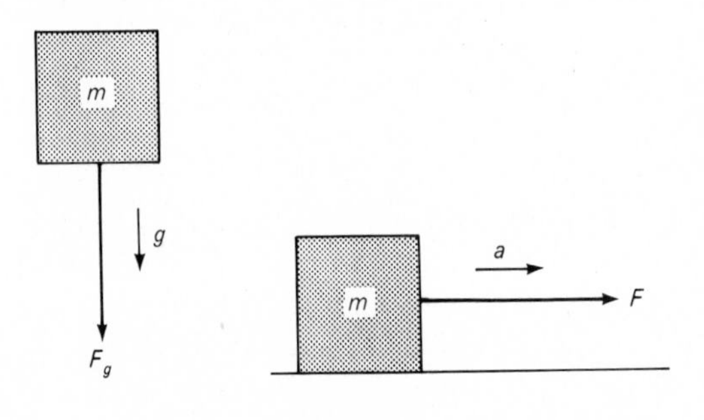

FIGURE A1 – 7
A mass m pulled by a force F has an acceleration a, and a mass m falling freely has an acceleration g.

Example

Find the weight of a mass of 6 kg.

Solution

Apply Equation (8) and obtain

$$F_g = mg = 6\ \text{kg} \times 9.80\ \frac{\text{m}}{\text{s}^2} = 58.8\ \frac{\text{kg} \cdot \text{m}}{\text{s}^2}$$

$$F_g = 58.8\ \text{N}$$

The difference between weight mg and mass m can be illustrated by imagining we carry a body into free space, away from all other bodies and their gravitational attraction. There a body will still have its same mass m but it will have *zero weight.* That a free body has its mass is illustrated by having another body bump into it. The larger the approaching mass, the greater will be the recoil of the first mass from the impact.

A1.10 Action and reaction

Of Newton's three laws of motion, the third law appears to be the simplest, yet it is often the most difficult to apply. The law may be stated simply as follows: Action and reaction forces are equal and oppositely directed.

Action and reaction forces will first be illustrated by applying Newton's law of gravitation to two neighboring bodies. "Any two bodies attract each other with a force proportional to the product of their masses and inversely proportional to the square of the distance between them." In algebraic symbols,

$$F = G\left(\frac{m_1 m_2}{d^2}\right) \qquad [9]$$

As illustrated in Figure A1 – 8, m_1 and m_2 are the masses in kilograms, d is the distance between their centers in meters, F is the force of attraction in newtons, and G is the proportionality constant,

$$G = 6.6732 \times 10^{-11}\ \frac{\text{m}^3}{\text{kg} \cdot \text{s}^2} \qquad [10]$$

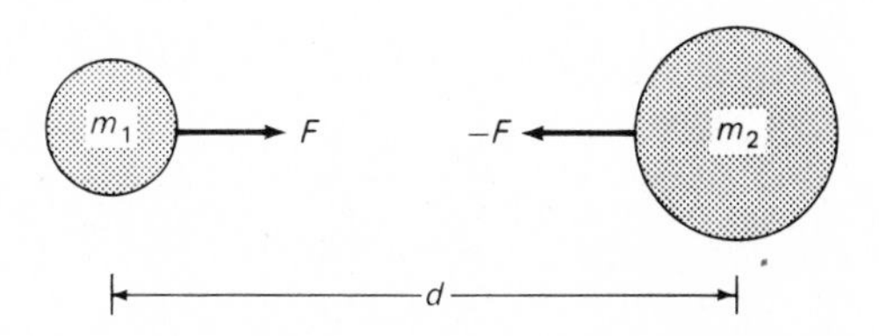

FIGURE A1 – 8 The gravitational attraction of any one body of mass m_1 for another body of mass m_2.

Mass m_2 pulls on m_1 to the right with a force F, and m_1 pulls to the left on m_2 with an equal but opposite force $-F$. These two are *action and reaction forces.*

Another illustration is shown in Figure A1–9—a bat striking a ball. At the time of impact, the action force F gives the ball an acceleration to the right, while reaction force $-F$ exerted on the bat by the ball slows it down (a negative acceleration).

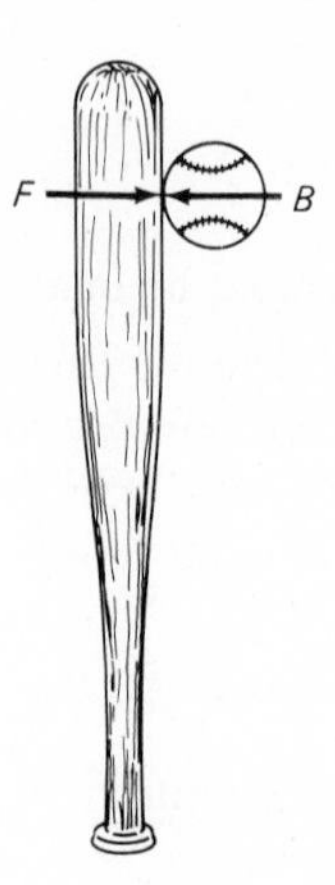

FIGURE A1–9
Action and reaction forces are always equal in magnitude and opposite in direction. Each force acts on a different body.

A third example is shown in Figure A1–10. A mass m_1 hanging by a cord from the end of a leaf spring comes to rest in the position shown. The weight of the block $-F_g$ is the downward pull of the earth on the block, and the equal but opposite force $-F_g$ is the upward force of the block on the earth. In addition to this pair of forces, the cord pulls on the block with the upward force $+F$, while the block pulls downward on the cord by the force $-F$. Note that there are two forces acting on the block. These are $+F$ pulling up and $-F_g$

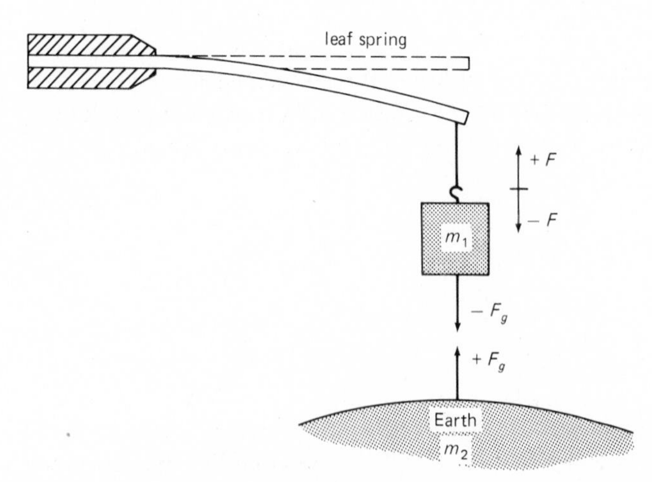

FIGURE A1–10
Two pairs of action and reaction forces.

pulling down. Although these are equal and opposite, they are not action and reaction; they are both action forces. Acting on the block in opposite directions, they keep the mass *m* from moving up or down.

It is important to note that any action and reaction pair of forces act on different bodies. Alternatively, whether a body is at rest or in uniform motion depends upon the forces acting on it and not upon any forces it exerts on some other body.

A1.11 Vector addition

Any measurable quantity that has magnitude and direction is a vector quantity. In mechanics, distance is often referred to as a displacement, and displacement is a vector quantity. If, for example, a man walks east from point A to point B, as shown in Figure A1 – 11, his displacement is the straight-line distance AB. Let this distance be 80 m. If the man now walks north from B to C, his displacement is the straight line BC. Let this distance be 60 m. We now ask two questions: (1) how far has the man walked and (b) how far is the man from his starting point?

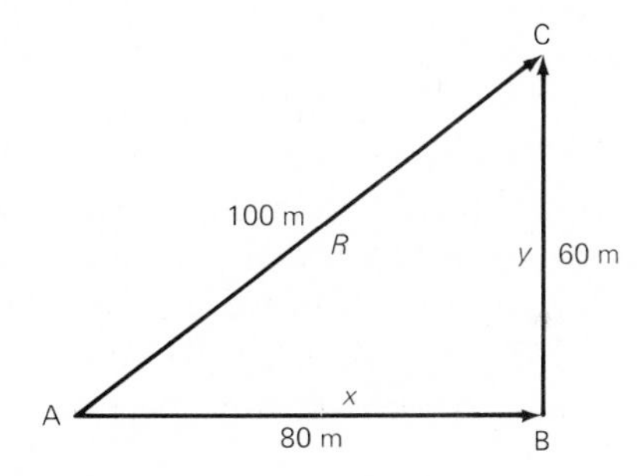

FIGURE A1 – 11
The process of vector addition applied to two displacements *x* and *y* to find their resultant *R*.

The answer to question (1) is simple; it is 80 m + 60 m = 140 m. The answer to question (2) requires the addition of the two displacements at right angles to each other. Joining A and C, we form a right triangle with sides *x, y,* and *R,* and the length of side *R* is the answer. To find this resultant *R,* we use the Pythagorean theorem:

$$R^2 = x^2 + y^2 \qquad [11]$$

Solving for the unknown *R,* we obtain

$$R = \sqrt{x^2 + y^2} = \sqrt{(80)^2 + (60)^2}$$

$$R = 100 \text{ m}$$

This process of adding displacements is called *vector addition.* Observe in Figure A1 – 12 that if the two displacements of 80 m and 60 m were in the same direction, the resultant *R* would be 140 m, whereas if they were in oppo-

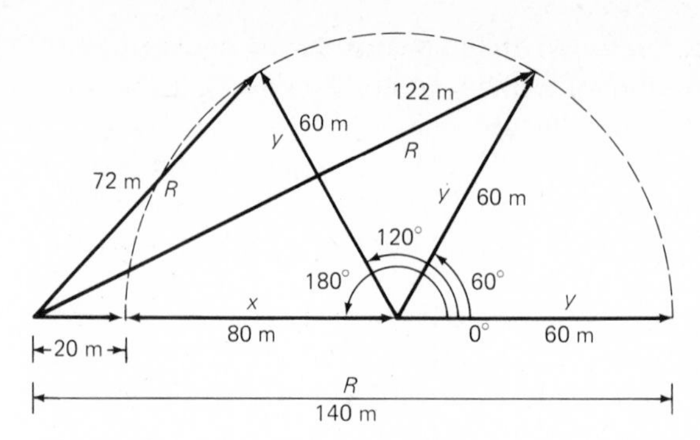

FIGURE A1–12 **Vector addition of the same two displacements at four different angles, 0°, 60°, 120° and 180°, give resultants *R* of 140 m, 122 m, 72 m and 20 m, respectively.**

site directions, *R* would be 20 m. At other angles, *R* must be between the two extremes of 20 m and 140 m. If *y* makes an angle of 60° with the direction of *x*, the resultant displacement will be 122 m, and if it makes an angle of 120°, the resultant will be 72 m.

If three displacements are to be added vectorially, the same procedure outlined above is followed. In Figure A1–13, let $a_1 = 6$ m in the $+x$ direction, $a_2 = 4$ m in a direction making 45° with the $+x$-axis, and $a_3 = 7$ m in the direction making 135° with the $+x$-axis. To find the resultant, we first add a_1 and a_2 vectorially and find the resultant R_1. To this resultant we add a_3 and obtain the final resultant *R*. If the lengths of the lines are drawn to a scale, with the lengths drawn proportionally to their magnitudes, and the angles are drawn using a protractor, *R* will be found to be 8.7 m.

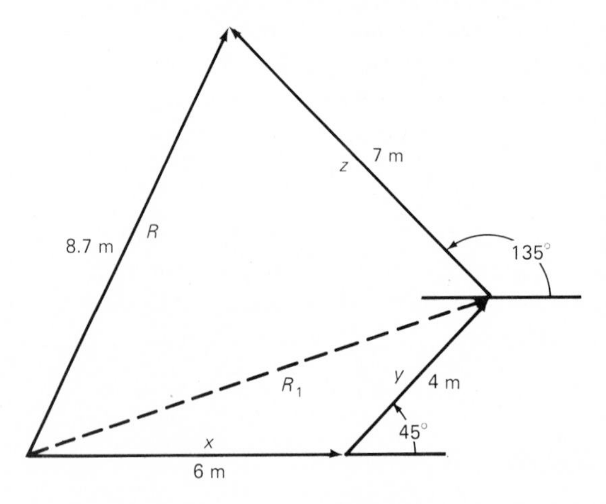

FIGURE A1–13 **Vector addition of three displacements, 6 m, 4 m, and 7 m, at the angles shown produce a resultant of 8.7 m.**

The procedure for adding additional vectors is simply an extension of this process. As strange as it may seem, the order in which any set of vectors is added makes no difference; the resultant *R* will always have the same magnitude and direction.

Velocities, accelerations, and forces are all vector quantities, and to add two or more such quantities of the same kind, we follow the procedure outlined above.

A1.12 Work

One of the most important concepts in nature is *energy.* Energy is important because it represents a fundamental entity common to all forms of matter in all parts of the known universe. Closely associated with energy is another concept, *work,* the transfer of energy. This is a term used in everyday life to describe the expenditure of one's stored up bodily energy. Energy is most easily described in terms of work.

In its simplest form, work is defined as force times the distance through which the force acts.

work = force × distance

$$W = F \times x \qquad [12]$$

Example

How much work is done in lifting a mass of 8 kg vertically upward a distance of 4.5 m? See Figure A1 – 14.

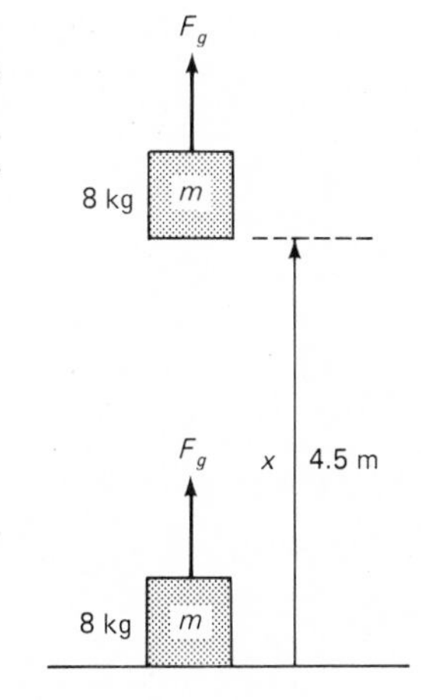

FIGURE A1 – 14
Work is defined as force, F_g, times distance, moved x.

Solution

The force required to lift a mass is given by Equation (8).

$$F_g = mg = 8 \text{ kg} \times 9.80 \frac{\text{m}}{\text{s}^2}$$

$$F_g = 78.4 \frac{\text{kg} \cdot \text{m}}{\text{s}^2}$$

$$F_g = 78.4 \text{ N}$$

Using Equation (12), we obtain

$$W = F_g \times x = 78.4 \frac{\text{kg} \cdot \text{m}}{\text{s}^2} \times 4.5 \text{ m}$$

$$W = 353 \frac{\text{kg} \cdot \text{m}^2}{\text{s}^2}$$

The units of work done are kilogram meter squared per second squared. Observe that they involve all three fundamental units of length, mass, and time and have the special name of joules (J). Since the units of kilogram meter per second squared are called newtons, the joule is equivalent to a force of 1 N acting for a distance of 1 m.

$$1 \text{ J} = 1 \text{ N} \times 1 \text{ m} \qquad [13]$$

A1.13 Potential energy

Mechanical energy is divided into two categories, potential energy E_p and kinetic energy E_k. A body is said to have potential energy if, by virtue of its position or state, it is able to do work. A wound clock spring, or a car at the top of a hill, has potential energy. The clock spring may keep a clock running for some time, while a car may coast downhill and travel some distance. Potential energy is measured by the amount of work that can be done, and it is measured in joules.

If a mass m is raised a distance x, it has, by virtue of its position above the ground level, a potential energy $F \times x$. The work done in lifting the mass has been stored up as potential energy in the mass. This energy can be regained by dropping the mass to the ground and in so doing have it perform work for us. By definition, then,

$$\text{potential energy} = F \times x$$
$$E_p = F \times x \qquad [14]$$

or

$$E_p = mgx$$

A1.14 Kinetic energy

A moving body of mass m has kinetic energy by virtue of its motion. An automobile moving along the highway has *kinetic energy of translation,* while a rotating wheel has *kinetic energy of rotation.* A given mass m moving with a constant velocity V has kinetic energy given by

$$E_k = \tfrac{1}{2}mV^2 \qquad [15]$$

See Figure A1 – 15.

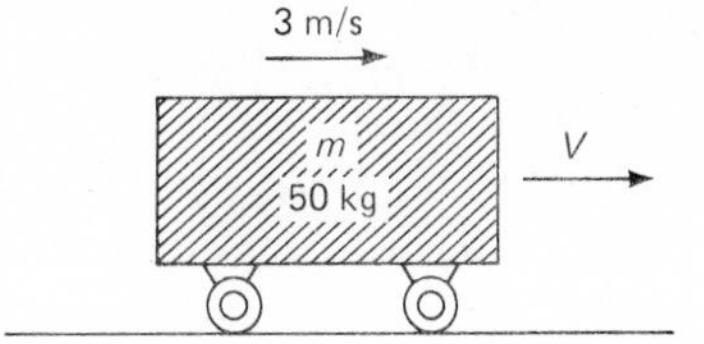

FIGURE A1 – 15
A mass m moving with a velocity V has kinetic energy $\frac{1}{2}mV^2$ by virtue of its motion.

Example

Find the kinetic energy of a 50-kg mass moving with a velocity of 3 m/s.

Solution

By direct substitution into Equation (15), we obtain

$$E_k = \frac{1}{2}\, mV^2 = \frac{1}{2}(50 \text{ kg}) \times \left(3\frac{\text{m}}{\text{s}}\right)^2$$

$$E_k = 225\, \frac{\text{kg} \cdot \text{m}^2}{\text{s}^2}$$

This answer has exactly the same fundamental units as work and potential energy, and it can be written in the same derived units:

$$E_k = 225 \text{ J}$$

By applying a constant force F on a body of mass m for a distance x, the body will acquire kinetic energy $\frac{1}{2}mV^2$.

$$F \times x = \frac{1}{2}\, mV^2 \qquad [16]$$

Conversely, a moving body with kinetic energy $\frac{1}{2}mV^2$, by being brought to rest, can exert a force F on some other system, and this force, acting through a distance x, can do work $F \times x$. This statement, expressed by Equation (16), is a special case of the *law of conservation of energy.* This basic law of physics may be stated as follows: in transforming energy from one form to another, energy is always conserved. In the treatment above, kinetic energy is converted to potential energy and vice versa. The work is equal to the energy transferred from one form to the other.

A1.15 Friction

In all the discussions above involving motion, we neglected the force of friction. All along we have assumed that friction was completely absent. We will now see how friction may be taken into account.

When we apply a force to pull a heavy box across the floor, friction plays an important role in the resultant motion, and the force equation must be modified. When one body slides over another, frictional forces always oppose the motion. See Figure A1 – 16. Such forces are largely due to the molecular attractive forces at the small areas of contact. The smoothness of the surfaces, within limits, does not greatly affect *f, the force of sliding friction.* If the surfaces are smooth, there will be many small areas in contact, and if they are rough, there will be fewer but larger areas.

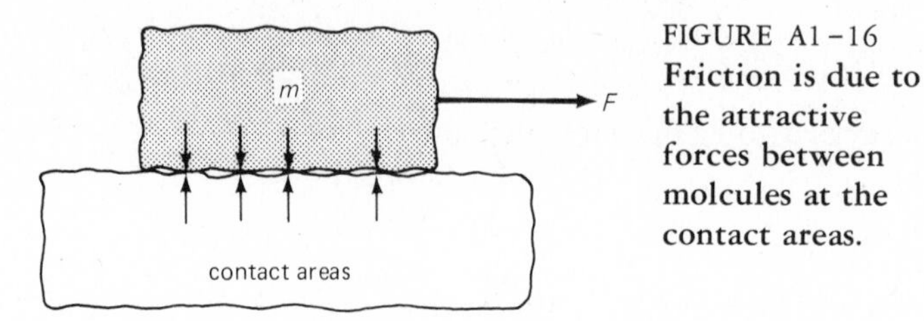

FIGURE A1 – 16
Friction is due to the attractive forces between molcules at the contact areas.

To start a body sliding requires a greater force than that required to keep it moving. Starting friction is greater than sliding friction. Once a body is moving, the force of sliding friction increases very little with speed and remains nearly constant over a wide range of speeds.

If the force F applied to a body is greater than the force of friction, the effective force producing acceleration is given by the difference,

$$F_{\text{eff}} = F - f \qquad [17]$$

Force is a vector quantity, and in Figure A1 – 17 it can be seen that F and f are oppositely directed. Hence Newton's second law of motion can be written as

$$F - f = ma \qquad [18]$$

The force to start a body moving, or to keep it moving with constant speed, can be measured experimentally as shown in Figure A1 – 17.

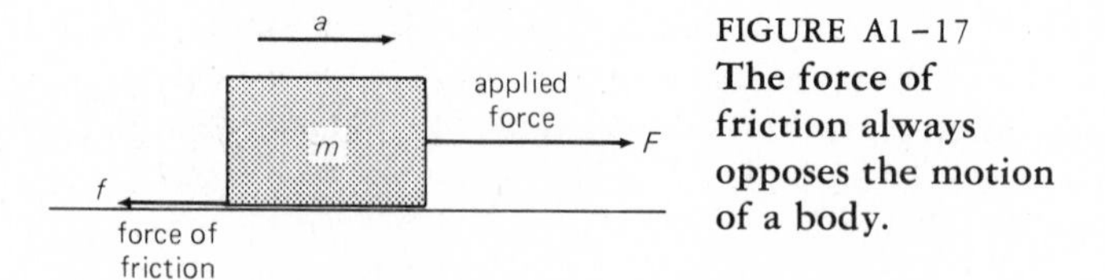

FIGURE A1 – 17
The force of friction always opposes the motion of a body.

Example

A heavy box with a mass of 60 kg requires a force of 250 N to move it across the floor at constant speed. Find the acceleration when a constant force of 400 N is applied to the box.

Solution

The known quantities are $f = 250$ N, $F = 400$ N, and $m = 60$ kg. By solving Equation (18) for the acceleration and substituting known quantities, we obtain

$$a = \frac{F - f}{m} = \frac{400\ \text{N} - 250\ \text{N}}{60\ \text{kg}} = 2.50\ \frac{\text{m}}{\text{s}^2}$$

In addition to sliding friction, there are fluid friction and rolling friction. The motion of a body through the air, or through water, is quite complex, but it may be taken into account in much the same way: by experimentally measuring the force of friction for each example.

A1.16 Power

Power is defined as the time rate at which energy is being expended.

$$\text{power} = \frac{\text{energy}}{\text{time}}$$

$$P = \frac{E}{t} \qquad [19]$$

The faster energy is expended, the greater is the power required to do the work that needs to be done. The longer the time t in the equation above, the smaller is the fraction E/t and the smaller is the required power.

In the MKS system of units, work is measured in joules, and power is measured in joules per second. One joule per second is called the watt (W), the unit of power.

$$1\ \text{watt} = \frac{1\ \text{joule}}{1\ \text{second}}$$

$$1\ \text{W} = \frac{1\ \text{J}}{1\ \text{s}} \qquad [20]$$

Example

Find the power required for a hoist if it is to raise 25 kg a distance of 18 m in the time of 12 s. See Figure A1 – 18.

Solution

The energy expended is given by Equation (14), and using Equation (19), we obtain

$$P = \frac{E}{t} = \frac{mgx}{t} = \frac{25\ \text{kg} \times 9.8\ \text{m/s}^2 \times 18\ \text{m}}{12\ \text{s}}$$

$$P = 367.5 \frac{\text{kg} \cdot \text{m}^2}{\text{s}^2} = 367.5 \text{ W}$$

The kilowatt (kW) is a larger unit of power and is equal to 1000 W.

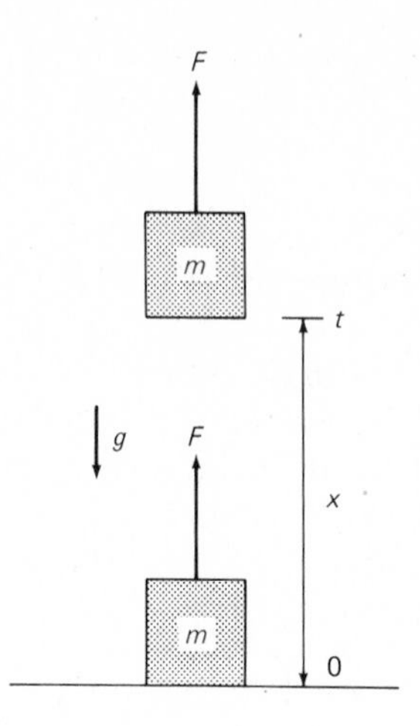

FIGURE A1-18
Power is defined as the time rate at which energy is being expended.

A1.17 Pressure

Pressure exists in solids, liquids, and gases and is best described in liquids and gases that are generally classified as fluids. Consider a vessel of water as shown in Figure A1-19. At a depth h_1, the pressure P_1 is given by the weight of the column of fluid of unit cross section and depth h_1. At a greater depth h_2, the pressure P_2 is given by the weight of a greater column of water. This can be demonstrated with a glass cylinder and a disk held over the lower end by a string. See diagram (b). If the cylinder is gradually filled with water, an increasing downward force is exerted on the disk. The instant the water inside reaches the level of the water outside, the disk drops off the cylinder, showing

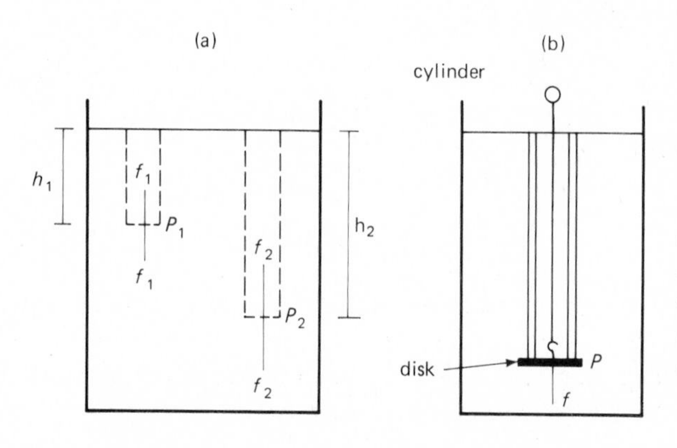

FIGURE A1-19
The pressure at any given depth in a liquid is equal to the weight per unit cross section of the liquid directly above.
(a) An illustration.
(b) An experimental demonstration.

that the downward force F inside and the upward force outside at the same level become equal. We can conclude from this experiment that the pressure at any given point in a fluid is given by the weight of the fluid above it, and that the pressure at one depth is equal to the pressure at the same depth anywhere in the fluid.

Pressure is defined as the normal (perpendicular) force exerted by a fluid per unit area, and it is usually written as

$$P = \frac{F}{A} \qquad [21]$$

where F is in newtons and A is in meters squared. Pressure is not a force; it is a force per unit area.

At any given depth in a liquid, the force exerted on an element of surface is perpendicular to the surface and is independent of its orientation. In other words, at any given depth, the pressure is the same in all directions. In Figure A1 – 20, for example, a hollow steel ball B is filled with water. By pushing down on the handle H, the plunger Q forces water out of several metal tubes T leading from the sides and bottom of the ball. Equal force in all directions is indicated by water jets rising to the same height.

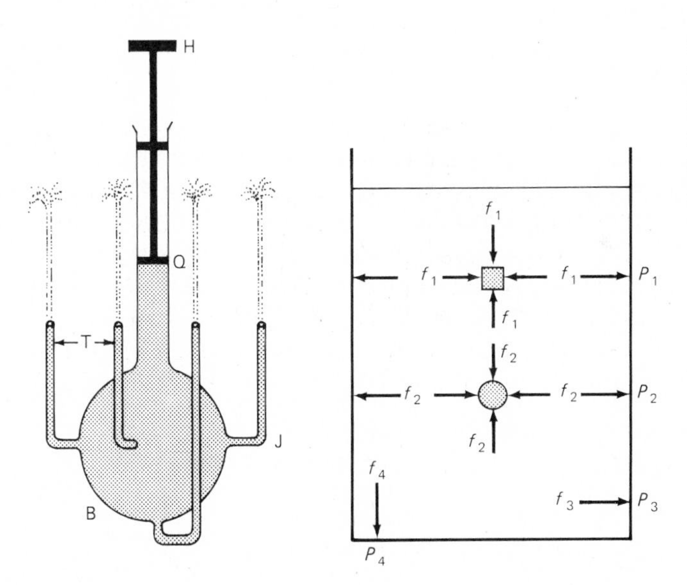

FIGURE A1 – 20
The pressure at any given depth in a liquid is the same in all directions.

A1.18 Atmospheric pressure

We on the earth are living in a great sea of air called the atmosphere. The air is a mixture of gases: about 77 percent nitrogen, 21 percent oxygen, and 1 percent argon. The remaining 1 percent includes small quantities of such gases as carbon dioxide, hydrogen, neon, krypton, helium, ozone, and xenon.

The atmosphere is most dense at sea level and extends upward to a height of 100 miles (mi) or more. The uncertainty as to the height is due to the fact that the higher one goes, the thinner the air gets, and finally it thins out into interplanetary space. Air at sea level has an average density of 1.293 kg/m³. By comparison, water has a density of 1000 kg/m³. Although the air is very light by comparison, it does extend to great heights and exerts a considerable pressure at sea level.

A column of air one square meter in cross section, and reaching to the top of the atmosphere, has a mass of approximately 10,340 kg. The weight of this air per square meter is what we call *atmospheric pressure.* By Equation (21), this amounts to

$$P = \frac{F}{A} = \frac{mg}{A} = \frac{10{,}340 \text{ kg} \times 9.8 \text{ m/s}^2}{1 \text{ m}^2}$$

$$P = 101{,}300 \frac{\text{N}}{\text{m}^2} \qquad [22]$$

This is called **standard atmospheric pressure** and is usually written as

$$P_0 = 1.013 \times 10^5 \frac{\text{N}}{\text{m}^2} \qquad [23]$$

A1.19 Radians

When a wheel or any object is in a state of rotation about any axis, the angle through which it turns is measured in *degrees* or in *radians* (rad), and the angular velocity or angular speed is measured in *radians per second* (rad/s). The radian is a unit of *angular measure* just as the meter is a unit of *linear measure.* It is defined as the angle subtended by the arc of a circle whose length is equal to the radius of the same circle. See Figure A1 – 21.

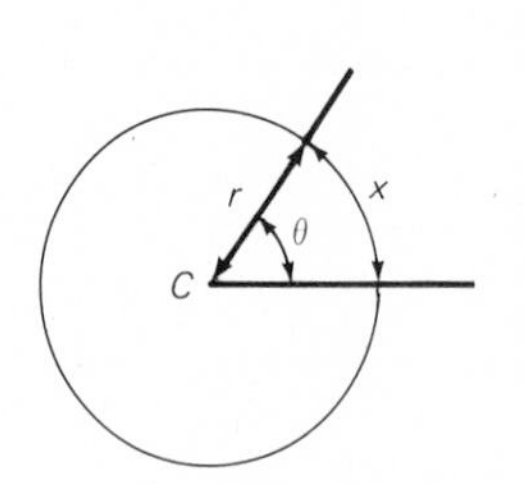

FIGURE A1 – 21
The radian is a unit of angular measure. When the length of the arc *x* equals the radius of the circle *r*, the angle *θ* equals 1 radian.

The circumference of a circle is given by 2π times the radius r.

$$\text{circumference} = 2\pi r \qquad [24]$$

There are therefore 2π radians in a complete circle:

$$2\pi \text{ rad} = 360° \qquad [25]$$

Since $\pi = 3.14159$, we obtain

$$1 \text{ rad} = \frac{360°}{2\pi} = 57.296° \qquad [26]$$

It follows from the relations above that any angle θ in radians is given by

$$\text{angle in radians} = \frac{\text{arc length}}{\text{radius}}$$

$$\theta = \frac{x}{r} \qquad [27]$$

If $x = r$, then $\theta = 1$ rad. If the arc x is twice as long, $x = 2r$, and $\theta = 2$ rad, and so on.

Here are some examples of angular measure:

$$30° = \frac{\pi}{6} \text{ rad} \qquad 90° = \frac{\pi}{2} \text{ rad} \qquad 180° = \pi \text{ rad}$$

$$45° = \frac{\pi}{4} \text{ rad} \qquad 120° = \frac{2\pi}{3} \text{ rad} \qquad 270° = \frac{3\pi}{2} \text{ rad}$$

$$60° = \frac{\pi}{3} \text{ rad} \qquad 150° = \frac{5\pi}{6} \text{ rad} \qquad 360° = 2\pi \text{ rad}$$

The reason for measuring angles in radians is that it simplifies all formulas for rotary motion. See Figure A1 – 22. As an illustration, the angular velocity of a body is defined as the angle turned through divided by the time:

$$\text{angular velocity} = \frac{\text{angle turned through}}{\text{elapsed time}}$$

$$\omega = \frac{\theta}{t} \qquad [28]$$

This formula should be compared with the corresponding definition of linear velocity. See Equation (2), which is repeated here:

$$V = \frac{x}{t} \qquad [29]$$

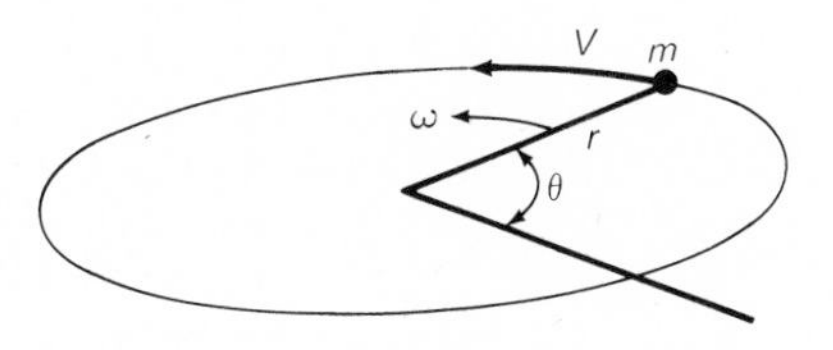

FIGURE A1 – 22
A mass *m*, whirled in a circle at the end of a string, has an angular velocity *ω*.

A1.20 Conversion factors

Occasionally it becomes necessary to convert measurements in the English system of units to the metric system, or vice versa. The following tables will serve in this capacity and are based upon the United States legal standard of length.

TABLE A1-1:
Conversion Factors for Units of Length

LENGTH	km	m	cm	in.	ft	mi
1 kilometer =	1	1,000	100,000	39,370	3,280.83	0.62137
1 meter =	0.00100	1	100	39.370	3.28083	6.21×10^{-4}
1 centimeter =	1.0×10^{-5}	0.0100	1	0.39370	0.032808	6.21×10^{-6}
1 inch =	2.54×10^{-5}	0.02540	2.5400	1	0.08333	1.58×10^{-5}
1 foot =	3.05×10^{-4}	0.30480	30.480	12	1	1.89×10^{-4}
1 mile =	1.60935	1,609.35	160,935	63,360	5,280	1

TABLE A1-2:
Conversion Factors for Speed and Velocity

VELOCITY	m/s	ft/s	km/h	mi/h	knots
1 m/s =	1	3.281	3.600	2.240	1.940
1 ft/s =	0.30480	1	1.0973	0.6818	0.5921
1 km/h =	0.27778	0.9113	1	0.6214	0.5396
1 mi/h =	0.44704	1.4667	1.6093	1	0.8694
1 knot =	0.51480	1.689	1.853	1.152	1

1 kg weighs 2.2046 pounds (lb)
453.6 g weighs 1 lb

As an illustration of how to use these tables, consider the following examples.

Example

Find the number of meters in 3.52 mi.

Solution

For the answer, we examine the rows and columns of the first table and find that 1 mi = 1609.35 m. The answer is therefore given by the product

$$3.52 \text{ mi} = 3.52 \times 1609.35 = 5664.91 \text{ m}$$

which, to four figures, is 5665 m.

Example

An automobile is traveling at 60 mi/h. Find its speed in (a) meters per second and (b) feet per second.

Solution

From the second table, we observe that 1 mi/h = 0.44704 m/s and 1.4667 ft/s. If we multiply both factors by 60, separately, we obtain

$$60 \text{ mi/h} = 60 \times 0.44704 = 26.82 \text{ m/s}$$
$$60 \text{ mi/h} = 60 \times 1.4667 = 88.0 \text{ ft/s}$$

PROBLEMS

1. A car traveling north averages 85 km/h. Find the average velocity in (a) meters per second and (b) miles per hour.

2. The speed of sound in air is measured and found to be 356 m/s. How far will it travel in 45 s? Find your answer in (a) kilometers and (b) miles.

3. A jet plane starts from rest on a runway and reaches a speed of 180 km/h in 42 s. Find its average acceleration in meters per second squared.

4. A mass of 75 kg is given an acceleration of 2.45 m/s^2. Find the magnitude of the applied force.

5. A constant force of 250 N is applied to a mass of 60 kg. Find the acceleration.

6. Find the weight of a person whose mass is 72 kg.

7. A boy walks south at 5 km/h for 1 h. He then walks east at 6 km/h for 2 h. Find (a) the total distance walked and (b) his distance from the starting point.

8. A hiker in climbing a mountain trail rises a distance of 520 m in 30 min. If his total mass is 86 kg, find (a) his stored potential energy in joules and (b) the power he develops in watts.

9. A 2000-kg car travels at 90 km/h along a highway. Find its kinetic energy in joules.

10. Neglecting friction, what force in newtons applied to a 3500-kg truck will give it a speed of 80 km/h in 200 m?

Appendix II

LOGARITHMS AND POWERS OF TEN

A2.1 Use of exponents

In recording the size and shape of an object, or in recording the time interval between the occurrence of two consecutive events, it is convenient to express very large numbers and very small decimals in an abbreviated form. This is done principally to conserve time and space. It is convenient for the astronomer in the study of stars on the one hand and the physicist in the study of atoms on the other. The abbreviations in common use today are based upon powers of ten, as shown in Table A2–1. In the table, the abbreviated form on the right side of each equation is mathematically correct. For example,

$$10^3 = 10 \times 10 \times 10 = 1000$$

and

$$10^{-3} = \frac{1}{10^3} = \frac{1}{1000} = 0.001$$

TABLE A2–1: Powers of Ten

$1 = 10^0$	$1 = 10^0$
$10 = 10^1$	$0.1 = 10^{-1}$
$100 = 10^2$	$0.01 = 10^{-2}$
$1{,}000 = 10^3$	$0.001 = 10^{-3}$
$10{,}000 = 10^4$	$0.0001 = 10^{-4}$
$100{,}000 = 10^5$	$0.00001 = 10^{-5}$
$1{,}000{,}000 = 10^6$	$0.000001 = 10^{-6}$

In every case, the exponent is seen to give directly the number of digits the decimal point is moved from unity, positive integers specifying the number of places the decimal point is moved to the right to make large numbers and negative integers specifying the number of places it is moved to the left to make fractions.

To illustrate the use of this system, let us suppose that a large jet plane has a mass of 600,000 kg. This can be written as

$$600{,}000 \text{ kg} = 6 \times 100{,}000 \text{ kg}$$
$$m = 6 \times 10^5 \text{ kg}$$

In the abbreviated notation, the mass is therefore written as 6×10^5 kg.

If more than one numeral occurs, any one of several abbreviations might be written. For example, in the case of large numbers,

$$240{,}000{,}000 = 24 \times 10{,}000{,}000 = 24 \times 10^7$$

or

$$240{,}000{,}000 = 2.4 \times 100{,}000{,}000 = 2.4 \times 10^8$$

In the case of small numbers, on the other hand,

$$0.0036 = 3.6 \times 10^{-3} \quad \text{or} \quad 36 \times 10^{-4}$$

To illustrate the advantages of this abbreviated notation, we can write the mass of the earth and the mass of an electron as follows:

$$\text{mass of the earth, } m = 5.97 \times 10^{24} \text{ kg}$$
$$\text{mass of an electron, } m = 9.11 \times 10^{-31} \text{ kg}$$

If these were written in complete decimal form, they would appear as follows:

$$\text{mass of the earth} = 5{,}970{,}000{,}000{,}000{,}000{,}000{,}000{,}000 \text{ kg}$$
$$\text{mass of an electron} = 0.000000000000000000000000000000911 \text{ kg}$$

At a meeting held by the International Union for Pure and Applied Physics a few years ago, the following symbolism was adopted for general use:

10^3	kilo	k	10^{-3}	milli	m
10^6	mega	M	10^{-6}	micro	μ
10^9	giga	G	10^{-9}	nano	n
10^{12}	tera	T	10^{-12}	pico	p
10^{15}	peta	P	10^{-15}	femto	f
10^{18}	exa	E	10^{-18}	atto	a

A2.2 Multiplication and division of large and small numbers

The multiplication and division of large and small numbers in the abbreviated notation involves the addition and subtraction of exponents by the following rules.

Rule 1. When a power number is changed from numerator to denominator, or vice versa, the sign of the exponent is reversed. For example,

$$\frac{5}{2 \times 10^{-6}} = \frac{5 \times 10^{6}}{2}$$

Rule 2. When two power numbers are multiplied, their exponents are added. For example,

$$3 \times 10^{5} \times 2 \times 10^{4} = 3 \times 2 \times 10^{5+4} = 6 \times 10^{9}$$

As anothèr example,

$$4 \times 10^{18} \times 2 \times 10^{-11} = 4 \times 2 \times 10^{18-11} = 8 \times 10^{7}$$

Rule 3. When one power number is divided by another, their exponents are subtracted. For example,

$$\frac{6 \times 10^{9}}{3 \times 10^{3}} = \frac{6 \times 10^{9-3}}{3} = 2 \times 10^{6}$$

As another example,

$$\frac{6 \times 10^{-8}}{2 \times 10^{-3}} = \frac{6 \times 10^{-8+3}}{2} = 3 \times 10^{-5}$$

A2.3 Logarithms

Observe in Table A2–1 that the exponents of powers of ten are all whole numbers. If we now introduce exponents that can take on any values whatever,

TABLE A2–2:
Logarithms

ROW	$\log x = a$	$10^{a} = x$	$\log x = a$
1	$\log 1 = 0$	$10^{0} = 1.0000$	$\log 1.0000 = 0$
2	$\log 10 = 1$	$10^{0.1} = 1.2589$	$\log 1.2589 = 0.1$
3	$\log 100 = 2$	$10^{0.2} = 1.5849$	$\log 1.5849 = 0.2$
4	$\log 1000 = 3$	$10^{0.3} = 1.9953$	$\log 1.9953 = 0.3$
5	$\log 10^{4} = 4$	$10^{0.4} = 2.5119$	$\log 2.5119 = 0.4$
6	$\log 10^{5} = 5$	$10^{0.5} = 3.1623$	$\log 3.1623 = 0.5$
7	$\log 10^{6} = 6$	$10^{0.6} = 3.9811$	$\log 3.9811 = 0.6$
8	$\log 10^{7} = 7$	$10^{0.7} = 5.0119$	$\log 5.0119 = 0.7$
9	$\log 10^{8} = 8$	$10^{0.8} = 6.3096$	$\log 6.3096 = 0.8$
10	$\log 10^{9} = 9$	$10^{0.9} = 7.9433$	$\log 7.9433 = 0.9$
11	$\log 10^{10} = 10$	$10^{1.0} = 10.0000$	$\log 10.000 = 1.0$
12	$\log 10^{11} = 11$	$10^{1.1} = 12.589$	$\log 12.589 = 1.1$
13	$\log 10^{12} = 12$	$10^{1.2} = 15.849$	$\log 15.849 = 1.2$
	etc.	etc.	etc.

including fractions or decimals as well as whole numbers, we can develop a system of numbers called **logarithms**.

We define a logarithm as a number a that is the power to which 10 must be raised to obtain that number. From this definition and Table A2–1, we construct Table A2–2.

Since the cost of a pocket calculator is about the same as that of a slide rule or a book of tables of logarithms, no student should be without one. It is recommended that the student should obtain a pocket calculator with the following keys or their equivalents: one or more storage keys, as well as x, x^2, $\sqrt[x]{y}$, 10^x, log, e^x, ln, y^x, arc, sin, cos, tan, 1/x, and π.

To verify the values in Table A2–2 with a calculator, consider, for example, all values given in row 6.

Column 2 Enter 100,000 and press the [log] key.* Read 5.
Column 3 Enter 10 and press the [y^x] key. Enter 0.5 and press the [=] key. Read 3.1622777, which to five figures is 3.1623.
Column 4 Enter 3.1623 and press the [log] key. Read 0.500.

The values in Table A2–2 will also verify the relationships given by the following Equations (1) and (2).

$$x = 10^a$$
$$\log x = \log 10^a$$
$$\log x = a \log 10 \qquad [1]$$

Since $\log_{10} 10 = 1$,

$$\log x = a \qquad [2]$$

To demonstrate the usefulness of these relations, suppose we select any two numbers x and y and multiply one by the other. Let

$$x = 10^a \quad \text{and} \quad y = 10^b$$

Multiplying one by the other gives

$$xy = 10^a \times 10^b \qquad [3]$$

Using Rule 2 above,

$$xy = 10^{a+b} \qquad [4]$$

Using Equations (1) and (2),

$$\log xy = \log 10^{a+b} \qquad [5]$$
$$\log xy = (a + b) \log 10$$
$$\log xy = a + b$$
$$\log xy = \log x + \log y \qquad [6]$$

*Since some calculators use a different logic, this sequence may need to be altered. In any case, read the instruction book.

This says that the logarithm of the product of two numbers is equal to the sum of the logarithms of these numbers.

From Rule 3 above, we follow the same procedure and find

$$\log \frac{x}{y} = \log x - \log y \qquad [7]$$

which says that the logarithm of the quotient of two numbers is equal to the difference between the logarithms of these numbers.

Equations (6) and (7) have been extremely useful for many years for the multiplication and division of large numbers, as well as for tabulating values of physical measurements made on certain natural phenomena. As examples, see the decibel scale of sound intensity in Section 9.4 and in Figure 9–7, as well as the cents scale used in expressing musical intervals on the tempered musical scale (see Section 14.8 and Figure 14–9).

Appendix III

ABSORPTION COEFFICIENTS

Absorption coefficients for some of the common building materials and furnishings are given in the following table. For more complete and detailed lists, see other books on architectural acoustics. The table here is useful in making simple calculations of reverberation times in large as well as small rooms.

MATERIALS	THICKNESS	125 Hz	250 Hz	500 Hz	1000 Hz	2000 Hz	4000 Hz
Fiberglass tile, type A[a]	2.5 cm	0.17	0.44	0.91	0.99	0.82	0.77
Acousti-Celotex, C–4[a]	3.0 cm	0.25	0.58	0.99	0.75	0.58	0.50
Fibretex[a]	2.0 cm	0.16	0.45	0.50	0.78	0.84	0.78
Cushion Tone, A–3[a]	2.8 cm	0.25	0.56	0.99	0.99	0.91	0.82
Rock wool	2.5 cm	0.35	0.49	0.63	0.80	0.83	0.80
Carpet on hair felt	10 cm	0.11	0.24	0.57	0.69	0.71	0.73
Plywood paneling	6.0 mm	0.28	0.22	0.17	0.09	0.10	0.10
Brick, unglazed		0.03	0.03	0.03	0.04	0.05	0.07
Plaster, gypsum or lime on lath		0.11	0.10	0.05	0.04	0.04	0.03
Draperies							
10 oz/yd^2, or 0.34 kg/m^3		0.04	0.05	0.11	0.18	0.30	0.44
14 oz/yd^2, or 0.48 kg/m^3		0.05	0.08	0.23	0.32	0.40	0.44
18 oz/yd^2, or 0.61 kg/m^3		0.05	0.12	0.35	0.45	0.50	0.44
Audience seated in upholstered seats (per unit)		0.60	0.74	0.88	0.96	0.93	0.85
Unoccupied upholstered seats (per unit)		0.49	0.66	0.80	0.88	0.82	0.70
Unoccupied wooden seats (per unit)		0.15	0.19	0.22	0.39	0.38	0.30
Occupied wooden seats (per unit)		0.25	0.35	0.40	0.65	0.62	0.42
Floors							
Concrete or terrazzo		0.01	0.01	0.015	0.02	0.02	0.02
Wood		0.15	0.11	0.10	0.07	0.06	0.07
Asphalt tile		0.02	0.02	0.03	0.10	0.20	0.15

[a]Nailed to wood furring strips 2.0 cm by 4.0 cm and 30.0 cm apart on centers.

BIBLIOGRAPHY

This list is meant to refer the reader to major works related to contemporary understanding of musical acoustics. It is not meant to be exhaustive, as many important books have been omitted. Those interested in a more extensive list, historic references, or special topics are referred to the lists at the ends of individual chapters. Also see Resource Letter MA–1: "Musical Acoustics," by Thomas D. Rossing, *American Journal of Physics* 43 (November 1975). A book containing this resource letter, together with selected reprints, is available from the American Association of Physics Teachers, SUNY, Stony Brook, NY 11794. Also see "The Physics of Music," selected reprints from *Sci. Am.* San Francisco: W. H. Freeman, 1978.

Backus, J. "Vibrations of the Reed and Air Column in the Clarinet." *J. Acoust. Soc. Am.* 23 (1961):806.

———. *The Acoustical Foundations of Music.* 2d ed. New York: Norton, 1977.

Békésy, G. von. *Experiments in Hearing.* New York: McGraw-Hill, 1960.

Benade, A. H. "The Physics of Woodwinds." *Sci. Am.* (Oct. 1960):144.

———. *Horns, Strings and Harmony.* Garden City, N.Y.: Doubleday (Anchor Books), 1960.

———. *Fundamentals of Musical Acoustics.* New York: Oxford University Press, 1960.

———. "The Physics of the Brasses." *Sci. Am.* (July 1973):24.

Beranek, L. L. *Music, Acoustics and Architecture.* New York: Wiley, 1962.

Blackham, E. D. "Physics of the Piano." *Sci. Am.* (Dec. 1965):88.

Culver, C. A. *Musical Acoustics.* New York: McGraw-Hill, 1956.

Denes, P. B., and Pinson, E. N. *The Speech Chain.* Garden City, N.Y.: Doubleday (Anchor Books), 1973.

Doelle, L. L. *Environmental Acoustics.* New York: McGraw-Hill, 1973.

Flanagan, J. L. *Speech Analysis, Synthesis, and Perception.* New York: Springer-Verlag, 1972.

Fletcher, H. *Speech and Hearing in Communication.* Huntington, N.Y.: Krieger, 1972.

Gerber, S. E. *Introductory Hearing Science.* Philadelphia: Saunders, 1974.

Helmholtz, H. *On the Sensations of Musical Tone.* New York: Dover, 1954.

Hutchins, C. M. "The Physics of Violins." *Sci. Am.* (Nov. 1962):78.

———. "Founding a Family of Fiddles." *Phys. Today* 20 (1967):23.

Jeans, Sir J. *Science and Music.* New York: Dover, 1968.

Johnson, K., and Walker, W. *The Science of High Fidelity.* Dubuque, Iowa: Kendell/Hunt, 1977.

Josephs, J. J. *The Physics of Musical Sound.* New York: Van Nostrand, 1967.

Kinsler, L. E., and Frey, A. R. *Fundamentals of Acoustics.* New York: Wiley, 1962.

Leipp, E. *The Violin—History, Aesthetics, Manufacture, and Acoustics.* University of Toronto Press, 1969.

Miller, D. C. *The Science of Musical Sounds.* New York: Macmillan 1928. A classic.

Olson, H. F. *Music, Physics and Engineering.* New York: Dover, 1967.

Rabinowicz, E. "Stick and Slip." *Sci. Am.* (May 1956):109.

Rendall, G. *The Clarinet.* London: Ernest Benn, 1971.

Rigden, J. S. *Physics and the Sound of Music.* New York: Wiley, 1977.

Roederer, J. G. *Introduction to the Physics and Psychophysics of Music.* 2d ed. New York: Springer-Verlag, 1975.

Rossing, T. D. "Acoustics of Percussion Instruments: Part I." *Phys. Teach.* 14 (Dec. 1967):546. "Part II." *Phys. Teach.* 15 (May 1977):278.

Sabine, W. C. *Collected Papers on Acoustics.* New York: Dover, 1967.

Savage, W. R. *Problems for Musical Acoustics.* New York: Oxford University Press, 1977.

Schnelleng, J. C. "Acoustical Effects of Violin Varnish." *J. Acoust. Soc. Am.* 44 (1968): 1175.

———. "The Physics of the Bowed String." *Sci. Am.* (Jan. 1974).

Schwartz, H. W. *The Story of Musical Instruments from Shepherds Pipe to Symphony.* New York: Doubleday, 1938.

Seashore, C. E. *Psychology of Music.* New York: Dover, 1967.

Stevens, S. S. *Handbook of Experimental Psychology.* New York: Wiley, 1951.

———, and Davis, H. *Hearing.* New York: Wiley, 1938.

———, and Warshofsky, F. *Sound and Hearing.* New York: Time Science Series, 1965.

Strong, W. J., and Plitnik, G. R. *Music, Speech and High Fidelity.* Provo, Utah: Brigham Young University Publications, 1977.

Taylor, C. A. *The Physics of Musical Sounds.* New York: Elsevier, 1965.

Taylor, R. *Noise.* Baltimore: Pelican/Penguin, 1970.

Van Bergeijk, W. A. *Waves and the Ear.* New York: Doubleday (Anchor Books), 1960.

White, H. E. *Modern College Physics.* 3d ed. New York: Van Nostrand, 1953. Chaps. 35–38.

Winckel, F. *Music, Sound and Sensation.* New York: Dover, 1967.

Wood, A., and Bowsher, J. M. *The Physics of Music.* New York: Halsted Press, 1975.

Yerges, L. *Sound, Noise and Vibration Control.* New York: Van Nostrand Reinhold, 1969.

ANSWERS TO ODD-NUMBERED PROBLEMS

Chapter 2

1. 0.002273 s
3. 0.313 cm

Chapter 3

1. (a) 0.61 m
   (b) 268.4 m/s

Chapter 4

1. (a) 3346 Hz
   (b) 1.240 m
   (c) 4149 m/s

Chapter 5

1. 351.0 m/s
3. (a) 349.8 m/s
   (b) 0.135 m
   (c) 2587 Hz
   (d) 2607 Hz

Chapter 6

1. (a) 0.628 s, 1.592 Hz
   (b) 1.987 s, 0.503 Hz
   (c) 6.28 s, 0.1592 Hz
3. (a) 426 Hz
   (b) 263 Hz

Chapter 7

1. (a) 17.5 cm
   (b) 0.219 rad
   (c) 12.5°
3. (a) 0.349 rad
   (b) 4.29 cm
   (c) 8132 Hz

Chapter 8

1. See art at right.
3. See art at right.
5.
   (b) $A = 35.4 \times 10^{-7}$ m and $\phi_0 =$ 75°
   (d) See art at right.
   (e) $T = 0.010$ s

Chapter 9

1. (a) 5.01×10^{-8} W/m²
   (b) 5.01×10^{4} pW/m²
3. (a) 12.57 B
   (b) 12.57 dB
5. (a) 1.58×10^{-8} W/m²
   (b) 5.0×10^{-2} W
   (c) 76.0 dB
7. 76.9 dB
9. 66.5 dB

Chapter 10

1. (a) 52 dB
   (b) 52 phons
   (c) 1.58×10^{-7} W/m²
   (d) 5.0×10^{-5} W
3. (a) 57, 63, and 66 phons, respectively
   (b) 3.3, 5.0, and 6.5 sones, respectively
   (c) 14.8 sones
   (d) 78 phons

Chapter 11

1. (a) 0.458 m
   (b) 6.272 N
   (c) 36.64 m/s
   (d) 0.004672 kg/m

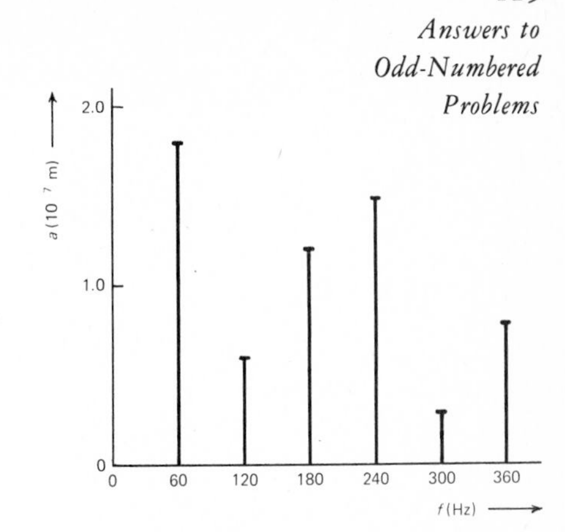

Chapter 8, 1.

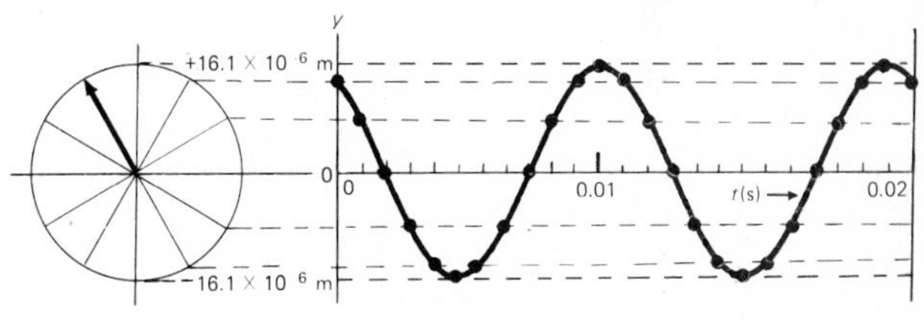

Chapter 8, 3.

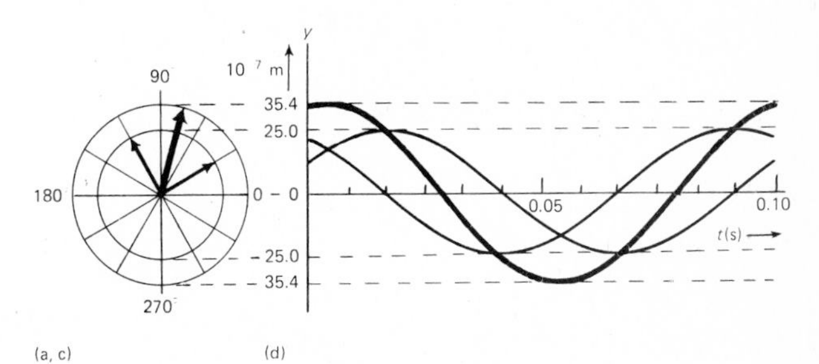

Chapter 8, 5d.

3. (a) a straight line
   (b) V is proportional to $\sqrt{F}$
   (c) $V = k\sqrt{F}$, where $k = 37.88$

Chapter 12

1. 969 m/s
3. (a) 169.9 m/s
   (b) 32.05 Hz
   (c) 128.2 Hz
5. (a) 331.5 m/s
   (b) 62.8 cm

Chapter 13

1. (a) 28.5 Hz
   (b) 26.0 Hz
   (c) 23.2 Hz
   (d) 21.2 Hz
   (e) 20.1 Hz
3. (a) 729 Hz
   (b) 1426 Hz

Chapter 14

1. (a) 70.7¢
   (b) 70.7¢
   (c) 203.9¢
   (d) 386.3¢
   (e) 315.6¢
   (f) 111.7¢
3. (a) 1.25
   (b) 1.875
   (c) 480 Hz

Chapter 15

1. (a) $f_{DT} = 260.7$ Hz,
   $f_{C_1} = 260.7$ Hz, and
   $f_{C_2} = 0$
   (b) C_4, C_4, and none
3. (a) 132 Hz, 264 Hz, 264 Hz, and 55 Hz
   (b) C_3, C_4, C_4, and A_1
5. (a) 88, 88, 88, 88, 176, 264 Hz
   (b) F_2, F_3, C_4
   (c) 1st, 2d, 3d, 4th, 5th, and 6th

Chapter 18

1. 59 cm

Chapter 19

1. 2.66 m

Chapter 21

1. (a) 24 V
   (b) 72 W
3. (a) 3 A
   (b) 24 V

Chapter 22

1. (a) See art at right.
   (b) See art at right.
3. See art, below.

Chapter 23

1. (a) 667 turns
   (b) About 400 m for a typical LP record
3. 250,000 cycles

Chapter 24

1. (a) 110 Hz and 770 Hz
   (b) A_2 and about G_5
3. Pink noise: constant dB/octave
   White noise: $3N$ dB/octave
   where N is the octave number

Chapter 25

0.41 ms

Chapter 27

1. 2.24 s
3. 245 m²

Chapter 28

1. (a) 29 ms
   (b) No. The sounds will blend for time separations less than 35 ms, although the precedence effect may cause the listener to identify the closer speaker as the sound source.
3. 17.4 m

Appendix 1

1. (a) 23.61 m/s
   (b) 52.8 mi/h
3. 1.190 m/s²

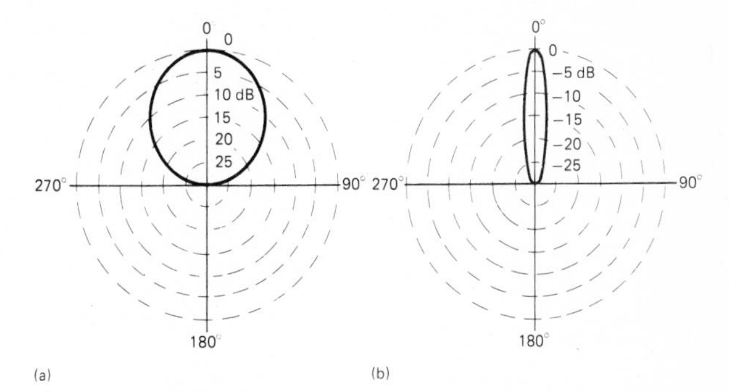

Chapter 22, 1a.

Chapter 22, 1b.

Chapter 22, 3.

5. 4.167 m/s^2
7. (a) 17 km
   (b) 13 km
9. 625,000 J

INDEX